Radio Frequency Integrated Circuit Design

For a listing of recent titles in the *Artech House Microwave Library*, turn to the back of this book.

Radio Frequency Integrated Circuit Design

John Rogers
Calvin Plett

Artech House
Boston • London
www.artechhouse.com

Library of Congress Cataloging-in-Publication Data
Rogers, John (John W. M.)
　　Radio frequency integrated circuit design / John Rogers, Calvin Plett.
　　　p. cm. — (Artech House microwave library)
　　Includes bibliographical references and index.
　　ISBN 1-58053-502-x (alk. paper)
　　1. Radio frequency integrated circuits—Design and construction.　2. Very high speed integrated circuits.　I. Plett, Calvin.　II. Title.　III. Series.
TK7874.78.R64　2003
621.3845—dc21
　　　　　　　　　　　　　　　　　　　　　　　　　　　　　　　　　　　2003041891

British Library Cataloguing in Publication Data
Rogers, John
　　Radio frequency integrated circuit design. — (Artech House microwave library)
　　1. Radio circuits—Design and construction　2. Linear integrated circuits—Design and construction　3. Microwave integrated circuits—Design and construction
　　4. Bipolar integrated circuits—Design and construction　I. Title　II. Plett, Calvin
　　621.3'812

　　ISBN 1-58053-502-x

Cover design by Igor Valdman

© 2003 ARTECH HOUSE, INC.
685 Canton Street
Norwood, MA 02062

All rights reserved. Printed and bound in the United States of America. No part of this book may be reproduced or utilized in any form or by any means, electronic or mechanical, including photocopying, recording, or by any information storage and retrieval system, without permission in writing from the publisher.
　　All terms mentioned in this book that are known to be trademarks or service marks have been appropriately capitalized. Artech House cannot attest to the accuracy of this information. Use of a term in this book should not be regarded as affecting the validity of any trademark or service mark.

International Standard Book Number: 1-58053-502-x
Library of Congress Catalog Card Number: 2003041891

10 9 8 7 6 5 4 3 2 1

Contents

	Foreword	**xv**
	Acknowledgments	**xix**
1	**Introduction to Communications Circuits**	**1**
1.1	Introduction	1
1.2	Lower Frequency Analog Design and Microwave Design Versus Radio Frequency Integrated Circuit Design	2
1.2.1	Impedance Levels for Microwave and Low-Frequency Analog Design	2
1.2.2	Units for Microwave and Low-Frequency Analog Design	3
1.3	Radio Frequency Integrated Circuits Used in a Communications Transceiver	4
1.4	Overview	6
	References	6
2	**Issues in RFIC Design, Noise, Linearity, and Filtering**	**9**
2.1	Introduction	9

v

2.2	Noise	9
2.2.1	Thermal Noise	10
2.2.2	Available Noise Power	11
2.2.3	Available Power from Antenna	11
2.2.4	The Concept of Noise Figure	13
2.2.5	The Noise Figure of an Amplifier Circuit	14
2.2.6	The Noise Figure of Components in Series	16
2.3	Linearity and Distortion in RF Circuits	23
2.3.1	Power Series Expansion	23
2.3.2	Third-Order Intercept Point	27
2.3.3	Second-Order Intercept Point	29
2.3.4	The 1-dB Compression Point	30
2.3.5	Relationships Between 1-dB Compression and IP3 Points	31
2.3.6	Broadband Measures of Linearity	32
2.4	Dynamic Range	35
2.5	Filtering Issues	37
2.5.1	Image Signals and Image Reject Filtering	37
2.5.2	Blockers and Blocker Filtering	39
	References	41
	Selected Bibliography	42

3	**A Brief Review of Technology**	**43**
3.1	Introduction	43
3.2	Bipolar Transistor Description	43
3.3	β Current Dependence	46
3.4	Small-Signal Model	47
3.5	Small-Signal Parameters	48
3.6	High-Frequency Effects	49
3.6.1	f_T as a Function of Current	51
3.7	Noise in Bipolar Transistors	53
3.7.1	Thermal Noise in Transistor Components	53
3.7.2	Shot Noise	53
3.7.3	$1/f$ Noise	54

3.8	Base Shot Noise Discussion	55
3.9	Noise Sources in the Transistor Model	55
3.10	Bipolar Transistor Design Considerations	56
3.11	CMOS Transistors	57
3.11.1	NMOS	58
3.11.2	PMOS	58
3.11.3	CMOS Small-Signal Model Including Noise	58
3.11.4	CMOS Square Law Equations	60
	References	61

4 Impedance Matching — 63

4.1	Introduction	63
4.2	Review of the Smith Chart	66
4.3	Impedance Matching	69
4.4	Conversions Between Series and Parallel Resistor-Inductor and Resistor-Capacitor Circuits	74
4.5	Tapped Capacitors and Inductors	76
4.6	The Concept of Mutual Inductance	78
4.7	Matching Using Transformers	81
4.8	Tuning a Transformer	82
4.9	The Bandwidth of an Impedance Transformation Network	83
4.10	Quality Factor of an LC Resonator	85
4.11	Transmission Lines	88
4.12	S, Y, and Z Parameters	89
	References	93

5 The Use and Design of Passive Circuit Elements in IC Technologies — 95

5.1	Introduction	95
5.2	The Technology Back End and Metallization in IC Technologies	95

5.3	Sheet Resistance and the Skin Effect	97
5.4	Parasitic Capacitance	100
5.5	Parasitic Inductance	101
5.6	Current Handling in Metal Lines	102
5.7	Poly Resistors and Diffusion Resistors	103
5.8	Metal-Insulator-Metal Capacitors and Poly Capacitors	103
5.9	Applications of On-Chip Spiral Inductors and Transformers	104
5.10	Design of Inductors and Transformers	106
5.11	Some Basic Lumped Models for Inductors	108
5.12	Calculating the Inductance of Spirals	110
5.13	Self-Resonance of Inductors	110
5.14	The Quality Factor of an Inductor	111
5.15	Characterization of an Inductor	115
5.16	Some Notes About the Proper Use of Inductors	117
5.17	Layout of Spiral Inductors	119
5.18	Isolating the Inductor	121
5.19	The Use of Slotted Ground Shields and Inductors	122
5.20	Basic Transformer Layouts in IC Technologies	122
5.21	Multilevel Inductors	124
5.22	Characterizing Transformers for Use in ICs	127
5.23	On-Chip Transmission Lines	129
5.23.1	Effect of Transmission Line	130
5.23.2	Transmission Line Examples	131
5.24	High-Frequency Measurement of On-Chip Passives and Some Common De-Embedding Techniques	134

5.25	Packaging	135
5.25.1	Other Packaging Techniques	138
	References	139
6	**LNA Design**	**141**
6.1	Introduction and Basic Amplifiers	141
6.1.1	Common-Emitter Amplifier (Driver)	141
6.1.2	Simplified Expressions for Widely Separated Poles	146
6.1.3	The Common-Base Amplifier (Cascode)	146
6.1.4	The Common-Collector Amplifier (Emitter Follower)	148
6.2	Amplifiers with Feedback	152
6.2.1	Common-Emitter with Series Feedback (Emitter Degeneration)	152
6.2.2	The Common-Emitter with Shunt Feedback	154
6.3	Noise in Amplifiers	158
6.3.1	Input-Referred Noise Model of the Bipolar Transistor	159
6.3.2	Noise Figure of the Common-Emitter Amplifier	161
6.3.3	Input Matching of LNAs for Low Noise	163
6.3.4	Relationship Between Noise Figure and Bias Current	169
6.3.5	Effect of the Cascode on Noise Figure	170
6.3.6	Noise in the Common-Collector Amplifier	171
6.4	Linearity in Amplifiers	172
6.4.1	Exponential Nonlinearity in the Bipolar Transistor	172
6.4.2	Nonlinearity in the Output Impedance of the Bipolar Transistor	180
6.4.3	High-Frequency Nonlinearity in the Bipolar Transistor	182
6.4.4	Linearity in Common-Collector Configuration	182
6.5	Differential Pair (Emitter-Coupled Pair) and Other Differential Amplifiers	183
6.6	Low-Voltage Topologies for LNAs and the Use of On-Chip Transformers	184

6.7	DC Bias Networks	187
6.7.1	Temperature Effects	189
6.8	Broadband LNA Design Example	189
	References	194
	Selected Bibliography	195

7 Mixers — 197

7.1	Introduction	197
7.2	Mixing with Nonlinearity	197
7.3	Basic Mixer Operation	198
7.4	Controlled Transconductance Mixer	198
7.5	Double-Balanced Mixer	200
7.6	Mixer with Switching of Upper Quad	202
7.6.1	Why LO Switching?	203
7.6.2	Picking the LO Level	204
7.6.3	Analysis of Switching Modulator	205
7.7	Mixer Noise	206
7.8	Linearity	215
7.8.1	Desired Nonlinearity	215
7.8.2	Undesired Nonlinearity	215
7.9	Improving Isolation	217
7.10	Image Reject and Single-Sideband Mixer	217
7.10.1	Alternative Single-Sideband Mixers	219
7.10.2	Generating 90° Phase Shift	220
7.10.3	Image Rejection with Amplitude and Phase Mismatch	224
7.11	Alternative Mixer Designs	227
7.11.1	The Moore Mixer	228
7.11.2	Mixers with Transformer Input	228
7.11.3	Mixer with Simultaneous Noise and Power Match	229
7.11.4	Mixers with Coupling Capacitors	230

7.12	General Design Comments	231
7.12.1	Sizing Transistors	232
7.12.2	Increasing Gain	232
7.12.3	Increasing IP3	232
7.12.4	Improving Noise Figure	233
7.12.5	Effect of Bond Pads and the Package	233
7.12.6	Matching, Bias Resistors, and Gain	234
7.13	CMOS Mixers	242
	References	244
	Selected Bibliography	244

8 Voltage-Controlled Oscillators 245

8.1	Introduction	245
8.2	Specification of Oscillator Properties	245
8.3	The LC Resonator	247
8.4	Adding Negative Resistance Through Feedback to the Resonator	248
8.5	Popular Implementations of Feedback to the Resonator	250
8.6	Configuration of the Amplifier (Colpitts or $-G_m$)	251
8.7	Analysis of an Oscillator as a Feedback System	252
8.7.1	Oscillator Closed-Loop Analysis	252
8.7.2	Capacitor Ratios with Colpitts Oscillators	255
8.7.3	Oscillator Open-Loop Analysis	258
8.7.4	Simplified Loop Gain Estimates	260
8.8	Negative Resistance Generated by the Amplifier	262
8.8.1	Negative Resistance of Colpitts Oscillator	262
8.8.2	Negative Resistance for Series and Parallel Circuits	263
8.8.3	Negative Resistance Analysis of $-G_m$ Oscillator	265
8.9	Comments on Oscillator Analysis	268
8.10	Basic Differential Oscillator Topologies	270

8.11	A Modified Common-Collector Colpitts Oscillator with Buffering	270
8.12	Several Refinements to the $-G_m$ Topology	270
8.13	The Effect of Parasitics on the Frequency of Oscillation	274
8.14	Large-Signal Nonlinearity in the Transistor	275
8.15	Bias Shifting During Startup	277
8.16	Oscillator Amplitude	277
8.17	Phase Noise	283
8.17.1	Linear or Additive Phase Noise and Leeson's Formula	283
8.17.2	Some Additional Notes About Low-Frequency Noise	291
8.17.3	Nonlinear Noise	292
8.18	Making the Oscillator Tunable	295
8.19	VCO Automatic-Amplitude Control Circuits	302
8.20	Other Oscillators	313
	References	316
	Selected Bibliography	317

9	**High-Frequency Filter Circuits**	**319**
9.1	Introduction	319
9.2	Second-Order Filters	320
9.3	Integrated RF Filters	321
9.3.1	A Simple Bandpass LC Filter	321
9.3.2	A Simple Bandstop Filter	322
9.3.3	An Alternative Bandstop Filter	323
9.4	Achieving Filters with Higher Q	327
9.4.1	Differential Bandpass LNA with Q-Tuned Load Resonator	327
9.4.2	A Bandstop Filter with Colpitts-Style Negative Resistance	329
9.4.3	Bandstop Filter with Transformer-Coupled $-G_m$ Negative Resistance	331

9.5	Some Simple Image Rejection Formulas	333
9.6	Linearity of the Negative Resistance Circuits	336
9.7	Noise Added Due to the Filter Circuitry	337
9.8	Automatic Q Tuning	339
9.9	Frequency Tuning	342
9.10	Higher-Order Filters	343
	References	346
	Selected Bibliography	347
10	**Power Amplifiers**	**349**
10.1	Introduction	349
10.2	Power Capability	350
10.3	Efficiency Calculations	350
10.4	Matching Considerations	351
10.4.1	Matching to S_{22}^* Versus Matching to Γ_{opt}	352
10.5	Class A, B, and C Amplifiers	353
10.5.1	Class A, B, and C Analysis	356
10.5.2	Class B Push-Pull Arrangements	362
10.5.3	Models for Transconductance	363
10.6	Class D Amplifiers	367
10.7	Class E Amplifiers	368
10.7.1	Analysis of Class E Amplifier	370
10.7.2	Class E Equations	371
10.7.3	Class E Equations for Finite Output Q	372
10.7.4	Saturation Voltage and Resistance	373
10.7.5	Transition Time	373
10.8	Class F Amplifiers	375
10.8.1	Variation on Class F: Second-Harmonic Peaking	379
10.8.2	Variation on Class F: Quarter-Wave Transmission Line	379
10.9	Class G and H Amplifiers	381
10.10	Class S Amplifiers	383

10.11	Summary of Amplifier Classes for RF Integrated Circuits	384
10.12	AC Load Line	385
10.13	Matching to Achieve Desired Power	385
10.14	Transistor Saturation	388
10.15	Current Limits	388
10.16	Current Limits in Integrated Inductors	390
10.17	Power Combining	390
10.18	Thermal Runaway—Ballasting	392
10.19	Breakdown Voltage	393
10.20	Packaging	394
10.21	Effects and Implications of Nonlinearity	394
10.21.1	Cross Modulation	395
10.21.2	AM-to-PM Conversion	395
10.21.3	Spectral Regrowth	395
10.21.4	Linearization Techniques	396
10.21.5	Feedforward	396
10.21.6	Feedback	397
10.22	CMOS Power Amplifier Example	398
	References	399

About the Authors   **401**

Index   **403**

Foreword

I enjoyed reading this book for a number of reasons. One reason is that it addresses high-speed analog design in the context of microwave issues. This is an advanced-level book, which should follow courses in basic circuits and transmission lines. Most analog integrated circuit designers in the past worked on applications at low enough frequency that microwave issues did not arise. As a consequence, they were adept at lumped parameter circuits and often not comfortable with circuits where waves travel in space. However, in order to design radio frequency (RF) communications integrated circuits (IC) in the gigahertz range, one must deal with transmission lines at chip interfaces and where interconnections on chip are far apart. Also, impedance matching is addressed, which is a topic that arises most often in microwave circuits. In my career, there has been a gap in comprehension between analog low-frequency designers and microwave designers. Often, similar issues were dealt with in two different languages. Although this book is more firmly based in lumped-element analog circuit design, it is nice to see that microwave knowledge is brought in where necessary.

Too many analog circuit books in the past have concentrated first on the circuit side rather than on basic theory behind their application in communications. The circuits usually used have evolved through experience, without a satisfying intellectual theme in describing them. Why a given circuit works best can be subtle, and often these circuits are chosen only through experience. For this reason, I am happy that the book begins first with topics that require an intellectual approach—noise, linearity and filtering, and technology issues. I am particularly happy with how linearity is introduced (power series). In the rest of the book it is then shown, with specific circuits and numerical examples, how linearity and noise issues arise.

In the latter part of the book, the RF circuits analyzed are ones that experience has shown to be good ones. Concentration is on bipolar circuits, not metal oxide semiconductors (MOS). Bipolar still has many advantages at high frequency. The depth with which design issues are addressed would not be possible if similar MOS coverage was attempted. However, there might be room for a similar book, which concentrates on MOS.

In this book there is a lot of detailed academic exploration of some important high-frequency RF bipolar ICs. One might ask if this is important in design for application, and the answer is yes. To understand why, one must appreciate the central role of analog circuit simulators in the design of such circuits. At the beginning of my career (around 1955–1960) discrete circuits were large enough that good circuit topologies could be picked out by breadboarding with the actual parts themselves. This worked fairly well with some analog circuits at audio frequencies, but failed completely in the progression to integrated circuits.

In high-speed IC design nowadays, the computer-based circuit simulator is crucial. Such simulation is important at four levels. The first level is the use of simplified models of the circuit elements (idealized transistors, capacitors, and inductors). The use of such models allows one to pick out good topologies and eliminate bad ones. This is not done well with just paper analysis because it will miss key factors, such as the complexities of the transistor, particularly nonlinearity and bias and signal interaction effects. Exploration of topologies with the aid of a circuit simulator is necessary. The simulator is useful for quick iteration of proposed circuits, with simplified models to show any fundamental problems with a proposed circuit. This brings out the influence of model parameters on circuit performance. This first level of simulation may be avoided if the best topology, known through experience, is picked at the start.

The second level of simulation is where the models are representative of the type of fabrication technology being used. However, we do not yet use specific numbers from the specific fabrication process and make an educated approximation to likely parasitic capacitances. Simulation at this level can be used to home in on good values for circuit parameters for a given topology before the final fabrication process is available. Before the simulation begins, detailed preliminary analysis at the level of this book is possible, and many parameters can be wisely chosen before simulation begins, greatly shortening the design process and the required number of iterations. Thus, the analysis should focus on topics that arise, given a typical fabrication process. I believe this has been done well here, and the authors, through scholarly work and real design experience, have chosen key circuits and topics.

The third level of design is where a link with a proprietary industrial process has been made, and good simulator models are supplied for the process. The circuit is laid out in the proprietary process and simulation is done, including estimates of parasitic capacitances from interconnections and detailed models of the elements used.

The incorporation of the proprietary models in the simulation of the circuit is necessary because when the IC is laid out in the actual process, fabrication of the result must be successful to the highest possible degree. This is because fabrication and testing is extremely expensive, and any failure can result in the necessity to change the design, requiring further fabrication and retesting, causing delay in getting the product to market.

The fourth design level is the comparison of the circuit behavior predicted from simulation with that of measurements of the actual circuit. Discrepancies must be explained. These may be from design errors or from inadequacies in the models, which are uncovered by the experimental result. These model inadequacies, when corrected, may result in further simulation, which causes the circuit design and layout to be refined with further fabrication.

This discussion has served to bring attention to the central role that computer simulation has in the design of integrated RF circuits, and the accompanying importance of circuit analysis such as presented in this book. Such detailed analysis may save money by facilitating the early success of applications. This book can be beneficial to designers, or by those less focused on specific design, for recognizing key constraints in the area, with faith justified, I believe, that the book is a correct picture of the reality of high-speed RF communications circuit design.

Miles A. Copeland
Fellow IEEE
Professor Emeritus
Carleton University Department of Electronics
Ottawa, Ontario, Canada
April 2003

Acknowledgments

This book has evolved out of a number of documents including technical papers, course notes, and various theses. We decided that we would organize some of the research we and many others had been doing and turn it into a manuscript that would serve as a comprehensive text for engineers interested in learning about radio frequency integrated circuits (RFIC). We have focused mainly on bipolar technology in the text, but since many techniques in RFICs are independent of technology, we hope that designers working with other technologies will also find much of the text useful. We have tried very hard to identify and exterminate bugs and errors from the text. Undoubtedly there are still many remaining, so we ask you, the reader, for your understanding. Please feel free to contact us with your comments. We hope that these pages add to your understanding of the subject.

Nobody undertakes a project like this without support on a number of levels, and there are many people that we need to thank. Professors Miles Copeland and Garry Tarr provided technical guidance and editing. We would like to thank David Moore for his input and consultation on many aspects of RFIC design. David, we have tried to add some of your wisdom to these pages. Thanks also go to Dave Rahn and Steve Kovacic, who have both contributed to our research efforts in a variety of ways. We would like to thank Sandi Plett who tirelessly edited chapters, provided formatting, and helped beat the word processor into submission. She did more than anybody except the authors to make this project happen. We would also like to thank a number of graduate students, alumni, and colleagues who have helped us with our understanding of RFICs over the years. This list includes but is not limited to Neric Fong,

Bill Toole, José Macedo, Sundus Kubba, Leonard Dauphinee, Rony Amaya, John J. Nisbet, Sorin Voinegescu, John Long, Tom Smy, Walt Bax, Brian Robar, Richard Griffith, Hugues Lafontaine, Ash Swaminathan, Jugnu Ojha, George Khoury, Mark Cloutier, John Peirce, Bill Bereza, and Martin Snelgrove.

1

Introduction to Communications Circuits

1.1 Introduction

Radio frequency integrated circuit (RFIC) design is an exciting area for research or product development. Technologies are constantly being improved, and as they are, circuits formerly implemented as discrete solutions can now be integrated onto a single chip. In addition to widely used applications such as cordless phones and cell phones, new applications continue to emerge. Examples of new products requiring RFICs are *wireless local-area networks* (WLAN), keyless entry for cars, wireless toll collection, *Global Positioning System* (GPS) navigation, remote tags, asset tracking, remote sensing, and tuners in cable modems. Thus, the market is expanding, and with each new application there are unique challenges for the designers to overcome. As a result, the field of RFIC design should have an abundance of products to keep designers entertained for years to come.

This huge increase in interest in *radio frequency* (RF) communications has resulted in an effort to provide components and complete systems on an *integrated circuit* (IC). In academia, there has been much research aimed at putting a complete radio on one chip. Since *complementary metal oxide semiconductor* (CMOS) is required for the *digital signal processing* (DSP) in the back end, much of this effort has been devoted to designing radios using CMOS technologies [1–3]. However, bipolar design continues to be the industry standard because it is a more developed technology and, in many cases, is better modeled. Major research is being done in this area as well. CMOS traditionally had the advantage of lower production cost, but as technology dimensions become smaller, this is becoming less true. Which will win? Who is to say? Ultimately, both will probably be replaced by radically different technologies. In any case, as long as people want to communicate, engineers will still be building radios. In this book we will focus on bipolar RF circuits, although CMOS circuits will also be discussed. Contrary to popular belief, most of the design concepts in RFIC design are applicable regardless of what technology is used to implement them.

The objective of a radio is to transmit or receive a signal between source and destination with acceptable quality and without incurring a high cost. From the user's point of view, quality can be perceived as information being passed from source to destination without the addition of noticeable noise or distortion. From a more technical point of view, quality is often measured in terms of bit error rate, and acceptable quality might be to experience less than one error in every million bits. Cost can be seen as the price of the communications equipment or the need to replace or recharge batteries. Low cost implies simple circuits to minimize circuit area, but also low power dissipation to maximize battery life.

1.2 Lower Frequency Analog Design and Microwave Design Versus Radio Frequency Integrated Circuit Design

RFIC design has borrowed from both analog design techniques, used at lower frequencies [4, 5], and high-frequency design techniques, making use of microwave theory [6, 7]. The most fundamental difference between low-frequency analog and microwave design is that in microwave design, transmission line concepts are important, while in low-frequency analog design, they are not. This will have implications for the choice of impedance levels, as well as how signal size, noise, and distortion are described.

On-chip dimensions are small, so even at RF frequencies (0.1–5 GHz), transistors and other devices may not need to be connected by transmission lines (i.e., the lengths of the interconnects may not be a significant fraction of a wavelength). However, at the chip boundaries, or when traversing a significant fraction of a wavelength on chip, transmission line theory becomes very important. Thus, on chip we can usually make use of analog design concepts, although, in practice, microwave design concepts are often used. At the chip interfaces with the outside world, we must treat it like a microwave circuit.

1.2.1 Impedance Levels for Microwave and Low-Frequency Analog Design

In low-frequency analog design, input impedance is usually very high (ideally infinity), while output impedance is low (ideally zero). For example, an operational amplifier can be used as a buffer because its high input impedance does not affect the circuit to which it is connected, and its low output impedance can drive a measurement device efficiently. The freedom to choose arbitrary impedance levels provides advantages in that circuits can drive or be driven by an impedance that best suits them. On the other hand, if circuits are connected using transmission lines, then these circuits are usually designed to have an input and output impedance that match the characteristic impedance of the transmission line.

1.2.2 Units for Microwave and Low-Frequency Analog Design

Signal, noise, and distortion levels are also described differently in low frequency analog versus microwave design. In microwave circuits, power is usually used to describe signals, noise, or distortion with the typical unit of measure being decibels above 1 milliwatt (dBm). However, in analog circuits, since infinite or zero impedance is allowed, power levels are meaningless, so voltages and current are usually chosen to describe the signal levels. Voltage and current are expressed as peak, peak-to-peak, or *root-mean-square* (rms). Power in dBm, P_{dBm}, can be related to the power in watts, P_{watt}, as shown in (1.1) and Table 1.1, where voltages are assumed to be across 50Ω.

$$P_{dBm} = 10 \log_{10}\left(\frac{P_{watt}}{1 \text{ mW}}\right) \quad (1.1)$$

Assuming a sinusoidal voltage waveform, P_{watt} is given by

$$P_{watt} = \frac{v_{rms}^2}{R} \quad (1.2)$$

where R is the resistance the voltage is developed across. Note also that v_{rms} can be related to the peak voltage v_{pp} by

Table 1.1
Power Relationships

v_{pp}	v_{rms}	P_{watt} (50Ω)	P_{dBm} (50Ω)
1 nV	0.3536 nV	2.5×10^{-21}	−176
1 μV	0.3536 μV	2.5×10^{-15}	−116
1 mV	353.6 μV	2.5 nW	−56
10 mV	3.536 mV	250 nW	−36
100 mV	35.36 mV	25 μW	−16
632.4 mV	223.6 mV	1 mW	0
1V	353.6 mV	2.5 mW	+4
10V	3.536V	250 mW	+24

$$v_{\text{rms}} = \frac{v_{\text{pp}}}{2\sqrt{2}} \tag{1.3}$$

Similarly, noise in analog signals is often defined in terms of volts or amperes, while in microwave it will be in terms of dBm. Noise is usually represented as noise density per hertz of bandwidth. In analog circuits, noise is specified as squared volts per hertz, or volts per square root of hertz. In microwave circuits, the usual measure of noise is dBm/Hz or noise figure, which is defined as the reduction in signal-to-noise ratio caused by the addition of the noise.

In both analog and microwave circuits, an effect of nonlinearity is the appearance of harmonic distortion or intermodulation distortion, often at new frequencies. In low-frequency analog circuits, this is often described by the ratio of the distortion components compared to the fundamental components. In microwave circuits, the tendency is to describe distortion by gain compression (power level where the gain is reduced due to nonlinearity) or third-order intercept point (IP3).

Noise and linearity are discussed in detail in Chapter 2. A summary of low-frequency analog and microwave design is shown in Table 1.2.

1.3 Radio Frequency Integrated Circuits Used in a Communications Transceiver

A typical block diagram of most of the major circuit blocks that make up a typical superheterodyne communications transceiver is shown in Figure 1.1. Many aspects of this transceiver are common to all transceivers.

Table 1.2
Comparison of Analog and Microwave Design

Parameter	Analog Design (most often used on chip)	Microwave Design (most often used at chip boundaries and pins)
Impedance	$Z_{in} \Rightarrow \infty$ $Z_{out} \Rightarrow 0$	$Z_{in} \Rightarrow 50\Omega$ $Z_{out} \Rightarrow 50\Omega$
Signals	Voltage, current, often peak or peak-to-peak	Power, often dBm
Noise	nV/$\sqrt{\text{Hz}}$	Noise factor F, noise figure NF
Nonlinearity	Harmonic distortion, intermodulation, clipping	Third-order intercept point IP3 1-dB compression

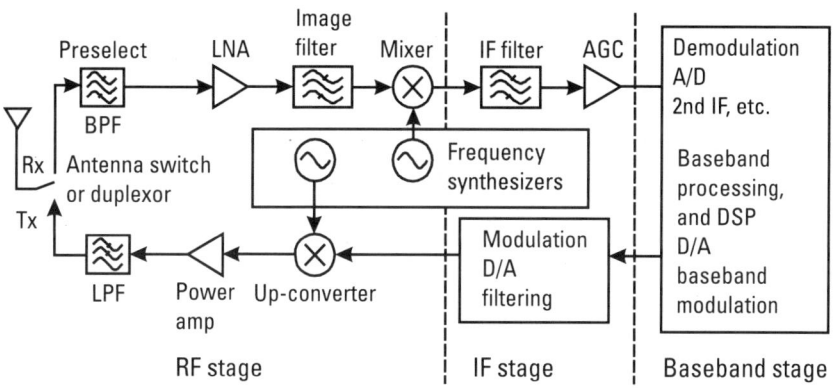

Figure 1.1 Typical transceiver block diagram.

This transceiver has a transmit side (Tx) and a receive side (Rx), which are connected to the antenna through a duplexer that can be realized as a switch or a filter, depending on the communications standard being followed. The input preselection filter takes the broad spectrum of signals coming from the antenna and removes the signals not in the band of interest. This may be required to prevent overloading of the *low-noise amplifier* (LNA) by out-of-band signals. The LNA amplifies the input signal without adding much noise. The input signal can be very weak, so the first thing to do is strengthen the signal without corrupting it. As a result, noise added in later stages will be of less importance. The image filter that follows the LNA removes out-of-band signals and noise (which will be discussed in detail in Chapter 2) before the signal enters the mixer. The mixer translates the input RF signal down to the intermediate frequency, since filtering, as well as circuit design, becomes much easier at lower frequencies for a multitude of reasons. The other input to the mixer is the *local oscillator* (LO) signal provided by a voltage-controlled oscillator inside a frequency synthesizer. The desired output of the mixer will be the difference between the LO frequency and the RF frequency.

At the input of the radio there may be many different channels or frequency bands. The LO frequency is adjusted so that the desired RF channel or frequency band is mixed down to the same *intermediate frequency* (IF) in all cases. The IF stage then provides channel filtering at this one frequency to remove the unwanted channels. The IF stage provides further amplification and *automatic gain control* (AGC) to bring the signal to a specific amplitude level before the signal is passed on to the back end of the receiver. It will ultimately be converted into bits (most modern communications systems use digital modulation schemes) that could represent, for example, voice, video, or data through the use of an analog-to-digital converter.

On the transmit side, the back-end digital signal is used to modulate the carrier in the IF stage. In the IF stage, there may be some filtering to remove unwanted signals generated by the baseband, and the signal may or may not be converted into an analog waveform before it is modulated onto the IF carrier. A mixer converts the modulated signal and IF carrier up to the desired RF frequency. A frequency synthesizer provides the other mixer input. Since the RF carrier and associated modulated data may have to be transmitted over large distances through lossy media (e.g., air, cable, and fiber), a *power amplifier* (PA) must be used to increase the signal power. Typically, the power level is increased from the milliwatt range to a level in the range of hundreds of milliwatts to watts, depending on the particular application. A lowpass filter after the PA removes any harmonics produced by the PA to prevent them from also being transmitted.

1.4 Overview

We will spend the rest of this book trying to convey the various design constraints of all the RF building blocks mentioned in the previous sections. Components are designed with the main concerns being frequency response, gain, stability, noise, distortion (nonlinearity), impedance matching, and power dissipation. Dealing with design constraints is what keeps the RFIC designer employed.

The focus of this book will be how to design and build the major circuit blocks that make up the RF portion of a radio using an IC technology. To that end, block level performance specifications are described in Chapter 2. A brief overview of IC technologies and transistor performance is given in Chapter 3. Various methods of matching impedances, which are very important at chip boundaries and for some interconnections of circuits on-chip, will be discussed in Chapter 4. The realization and limitations of passive circuit components in an IC technology will be discussed in Chapter 5. Chapters 6 through 10 will be devoted to individual circuit blocks such as LNAs, mixers, *voltage-controlled oscillators* (VCOs), filters, and power amplifiers. However, the design of complete synthesizers is beyond the scope of this book. The interested reader is referred to [8–10].

References

[1] Lee, T. H., *The Design of CMOS Radio Frequency Integrated Circuits*, Cambridge, England: Cambridge University Press, 1998.

[2] Razavi, B., *RF Microelectronics*, Upper Saddle River, NJ: Prentice Hall, 1998.

[3] Crols, J., and M. Steyaert, *CMOS Wireless Transceiver Design,* Dordrecht, the Netherlands: Kluwer Academic Publishers, 1997.

[4] Gray, P. R., et al., *Analysis and Design of Analog Integrated Circuits,* 4th ed., New York: John Wiley & Sons, 2001.

[5] Johns, D. A., and K. Martin, *Analog Integrated Circuit Design,* New York: John Wiley & Sons, 1997.

[6] Gonzalez, G., *Microwave Transistor Amplifiers Analysis and Design,* 2nd ed., Upper Saddle River, NJ: Prentice Hall, 1997.

[7] Pozar, D. M., *Microwave Engineering,* 2nd ed., New York: John Wiley & Sons, 1998.

[8] Crawford, J. A., *Frequency Synthesizer Design Handbook,* Norwood, MA: Artech House, 1994.

[9] Wolaver, D. H., *Phase-Locked Loop Circuit Design,* Englewood Cliffs, NJ: Prentice Hall, 1991.

[10] Razavi, B., (ed.), *Monolithic Phase-Locked Loops and Clock Recovery Circuits: Theory and Design,* New York: IEEE Press, 1996.

2

Issues in RFIC Design, Noise, Linearity, and Filtering

2.1 Introduction

In this chapter we will have a brief look at some general issues in RF circuit design. Nonidealities we will consider include noise and nonlinearity. We will also consider the effect of filtering. An ideal circuit, such as an amplifier, produces a perfect copy of the input signal at the output. In a real circuit, the amplifier will introduce both noise and distortion to that waveform. Noise, which is present in all resistors and active devices, limits the minimum detectable signal in a radio. At the other amplitude extreme, nonlinearities in the circuit blocks will cause the output signal to become distorted, limiting the maximum signal amplitude.

At the system level, specifications for linearity and noise as well as many other parameters must be determined before the circuit can be designed. In this chapter, before we look at circuit details, we will look at some of these system issues in more detail. In order to design radio frequency integrated circuits with realistic specifications, we need to understand the impact of noise on minimum detectable signals and the effect of nonlinearity on distortion. Knowledge of noise floors and distortion will be used to understand the requirements for circuit parameters.

2.2 Noise

Signal detection is more difficult in the presence of noise. In addition to the desired signal, the receiver is also picking up noise from the rest of the universe.

Any matter above 0K contains thermal energy. This thermal energy moves atoms and electrons around in a random way, leading to random currents in circuits, which are also noise. Noise can also come from man-made sources such as microwave ovens, cell phones, pagers, and radio antennas. Circuit designers are mostly concerned with how much noise is being added by the circuits in the transceiver. At the input to the receiver, there will be some noise power present that defines the noise floor. The minimum detectable signal must be higher than the noise floor by some *signal-to-noise ratio* (SNR) to detect signals reliably and to compensate for additional noise added by circuitry. These concepts will be described in the following sections.

We note that to find the total noise due to a number of sources, the relationship of the sources with each other has to be considered. The most common assumption is that all noise sources are random and have no relationship with each other, so they are said to be uncorrelated. In such a case, noise power is added instead of noise voltage. Similarly, if noise at different frequencies is uncorrelated, noise power is added. We note that signals, like noise, can also be uncorrelated, such as signals at different unrelated frequencies. In such a case, one finds the total output signal by adding the powers. On the other hand, if two sources are correlated, the voltages can be added. As an example, correlated noise is seen at the outputs of two separate paths that have the same origin.

2.2.1 Thermal Noise

One of the most common noise sources in a circuit is a resistor. Noise in resistors is generated by thermal energy causing random electron motion [1–3]. The thermal noise spectral density in a resistor is given by

$$N_{\text{resistor}} = 4kTR \tag{2.1}$$

where T is the Kelvin temperature of the resistor, k is Boltzmann's constant (1.38×10^{-23} J/K), and R is the value of the resistor. Noise power spectral density is expressed using volts squared per hertz (power spectral density). In order to find out how much power a resistor produces in a finite bandwidth, simply multiply (2.1) by the bandwidth of interest Δf:

$$v_n^2 = 4kTR\Delta f \tag{2.2}$$

where v_n is the rms value of the noise voltage in the bandwidth Δf. This can also be written equivalently as a noise current rather than a noise voltage:

$$i_n^2 = \frac{4kT\Delta f}{R} \tag{2.3}$$

Thermal noise is white noise, meaning it has a constant power spectral density with respect to frequency (valid up to approximately 6,000 GHz) [4]. The model for noise in a resistor is shown in Figure 2.1.

2.2.2 Available Noise Power

Maximum power is transferred to the load when R_{LOAD} is equal to R. Then v_o is equal to $v_n/2$. The output power spectral density P_o is then given by

$$P_o = \frac{v_o^2}{R} = \frac{v_n^2}{4R} = kT \quad (2.4)$$

Thus, available power is kT, independent of resistor size. Note that kT is in watts per hertz, which is a power density. To get total power out P_{out} in watts, multiply by the bandwidth, with the result that

$$P_{out} = kTB \quad (2.5)$$

2.2.3 Available Power from Antenna

The noise from an antenna can be modeled as a resistor [5]. Thus, as in the previous section, the available power from an antenna is given by

$$P_{available} = kT = 4 \times 10^{-21} \text{ W/Hz} \quad (2.6)$$

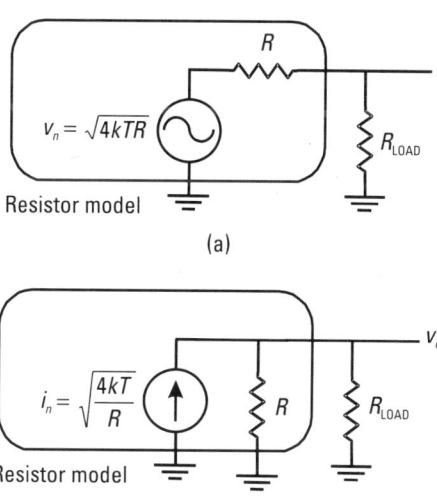

Figure 2.1 Resistor noise model: (a) with a voltage source, and (b) with a current source.

at $T = 290K$, or in dBm per hertz,

$$P_{available} = 10 \log_{10}\left(\frac{4 \times 10^{-21}}{1 \times 10^{-3}}\right) = -174 \text{ dBm/Hz} \quad (2.7)$$

Note that using 290K as the temperature of the resistor modeling the antenna is appropriate for cell phone applications where the antenna is pointed at the horizon. However, if the antenna were pointed at the sky, the equivalent noise temperature would be much lower, more typically 50K [6].

For any receiver required to receive a given signal bandwidth, the minimum detectable signal can now be determined. As can be seen from (2.5), the noise floor depends on the bandwidth. For example, with a bandwidth of 200 kHz, the noise floor is

$$\text{Noise floor} = kTB = 4 \times 10^{-21} \times 200,000 = 8 \times 10^{-16} \quad (2.8)$$

More commonly, the noise floor would be expressed in dBm, as in the following for the example shown above:

$$\text{Noise floor} = -174 \text{ dBm/Hz} + 10 \log_{10}(200,000) = -121 \text{ dBm} \quad (2.9)$$

Thus, we can now also formally define signal-to-noise ratio. If the signal has a power of S, then the SNR is

$$\text{SNR} = \frac{S}{\text{Noise floor}} \quad (2.10)$$

Thus, if the electronics added no noise and if the detector required a signal-to-noise ratio of 0 dB, then a signal at −121 dBm could just be detected. The minimum detectable signal in a receiver is also referred to as the receiver sensitivity. However, the SNR required to detect bits reliably (e.g., bit error rate (BER) = 10^{-3}) is typically not 0 dB. The actual required SNR depends on a variety of factors, such as bit rate, energy per bit, IF filter bandwidth, detection method (e.g., synchronous or not), and interference levels. Such calculations are the topics for a digital communications course [6, 7] and will not be discussed further here. But typical results for a bit error rate of 10^{-3} is about 7 dB for *quadrature phase shift keying* (QPSK), about 12 dB for 16 *quadrature amplitude modulation* (QAM), and about 17 dB for 64 QAM, though often higher numbers are quoted to leave a safety margin. It should be noted that for data transmission, lower BER is often required (e.g., 10^{-6}), resulting in an SNR requirement of 11 dB or more for QPSK. Thus, the input signal level must be above the noise floor level by at least this amount. Consequently, the minimum detectable signal level in a 200-kHz bandwidth is more like −114 dBm (assuming no noise is added by the electronics).

2.2.4 The Concept of Noise Figure

Noise added by electronics will be directly added to the noise from the input. Thus, for reliable detection, the previously calculated minimum detectable signal level must be modified to include the noise from the active circuitry. Noise from the electronics is described by noise factor F, which is a measure of how much the signal-to-noise ratio is degraded through the system. We note that

$$S_o = G \cdot S_i \quad (2.11)$$

where S_i is the input signal power, S_o is the output signal power, and G is the power gain S_o/S_i. We derive the following equation for noise factor:

$$F = \frac{\text{SNR}_i}{\text{SNR}_o} = \frac{S_i/N_{i(\text{source})}}{S_o/N_{o(\text{total})}} = \frac{S_i/N_{i(\text{source})}}{(S_i \cdot G)/N_{o(\text{total})}} = \frac{N_{o(\text{total})}}{G \cdot N_{i(\text{source})}} \quad (2.12)$$

where $N_{o(\text{total})}$ is the total noise at the output. If $N_{o(\text{source})}$ is the noise at the output originating at the source, and $N_{o(\text{added})}$ is the noise at the output added by the electronic circuitry, then we can write:

$$N_{o(\text{total})} = N_{o(\text{source})} + N_{o(\text{added})} \quad (2.13)$$

Noise factor can be written in several useful alternative forms:

$$F = \frac{N_{o(\text{total})}}{G \cdot N_{i(\text{source})}} = \frac{N_{o(\text{total})}}{N_{o(\text{source})}} = \frac{N_{o(\text{source})} + N_{o(\text{added})}}{N_{o(\text{source})}} = 1 + \frac{N_{o(\text{added})}}{N_{o(\text{source})}} \quad (2.14)$$

This shows that the minimum possible noise factor, which occurs if the electronics adds no noise, is equal to 1. Noise figure NF is related to noise factor F by

$$\text{NF} = 10 \log_{10} F \quad (2.15)$$

Thus, while noise factor is at least 1, noise figure is at least 0 dB. In other words, an electronic system that adds no noise has a noise figure of 0 dB.

In the receiver chain, for components with loss (such as switches and filters), the noise figure is equal to the attenuation of the signal. For example, a filter with 3 dB of loss has a noise figure of 3 dB. This is explained by noting that output noise is approximately equal to input noise, but signal is attenuated by 3 dB. Thus, there has been a degradation of SNR by 3 dB.

2.2.5 The Noise Figure of an Amplifier Circuit

We can now make use of the definition of noise figure just developed and apply it to an amplifier circuit [8]. For the purposes of developing (2.14) into a more useful form, it is assumed that all practical amplifiers can be characterized by an input-referred noise model, such as the one shown in Figure 2.2, where the amplifier is characterized with current gain A_i. (It will be shown in later chapters how to take a practical amplifier and make it fit this model.) In this model, all noise sources in the circuit are lumped into a series noise voltage source v_n and a parallel current noise source i_n placed in front of a noiseless transfer function.

If the amplifier has finite input impedance, then the input current will be split by some ratio α between the amplifier and the source admittance Y_s:

$$\text{SNR}_{in} = \frac{\alpha^2 i_{in}^2}{\alpha^2 i_{ns}^2} \tag{2.16}$$

Assuming that the input-referred noise sources are correlated, the output signal-to-noise ratio is

$$\text{SNR}_{out} = \frac{\alpha^2 A_i^2 i_{in}^2}{\alpha^2 A_i^2 \left(i_{ns}^2 + |i_n + v_n Y_s|^2\right)} \tag{2.17}$$

Thus, the noise factor can now be written in terms of the preceding two equations:

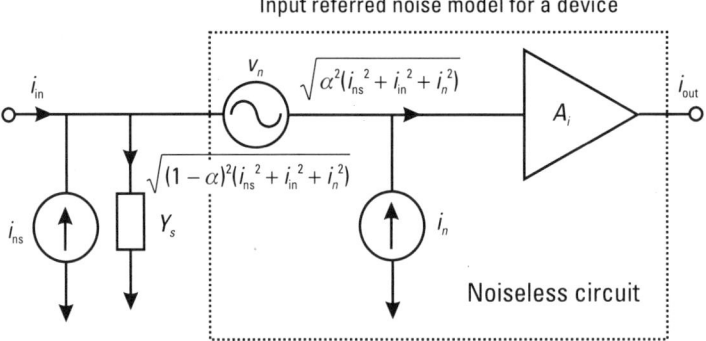

Figure 2.2 Input-referred noise model for a device.

$$F = \frac{i_{ns}^2 + |i_n + v_n Y_s|^2}{i_{ns}^2} = \frac{N_{o(\text{total})}}{N_{o(\text{source})}} \qquad (2.18)$$

This can also be interpreted as the ratio of the total output noise to the total output noise due to the source admittance.

In (2.17), it was assumed that the two input noise sources were correlated with each other. In general, they will not be correlated with each other, but rather the current i_n will be partially correlated with v_n and partially uncorrelated. We can expand both current and voltage into these two explicit parts:

$$i_n = i_c + i_u \qquad (2.19)$$

$$v_n = v_c + v_u \qquad (2.20)$$

In addition, the correlated components will be related by the ratio

$$i_c = Y_c v_c \qquad (2.21)$$

where Y_c is the correlation admittance.

The noise figure can now be written as

$$\text{NF} = 1 + \frac{i_u^2 + |Y_c + Y_s|^2 v_c^2 + v_u^2 |Y_s|^2}{i_{ns}^2} \qquad (2.22)$$

The noise currents and voltages can also be written in terms of equivalent resistance and admittance (these resistors would have the same noise behavior):

$$R_c = \frac{v_c^2}{4kT\Delta f} \qquad (2.23)$$

$$R_u = \frac{v_u^2}{4kT\Delta f} \qquad (2.24)$$

$$G_u = \frac{i_u^2}{4kT\Delta f} \qquad (2.25)$$

$$G_s = \frac{i_{ns}^2}{4kT\Delta f} \qquad (2.26)$$

Thus, the noise figure is now written in terms of these parameters:

$$\text{NF} = 1 + \frac{G_u + |Y_c + Y_s|^2 R_c + |Y_s|^2 R_u}{G_s} \quad (2.27)$$

$$\text{NF} = 1 + \frac{G_u + [(G_c + G_s)^2 + (B_c + B_s)^2] R_c + (G_s^2 + B_s^2) R_u}{G_s} \quad (2.28)$$

It can be seen from this equation that NF is dependent on the equivalent source impedance.

Equation (2.28) can be used not only to determine the noise figure, but also to determine the source loading conditions that will minimize the noise figure. Differentiating with respect to G_s and B_s and setting the derivative to zero yields the following two conditions for minimum noise (G_{opt} and B_{opt}) after several pages of math:

$$G_{\text{opt}} = \sqrt{\frac{G_u + R_u \left(\frac{R_c B_c}{R_c + R_u}\right)^2 + G_c^2 R_c + \left(B_c - \frac{R_c B_c}{R_c + R_u}\right)^2 R_c}{R_c + R_u}} \quad (2.29)$$

$$B_{\text{opt}} = \frac{-R_c B_c}{R_c + R_u} \quad (2.30)$$

2.2.6 The Noise Figure of Components in Series

For components in series, as shown in Figure 2.3, one can calculate the total output noise ($N_{o(\text{total})}$) and output noise due to the source ($N_{o(\text{source})}$) to determine the noise figure.

The output signal S_o is given by

$$S_o = S_i \cdot G_1 \cdot G_2 \cdot G_3 \quad (2.31)$$

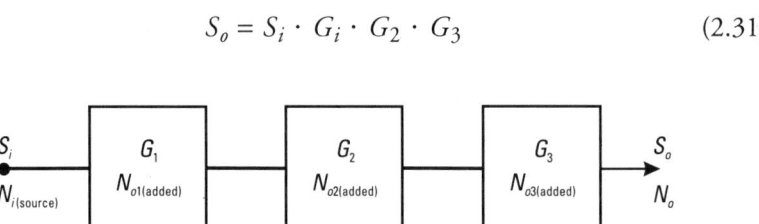

Figure 2.3 Noise figure in cascaded circuits with gain and noise added shown in each.

The input noise is

$$N_{i(\text{source})} = kT \tag{2.32}$$

The total output noise is

$$N_{o(\text{total})} = N_{i(\text{source})} G_1 G_2 G_3 + N_{o1(\text{added})} G_2 G_3 \\ + N_{o2(\text{added})} G_3 + N_{o3(\text{added})} \tag{2.33}$$

The output noise due to the source is

$$N_{o(\text{source})} = N_{i(\text{source})} G_1 G_2 G_3 \tag{2.34}$$

Finally, the noise factor can be determined as

$$F = \frac{N_{o(\text{total})}}{N_{o(\text{source})}} = 1 + \frac{N_{o1(\text{added})}}{N_{i(\text{source})} G_1} + \frac{N_{o2(\text{added})}}{N_{i(\text{source})} G_1 G_2} + \frac{N_{o3(\text{added})}}{N_{i(\text{source})} G_1 G_2 G_3} \\ = F_1 + \frac{F_2 - 1}{G_1} + \frac{F_3 - 1}{G_1 G_2} \tag{2.35}$$

The above formula shows how the presence of gain preceding a stage causes the effective noise figure to be reduced compared to the measured noise figure of a stage by itself. For this reason, we typically design systems with a low-noise amplifier at the front of the system. We note that the noise figure of each block is typically determined for the case in which a standard input source (e.g., 50Ω) is connected. The above formula can also be used to derive an equivalent model of each block as shown in Figure 2.4. If the input noise when measuring noise figure is

$$N_{i(\text{source})} = kT \tag{2.36}$$

and noting from manipulation of (2.14) that

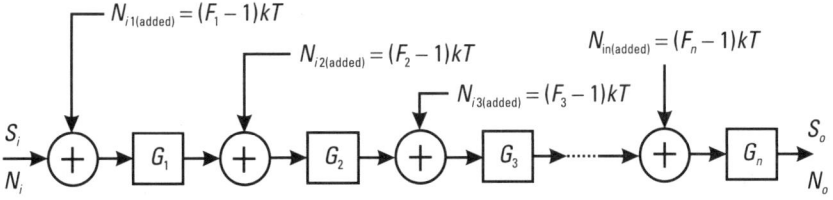

Figure 2.4 Equivalent noise model of a circuit.

$$N_{o1(\text{added})} = (F-1)N_{o(\text{source})} \qquad (2.37)$$

Now dividing both sides of (2.37) by G_1,

$$N_{i(\text{added})} = (F-1)\frac{N_{o(\text{source})}}{G_1} = (F-1)N_{i(\text{source})} = (F-1)kT \qquad (2.38)$$

Then the total input-referred noise to the first stage is

$$N_{i1} = N_{i(\text{source})} + (F_1-1)kT = kT + (F_1-1)kT = kTF_1 \qquad (2.39)$$

Thus, the input-referred noise model for cascaded stages as shown in Figure 2.4 can be derived.

Example 2.1 Noise Calculations

Figure 2.5 shows a 50-Ω source resistance loaded with 50Ω. Determine how much noise voltage per unit bandwidth is present at the output. Then, for any R_L, what is the maximum noise power that this source can deliver to any load? Also find the noise factor, assuming that R_L does not contribute to noise factor, and compare to the case where R_L does contribute to noise factor.

Solution

The noise from the 50Ω source is $\sqrt{4kTR} \approx 0.9$ nV/$\sqrt{\text{Hz}}$ at a temperature of 290K, which, after the voltage divider, becomes one half of this value, or $v_o = 0.45$ nV/$\sqrt{\text{Hz}}$.

Now, for maximum power transfer, the load must remain matched, so $R_L = R_S = 50\Omega$. Then the complete available power from the source is delivered to the load. In this case,

$$P_o = \frac{v_o^2}{4R_L} = P_{\text{in(available)}}$$

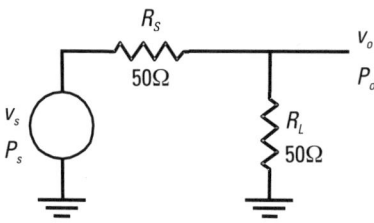

Figure 2.5 Simple circuit used for noise calculations.

$$P_{in(available)} = \frac{v_o^2}{4R_L} = \frac{4kTR_S}{4R_L} = kT = 4 \times 10^{-21}$$

At the output, the complete noise power (available) appears, and so if R_L is noiseless, the noise factor = 1. However, if R_L has noise of $\sqrt{4kTR_L}$ V/$\sqrt{\text{Hz}}$, then at the output, the total noise power is $2kT$, where kT is from R_S and kT is from R_L. Therefore, for a resistively matched circuit, the noise figure is 3 dB. Note that the output noise voltage is 0.45 nV/$\sqrt{\text{Hz}}$ from each resistor for a total of $\sqrt{2} \cdot 0.45$ nV/$\sqrt{\text{Hz}}$ = 0.636 nV/$\sqrt{\text{Hz}}$ (with noise the power adds because the noise voltage is uncorrelated).

Example 2.2 Noise Calculation with Gain Stages

In this example, Figure 2.6, a voltage gain of 20 has been added to the original circuit of Figure 2.5. All resistor values are still 50Ω. Determine the noise at the output of the circuit due to all resistors and then determine the circuit noise figure and signal-to-noise ratio assuming a 1-MHz bandwidth and the input is a 1-V sine wave.

Solution

In this example, at v_x the noise is still due to only R_S and R_2. As before, the noise at this point is 0.636 nV/$\sqrt{\text{Hz}}$. The signal at this point is 0.5V, thus at point v_y the signal is 10V and the noise due to the two input resistors R_S and R_2 is 0.636 · 20 = 12.72 nV/$\sqrt{\text{Hz}}$. At the output, the signal and noise from the input sources, as well as the noise from the two output resistors, all see a voltage divider. Thus, one can calculate the individual components. For the combination of R_S and R_2, one obtains

$$v_{R_S+R_2} = 0.5 \times 12.72 = 6.36 \text{ nV}/\sqrt{\text{Hz}}$$

The noise from the source can be determined from this equation:

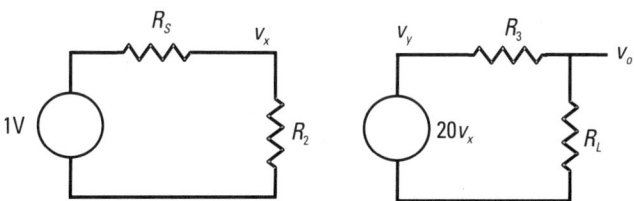

Figure 2.6 Noise calculation with a gain stage.

$$v_{R_S} = \frac{6.36 \text{ nV}/\sqrt{\text{Hz}}}{\sqrt{2}} = 4.5 \text{ nV}/\sqrt{\text{Hz}}$$

For the other resistors, the voltage is

$$v_{R_S} = 0.5 \cdot 0.9 = 0.45 \text{ nV}/\sqrt{\text{Hz}}$$

$$v_{R_L} = 0.5 \cdot 0.9 = 0.45 \text{ nV}/\sqrt{\text{Hz}}$$

Total output noise is given by

$$v_{no(total)} = \sqrt{v_{(R_S+R_L)}^2 + v_{R_S}^2 + v_{R_L}^2} = \sqrt{6.36^2 + 0.45^2 + 0.45^2}$$
$$= 6.392 \text{ nV}/\sqrt{\text{Hz}}$$

Therefore, the noise figure can now be determined:

$$\text{Noise factor} = F = \frac{N_{o(total)}}{N_{o(source)}} = \left(\frac{6.392}{4.5}\right)^2 = (1.417)^2 = 2.018$$

$$\text{NF} = 10 \log_{10} F = 10 \log_{10} 2.018 = 3.05 \text{ dB}$$

Since the output voltage also sees a voltage divider of 1/2, it has a value of 5V. Thus, the signal-to-noise ratio is

$$\frac{S}{N} = 20 \log\left(\frac{5}{\frac{6.392 \text{ nV}}{\sqrt{\text{Hz}}} \cdot \sqrt{1 \text{ MHz}}}\right) = 117.9 \text{ dB}$$

This example illustrates that noise from the source and amplifier input resistance are the dominant noise sources in the circuit. Each resistor at the input provides 4.5 nV/$\sqrt{\text{Hz}}$, while the two resistors behind the amplifier each only contribute 0.45 nV/$\sqrt{\text{Hz}}$. Thus, as explained earlier, after a gain stage, noise is less important.

Example 2.3 Effect of Impedance Mismatch on Noise Figure

Find the noise figure of Example 2.2 again, but now assume that $R_2 = 500\Omega$.

Solution

As before, the output noise due to the resistors is as follows:

$$v_{no(R_S)} = 0.9 \cdot \frac{500}{550} \cdot 20 \cdot 0.5 = 8.181 \text{ nV}/\sqrt{\text{Hz}}$$

where 500/550 accounts for the voltage division from the noise source to the node v_x.

$$v_{no(R_2)} = 0.9 \cdot \sqrt{10} \cdot \frac{50}{550} \cdot 20 \cdot 0.5 = 2.587 \text{ nV}/\sqrt{\text{Hz}}$$

where the $\sqrt{10}$ accounts for the higher noise in a 500-Ω resistor compared to a 50-Ω resistor.

$$v_{no(R_3)} = 0.9 \cdot 0.5 = 0.45 \text{ nV}/\sqrt{\text{Hz}}$$

$$v_{no(R_L)} = 0.9 \cdot 0.5 = 0.45 \text{ nV}/\sqrt{\text{Hz}}$$

The total output noise voltage is

$$v_{no(total)} = \sqrt{v_{R_S}^2 + v_{R_2}^2 + v_{R_3}^2 + v_{R_L}^2} = \sqrt{8.181^2 + 2.587^2 + 0.45^2 + 0.45^2}$$
$$= 8.604 \text{ nV}/\sqrt{\text{Hz}}$$

$$\text{Noise factor} = F = \frac{N_{o\,(total)}}{N_{o\,(source)}} = \left(\frac{8.604}{8.181}\right)^2 = 1.106$$

$$\text{NF} = 10 \log_{10} F = 10 \log_{10} 1.106 = 0.438 \text{ dB}$$

Note: This circuit is unmatched at the input. This example illustrates that a mismatched circuit may have better noise performance than a matched one. However, this assumes that it is possible to build a voltage amplifier that requires little power at the input. This may be possible on an IC. However, if transmission lines are included, power transfer will suffer. A matching circuit may need to be added.

Example 2.4 Cascaded Noise Figure and Sensitivity Calculation

Find the effective noise figure and noise floor of the system shown in Figure 2.7. The system consists of a filter with 3-dB loss, followed by a switch with 1-dB loss, an LNA, and a mixer. Assume the system needs an SNR of 7 dB for a bit error rate of 10^{-3}. Also assume that the system bandwidth is 200 kHz.

Solution

Since the bandwidth of the system has been given as 200 kHz, the noise floor of the system can be determined:

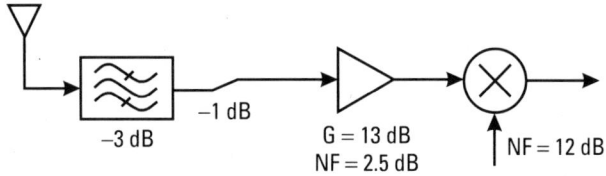

Figure 2.7 System for performance calculation.

$$\text{Noise floor} = -174 \text{ dBm} + 10 \log_{10}(200{,}000) = -121 \text{ dBm}$$

We make use of the cascaded noise figure equation and determine that the overall system noise figure is given by

$$\text{NF}_{\text{TOTAL}} = 3 \text{ dB} + 1 \text{ dB} + 10 \log_{10}\left[1.78 + \frac{15.84 - 1}{20}\right] \approx 8 \text{ dB}$$

Note that the LNA noise figure of 2.5 dB corresponds to a noise factor of 1.78 and the gain of 13 dB corresponds to a power gain of 20. Furthermore, the noise figure of 12 dB corresponds to a noise factor of 15.84.

Note that if the mixer also has gain, then possibly the noise due to the IF stage may be ignored. In a real system this would have to be checked, but here we will ignore noise in the IF stage.

Since it was stated that the system requires an SNR of 7 dB, the sensitivity of the system can now be determined:

$$\text{Sensitivity} = -121 \text{ dBm} + 7 \text{ dB} + 8 \text{ dB} = -106 \text{ dBm}$$

Thus, the smallest allowable input signal is −106 dBm. If this is not adequate for a given application, then a number of things can be done to improve this:

1. A smaller bandwidth could be used. This is usually fixed by IF requirements.
2. The loss in the preselect filter or switch could be reduced. For example, the LNA could be placed in front of one or both of these components.
3. The noise figure of the LNA could be improved.
4. The LNA gain could be increased reducing the effect of the mixer on the system NF.
5. A lower NF in the mixer would also improve the system NF.
6. If a lower SNR for the required BER could be tolerated, then this would also help.

2.3 Linearity and Distortion in RF Circuits

In an ideal system, the output is linearly related to the input. However, in any real device the transfer function is usually a lot more complicated. This can be due to active or passive devices in the circuit or the signal swing being limited by the power supply rails. Unavoidably, the gain curve for any component is never a perfectly straight line, as illustrated in Figure 2.8.

The resulting waveforms can appear as shown in Figure 2.9. For amplifier saturation, typically the top and bottom portions of the waveform are clipped equally, as shown in Figure 2.9(b). However, if the circuit is not biased between the two clipping levels, then clipping can be nonsymmetrical as shown in Figure 2.9(c).

2.3.1 Power Series Expansion

Mathematically, any nonlinear transfer function can be written as a series expansion of power terms unless the system contains memory, in which case a Volterra series is required [9, 10]:

$$v_{out} = k_0 + k_1 v_{in} + k_2 v_{in}^2 + k_3 v_{in}^3 + \ldots \quad (2.40)$$

To describe the nonlinearity perfectly, an infinite number of terms is required; however, in many practical circuits, the first three terms are sufficient to characterize the circuit with a fair degree of accuracy.

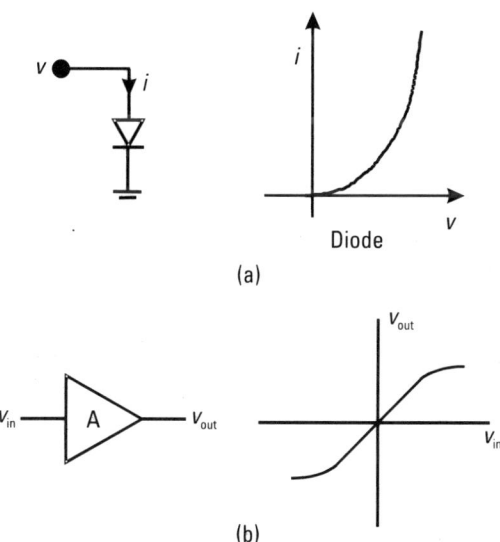

Figure 2.8 Illustration of the nonlinearity in (a) a diode, and (b) an amplifier.

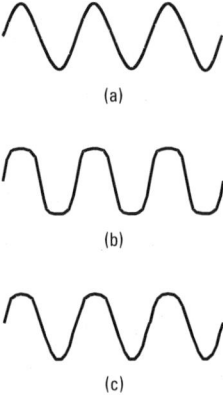

Figure 2.9 Distorted output waveforms: (a) input; (b) output, clipping; and (c) output, bias wrong.

Symmetrical saturation as shown in Figure 2.8(b) can be modeled with odd order terms; for example,

$$y = x - \frac{1}{10}x^3 \quad (2.41)$$

looks like Figure 2.10. In another example, an exponential nonlinearity as shown in Figure 2.8(a) has the form

$$x + \frac{x^2}{2!} + \frac{x^3}{3!} + \ldots \quad (2.42)$$

which contains both even and odd power terms because it does not have symmetry about the y-axis. Real circuits will have more complex power series expansions.

One common way of characterizing the linearity of a circuit is called the two-tone test. In this test, an input consisting of two sine waves is applied to the circuit.

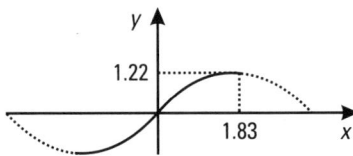

Figure 2.10 Example of output or input nonlinearity with first- and third-order terms.

$$v_{in} = v_1 \cos \omega_1 t + v_2 \cos \omega_2 t = X_1 + X_2 \tag{2.43}$$

When this tone is applied to the transfer function given in (2.40), the result is a number of terms:

$$v_0 = k_0 + \underbrace{k_1(X_1 + X_2)}_{\text{desired}} + \underbrace{k_2(X_1 + X_2)^2}_{\text{second order}} + \underbrace{k_3(X_1 + X_2)^3}_{\text{third order}} \tag{2.44}$$

$$v_0 = k_0 + k_1(X_1 + X_2) + k_2(X_1^2 + 2X_1X_2 + X_2^2) \tag{2.45}$$
$$+ k_3(X_1^3 + 3X_1^2 X_2 + 3X_1 X_2^2 + X_1^3)$$

These terms can be further broken down into various frequency components. For instance, the X_1^2 term has a zero frequency (dc) component and another at the second harmonic of the input:

$$X_1^2 = (v_1 \cos \omega_1 t)^2 = \frac{v_1^2}{2}(1 + \cos 2\omega_1 t) \tag{2.46}$$

The second-order terms can be expanded as follows:

$$(X_1 + X_2)^2 = \underbrace{X_1^2}_{\substack{\text{dc +} \\ \text{HD2}}} + \underbrace{2X_1 X_2}_{\text{MIX}} + \underbrace{X_2^2}_{\substack{\text{dc +} \\ \text{HD2}}} \tag{2.47}$$

where second-order terms are composed of second harmonics HD2, and mixing components, here labeled MIX but sometimes labeled IM2 for second-order intermodulation. The mixing components will appear at the sum and difference frequencies of the two input signals. Note also that second-order terms cause an additional dc term to appear.

The third-order terms can be expanded as follows:

$$(X_1 + X_2)^3 = \underbrace{X_1^3}_{\substack{\text{FUND} \\ + \text{HD3}}} + \underbrace{3X_1^2 X_2}_{\substack{\text{IM3 +} \\ \text{FUND}}} + \underbrace{3X_1 X_2^2}_{\substack{\text{IM3 +} \\ \text{FUND}}} + \underbrace{X_2^3}_{\substack{\text{FUND} \\ + \text{HD3}}} \tag{2.48}$$

Third-order nonlinearity results in third harmonics HD3 and third-order intermodulation IM3. Expansion of both the HD3 and IM3 terms shows output signals appearing at the input frequencies. The effect is that third-order nonlinearity can change the gain, which is seen as gain compression. This is summarized in Table 2.1.

Table 2.1
Summary of Distortion Components

Frequency	Component Amplitude
dc	$k_0 + \frac{k_2}{2}(v_1^2 + v_2^2)$
ω_1	$k_1 v_1 + k_3 v_1 \left(\frac{3}{4} v_1^2 + \frac{3}{2} v_2^2\right)$
ω_2	$k_1 v_2 + k_3 v_2 \left(\frac{3}{4} v_2^2 + \frac{3}{2} v_1^2\right)$
$2\omega_1$	$\frac{k_2 v_1^2}{2}$
$2\omega_2$	$\frac{k_2 v_2^2}{2}$
$\omega_1 \pm \omega_2$	$k_2 v_1 v_2$
$\omega_2 \pm \omega_1$	$k_2 v_1 v_2$
$3\omega_1$	$\frac{k_3 v_1^3}{4}$
$3\omega_2$	$\frac{k_3 v_2^3}{4}$
$2\omega_1 \pm \omega_2$	$\frac{3}{4} k_3 v_1^2 v_2$
$2\omega_2 \pm \omega_1$	$\frac{3}{4} k_3 v_1 v_2^2$

Note that in the case of an amplifier, only the terms at the input frequency are desired. Of all the unwanted terms, the last two at frequencies $2\omega_1 - \omega_2$ and $2\omega_2 - \omega_1$ are the most troublesome, since they can fall in the band of the desired outputs if ω_1 is close in frequency to ω_2 and therefore cannot be easily filtered out. These two tones are usually referred to as third-order intermodulation terms (IM3 products).

Example 2.5 Determination of Frequency Components Generated in a Nonlinear System

Consider a nonlinear circuit with 7- and 8-MHz tones applied at the input. Determine all output frequency components, assuming distortion components up to the third order.

Solution
Table 2.2 and Figure 2.11 show the outputs.

It is apparent that harmonics can be filtered out easily, while the third-order intermodulation terms, being close to the desired tones, may be difficult to filter.

Issues in RFIC Design, Noise, Linearity, and Filtering

Table 2.2
Outputs from Nonlinear Circuits with Inputs at $f_1 = 7$, $f_2 = 8$ MHz

	Symbolic Frequency	Example Frequency	Name	Comment
First order	f_1, f_2	7, 8	Fundamental	Desired output
Second order	$2f_1, 2f_2$	14, 16	HD2 (harmonics)	Can filter
	$f_2 - f_1, f_2 + f_1$	2, 15	IM2 (mixing)	Can filter
Third order	$3f_1, 3f_2$	21, 24	HD3 (harmonic)	Can filter harmonics
	$2f_1 - f_2,$	6	IM3 (intermod)	Close to fundamental,
	$2f_2 - f_1$	9	IM3 (intermod)	difficult to filter

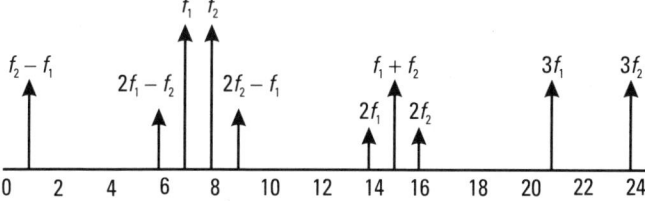

Figure 2.11 Output spectrum with inputs at 7 and 8 MHz.

2.3.2 Third-Order Intercept Point

One of the most common ways to test the linearity of a circuit is to apply two signals at the input, having equal amplitude and offset by some frequency, and plot fundamental output and intermodulation output power as a function of input power as shown in Figure 2.12. From the plot, the *third-order intercept point* (IP3) is determined. The third-order intercept point is a theoretical point where the amplitudes of the intermodulation tones at $2\omega_1 - \omega_2$ and $2\omega_2 - \omega_1$ are equal to the amplitudes of the fundamental tones at ω_1 and ω_2.

From Table 2.1, if $v_1 = v_2 = v_i$, then the fundamental is given by

$$\text{fund} = k_1 v_i + \frac{9}{4} k_3 v_i^3 \tag{2.49}$$

The linear component of (2.49) given by

$$\text{fund} = k_1 v_i \tag{2.50}$$

can be compared to the third-order intermodulation term given by

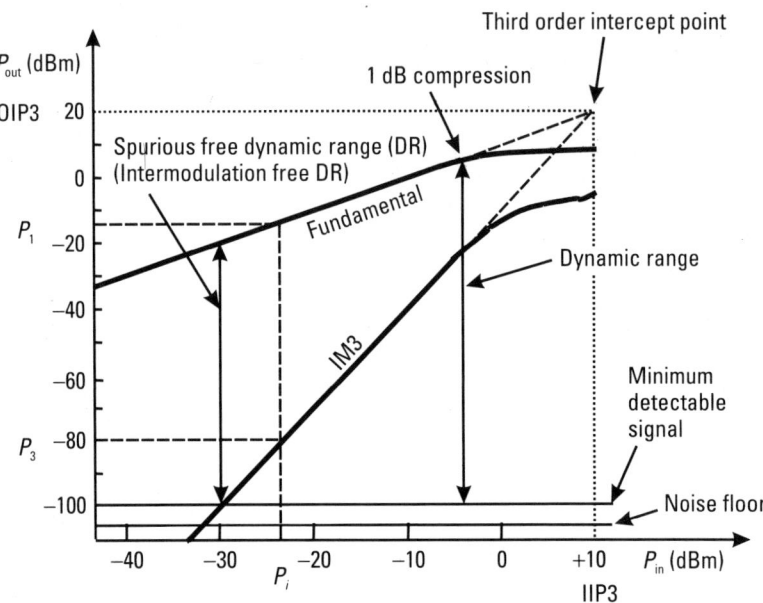

Figure 2.12 Plot of input output power of fundamental and IM3 versus input power.

$$\text{IM3} = \frac{3}{4} k_3 v_i^3 \quad (2.51)$$

Note that for small v_i, the fundamental rises linearly (20 dB/decade) and that the IM3 terms rise as the cube of the input (60 dB/decade). A theoretical voltage at which these two tones will be equal can be defined:

$$\frac{\frac{3}{4} k_3 v_{\text{IP3}}^3}{k_1 v_{\text{IP3}}} = 1 \quad (2.52)$$

This can be solved for v_{IP3}:

$$v_{\text{IP3}} = 2\sqrt{\frac{k_1}{3k_3}} \quad (2.53)$$

Note that (2.53) gives the input voltage at the third-order intercept point. The input power at this point is called the *input third-order intercept point* (IIP3). If IP3 is specified at the output, it is called the *output third-order intercept point* (OIP3).

Of course, the third-order intercept point cannot actually be measured directly, since by the time the amplifier reached this point, it would be heavily overloaded. Therefore, it is useful to describe a quick way to extrapolate it at a given power level. Assume that a device with power gain G has been measured to have an output power of P_1 at the fundamental frequency and a power of P_3 at the IM3 frequency for a given input power of P_i, as illustrated in Figure 2.12. Now, on a log plot (for example, when power is in dBm) of P_3 and P_1 versus P_i, the IM3 terms have a slope of 3 and the fundamental terms have a slope of 1. Therefore,

$$\frac{\text{OIP3} - P_1}{\text{IIP3} - P_i} = 1 \qquad (2.54)$$

$$\frac{\text{OIP3} - P_3}{\text{IIP3} - P_i} = 3 \qquad (2.55)$$

since subtraction on a log scale amounts to division of power.

Also note that

$$G = \text{OIP3} - \text{IIP3} = P_1 - P_i \qquad (2.56)$$

These equations can be solved to give

$$\text{IIP3} = P_1 + \frac{1}{2}[P_1 - P_3] - G = P_i + \frac{1}{2}[P_1 - P_3] \qquad (2.57)$$

2.3.3 Second-Order Intercept Point

A *second-order intercept point* (IP2) can be defined that is similar to the third-order intercept point. Which one is used depends largely on which is more important in the system of interest; for example, second-order distortion is particularly important in direct downconversion receivers.

If two tones are present at the input, then the second-order output is given by

$$v_{\text{IM2}} = k_2 v_i^2 \qquad (2.58)$$

Note that in this case, the IM2 terms rise at 40 dB/decade rather than at 60 dB/decade, as in the case of the IM3 terms.

The theoretical voltage at which the IM2 term will be equal to the fundamental term given in (2.50) can be defined:

$$\frac{k_2 v_{\text{IP2}}^2}{k_1 v_{\text{IP2}}} = 1 \tag{2.59}$$

This can be solved for v_{IP2}:

$$v_{\text{IP2}} = \frac{k_1}{k_2} \tag{2.60}$$

2.3.4 The 1-dB Compression Point

In addition to measuring the IP3 or IP2 of a circuit, the 1-dB compression point is another common way to measure linearity. This point is more directly measurable than IP3 and requires only one tone rather than two (although any number of tones can be used). The 1-dB compression point is simply the power level, specified at either the input or the output, where the output power is 1 dB less than it would have been in an ideally linear device. It is also marked in Figure 2.12.

We first note that at 1-dB compression, the ratio of the actual output voltage v_o to the ideal output voltage v_{oi} is

$$20 \log_{10}\left(\frac{v_o}{v_{oi}}\right) = -1 \text{ dB} \tag{2.61}$$

or

$$\frac{v_o}{v_{oi}} = 0.89125 \tag{2.62}$$

Now referring again to Table 2.1, we note that the actual output voltage for a single tone is

$$v_o = k_1 v_i + \frac{3}{4} k_3 v_i^3 \tag{2.63}$$

for an input voltage v_i. The ideal output voltage is given by

$$v_{oi} = k_1 v_i \tag{2.64}$$

Thus, the 1-dB compression point can be found by substituting (2.63) and (2.64) into (2.62):

$$\frac{k_1 v_{1dB} + \frac{3}{4} k_3 v_{1dB}^3}{k_1 v_{1dB}} = 0.89125 \tag{2.65}$$

Note that for a nonlinearity that causes compression, rather than one that causes expansion, k_3 has to be negative. Solving (2.65) for v_{1dB} gives

$$v_{1dB} = 0.38 \sqrt{\frac{k_1}{k_3}} \tag{2.66}$$

If more than one tone is applied, the 1-dB compression point will occur for a lower input voltage. In the case of two equal amplitude tones applied to the system, the actual output power for one frequency is

$$v_o = k_1 v_i + \frac{9}{4} k_3 v_i^3 \tag{2.67}$$

The ideal output voltage is still given by (2.64). So now the ratio is

$$\frac{k_1 v_{1dB} + \frac{9}{4} k_3 v_{1dB}^3}{k_1 v_{1dB}} = 0.89125 \tag{2.68}$$

Therefore, the 1-dB compression voltage is now

$$v_{1dB} = 0.22 \sqrt{\frac{k_1}{k_3}} \tag{2.69}$$

Thus, as more tones are added, this voltage will continue to get lower.

2.3.5 Relationships Between 1-dB Compression and IP3 Points

In the last two sections, formulas for the IP3 and the 1-dB compression point have been derived. Since we now have expressions for both these values, we can find a relationship between these two points. Taking the ratio of (2.53) and (2.66) gives

$$\frac{v_{IP3}}{v_{1dB}} = \frac{2\sqrt{\frac{k_1}{3k_3}}}{0.38 \sqrt{\frac{k_1}{k_3}}} = 3.04 \tag{2.70}$$

Thus, these voltages are related by a factor of 3.04, or about 9.66 dB, independent of the particulars of the nonlinearity in question. In the case of the 1-dB compression point with two tones applied, the ratio is larger. In this case,

$$\frac{v_{IP3}}{v_{1dB}} = \frac{2\sqrt{\frac{k_1}{3k_3}}}{0.22\sqrt{\frac{k_1}{k_3}}} = 5.25 \quad (2.71)$$

Thus, these voltages are related by a factor of 5.25 or about 14.4 dB.

Thus, one can estimate that for a single tone, the compression point is about 10 dB below the intercept point, while for two tones, the 1-dB compression point is close to 15 dB below the intercept point. The difference between these two numbers is just the factor of three (4.77 dB) resulting from the second tone.

Note that this analysis is valid for third-order nonlinearity. For stronger nonlinearity (i.e., containing fifth-order terms), additional components are found at the fundamental as well as at the intermodulation frequencies. Nevertheless, the above is a good estimate of performance.

Example 2.6 Determining IIP3 and 1-dB Compression Point from Measurement Data

An amplifier designed to operate at 2 GHz with a gain of 10 dB has two signals of equal power applied at the input. One is at a frequency of 2.0 GHz and another at a frequency of 2.01 GHz. At the output, four tones are observed at 1.99, 2.0, 2.01, and 2.02 GHz. The power levels of the tones are −70, −20, −20, and −70 dBm, respectively. Determine the IIP3 and 1-dB compression point for this amplifier.

Solution

The tones at 1.99 and 2.02 GHz are the IP3 tones. We can use (2.57) directly to find the IIP3:

$$\text{IIP3} = P_1 + \frac{1}{2}[P_1 - P_3] - G = -20 + \frac{1}{2}[-20 + 70] - 10 = -5 \text{ dBm}$$

The 1-dB compression point for a signal tone is 9.66 dB lower than this value, about −14.7 dBm at the input.

2.3.6 Broadband Measures of Linearity

Intercept and 1-dB compression points are two common measures of linearity, but they are by no means the only ones. Many others exist and, in fact, more could be defined. Two other measures of linearity that are common in wideband systems handling many signals simultaneously are called *composite triple-order beat* (CTB) and *composite second-order beat* (CSO) [11, 12]. In these tests of linearity, N signals of voltage v_i are applied to the circuit equally spaced in frequency, as shown in Figure 2.13. Note here that, as an example, the tones are spaced 6 MHz apart (this is the spacing for a cable television system for which this is a popular way to characterize linearity). Note also that the tones are never placed at a frequency that is an exact multiple of the spacing (in this case, 6 MHz). This is done so that third-order terms and second-order terms fall at different frequencies. This will be clarified shortly.

If we take three of these signals, then the third-order nonlinearity gets a little more complicated than before:

$$(x_1 + x_2 + x_3)^3 = \underbrace{x_1^3 + x_2^3 + x_3^3}_{\text{HD3}}$$

$$+ \underbrace{3x_1^2 x_2 + 3x_1^2 x_3 + 3x_2^2 x_1 + 3x_3^2 x_1 + 3x_2^2 x_3 + 3x_3^2 x_2}_{\text{IM3}}$$

$$+ \underbrace{6x_1 x_2 x_3}_{\text{TB}} \tag{2.72}$$

The last term in the expression causes CTB in that it creates terms at frequencies $\omega_1 \pm \omega_2 \pm \omega_3$ of magnitude $1.5 k_3 v_i$ where $\omega_1 < \omega_2 < \omega_3$. This is twice as large as the IM3 products. Note that, except for the case where all three are added ($\omega_1 + \omega_2 + \omega_3$), these tones can fall into any of the channels being used and many will fall into the same channel. For instance, in Figure

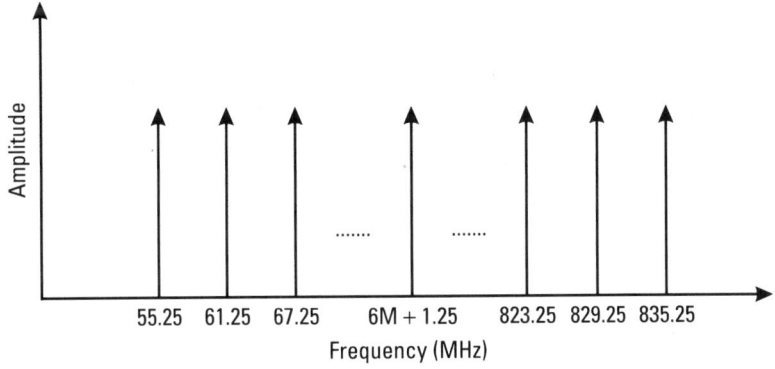

Figure 2.13 Equally spaced tones entering a broadband circuit.

2.13, 67.25 − 73.25 + 79.25 = 73.25 MHz and 49.25 − 55.25 + 79.25 = 73.25 MHz will both fall on the 73.25-MHz frequency. In fact, there will be many more *triple-beat* (TB) products than IM3 products. Thus, these terms become more important in a wide-band system. It can be shown that the maximum number of terms will fall on the tone at the middle of the band. With N tones, it can be shown that the number of tones falling there will be

$$\text{Tones} = \frac{3}{8} N^2 \tag{2.73}$$

We have already said that the voltage of these tones is twice that of the IP3 tones. We also note here that if the signal power is backed off from the IP3 power by some amount, the power in the IP3 tones will be backed off three times as much (calculated on a logarithmic scale). Therefore, if each fundamental tone is at a power level of P_s, then the power of the TB tones will be

$$\text{TB (dBm)} = P_{\text{IP3}} - 3(P_{\text{IP3}} - P_s) + 6 \tag{2.74}$$

where P_{IP3} is the IP3 power level for the given circuit.

Now, assuming that all tones add as power rather than voltage, and noting that CTB is usually specified as so many decibels down from the signal power,

$$\text{CTB (dB)} = P_s - \left[P_{\text{IP3}} - 3(P_{\text{IP3}} - P_s) + 6 + 10 \log\left(\frac{3}{8} N^2\right) \right] \tag{2.75}$$

Note that CTB could be found using either input- or output-referred power levels.

Similar to the CTB is the CSO, which can also be used to measure the linearity of a broadband system. Again, if we have N signals all at the same power level, we now consider the second-order distortion products of each pair of signals falling at frequencies $\omega_1 \pm \omega_2$. In this case, the signals fall at frequencies either above or below the carriers rather than right on top of them, as in the case of the triple-beat terms, provided that the carriers are not some even multiple of the channel spacing. For example, in Figure 2.13, 49.25 + 55.25 = 104.5 MHz. This is 1.25 MHz above the closest carrier at 103.25 MHz. All the sum terms will fall 1.25 MHz above the closest carrier, while the difference terms such as 763.25 − 841.25 = 78, will fall 1.25 MHz below the closest carrier at 79.25 MHz. Thus, the second-order and third-order terms can be measured separately. The number of terms that fall next to any given carrier will vary. Some of the $\omega_1 + \omega_2$ terms will fall out of band and the maximum number in band will fall next to the highest frequency carrier. The number of second-order beats above any given carrier is given by

$$N_B = (N-1)\frac{f - 2f_L + d}{2(f_H - f_L)} \tag{2.76}$$

where N is the number of carriers, f is the frequency of the measurement channel, f_L is the frequency of the lowest channel, f_H is the frequency of the highest channel, and d is the frequency offset from a multiple of the channel spacing (1.25 MHz in Figure 2.13).

For the case of the difference frequency second-order beats, there are more of these at lower frequencies, and the maximum number will be next to the lowest frequency carrier. In this case, the number of second-order products next to any carrier can be approximated by

$$N_B = (N-1)\left(1 - \frac{f - d}{f_H - f_L}\right) \tag{2.77}$$

Each of the second-order beats is an IP2 tone. Therefore, if each fundamental tone is at a power level of P_s, then the power of the *second-order beat* (SO) tones will be

$$\text{SO (dBm)} = P_{\text{IP2}} - 2(P_{\text{IP2}} - P_s) \tag{2.78}$$

Thus, the composite second-order beat product will be given by

$$\text{CSO (dB)} = P_s - [P_{\text{IP2}} - 2(P_{\text{IP2}} - P_s) + 10 \log(N_B)] \tag{2.79}$$

2.4 Dynamic Range

So far, we have discussed noise and linearity in circuits. Noise determines how small a signal a receiver can handle, while linearity determines how large a signal a receiver can handle. If operation up to the 1-dB compression point is allowed (for about 10% distortion, or IM3 is about −20 dB with respect to the desired output), then the dynamic range is from the minimum detectable signal to this point. This is illustrated in Figure 2.12. In this figure, intermodulation components are above the minimum detectable signal for $P_{\text{in}} > -30$ dBm, for which $P_{\text{out}} = -20$ dBm. Thus, for any P_{out} between the minimum detectable signal of −100 dBm and −20 dBm, no intermodulation components can be seen, so the spurious free dynamic range is 80 dB.

Example 2.7 Determining Dynamic Range

In Example 2.4 we determined the sensitivity of a receiver system. Figure 2.14 shows this receiver again with the linearity of the mixer and LNA specified. Determine the dynamic range of this receiver.

Solution

The overall receiver has a gain of 19 dB. The minimum detectable signal from Example 2.4 is −106 dBm or −87 dBm at the output. The IIP3 of the LNA referred to the input is −5 dBm + 4 = −1 dBm. The IIP3 of the mixer referred to the input is 0 − 13 + 4 = −9 dBm. Therefore, the mixer dominates the IIP3 for the receiver. The 1-dB compression point will be 9.6 dB lower than this, or −18.6 dBm. Thus, the dynamic range of the system will be −18.6 + 106 = 87.4 dB.

Example 2.8 Effect of Bandwidth on Dynamic Range

The data transfer rate of the previous receiver can be greatly improved if we use a bandwidth of 80 MHz rather than 200 kHz. What does this do to the dynamic range of the receiver?

Solution

This system is the same as the last one except that now the bandwidth is 80 MHz. Thus, the noise floor is now

$$\text{Noise floor} = -174 \text{ dBm} + 10 \log_{10}(80 \times 10^6) = -95 \text{ dBm}$$

Assuming that the same signal-to-noise ratio is required:

$$\text{Sensitivity} = -95 \text{ dBm} + 7 \text{ dB} + 8 \text{ dB} = -80 \text{ dBm}$$

Thus, the dynamic range is now −15.6 + 80 = 64.4 dB. In order to get this back to the value in the previous system, we would need to increase the linearity of the receiver by 25.3 dB. As we will see in future chapters, this would be no easy task.

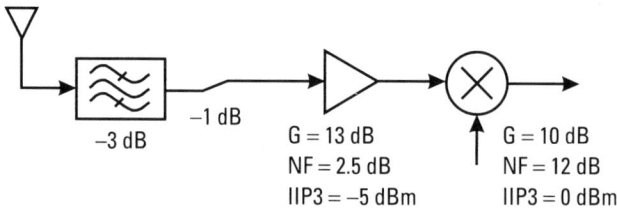

Figure 2.14 Circuit for system example.

2.5 Filtering Issues

To determine noise floor, the system bandwidth has to be known. The system bandwidth is set by filters, so it becomes necessary to discuss some of the filtering issues. There are additional reasons for needing filtering. The receiver must be able to maintain operation and to detect the desired signal in the presence of other signals often referred to as blocking signals. These other signals could be of large amplitude and could be close by in frequency. Such signals must be removed by filters, so a very general discussion of filters is in order. Actual monolithic filter circuits will be discussed in a later chapter.

2.5.1 Image Signals and Image Reject Filtering

The task of the receiver front end is to take the RF input and mix it either to baseband or to some IF where it can be more easily processed. A receiver in which the signal is taken directly to base band is called a *homodyne* or *direct-conversion receiver*. Although simpler than a receiver that takes the signal to some IF first (called a *superheterodyne receiver*), direct-conversion receivers suffer from numerous problems, including dc offsets, because much of the information is close to dc and also because of LO self-mixing [13]. A typical superheterodyne receiver front end consists of an LNA, an image filter, a mixer, and a VCO, as shown in Figure 2.15. An alternative to the image filter is to use an image reject mixer, which will be discussed in detail in Chapter 7. The image filter is required to suppress the unwanted image frequency, which is located a distance of two IFs away from the desired radio frequency [14]. Also, the image filter must prevent noise at the image frequency from mixing down to the IF and increasing the noise figure.

A superheterodyne receiver takes the desired RF input signal and mixes it with some reference signal to extract the difference frequency, as shown in Figure 2.16. The LO reference is mixed with the input to produce a signal at the difference frequency of the LO and RF. The problem is that a signal on the other side of the LO at the same distance from the LO will also mix down

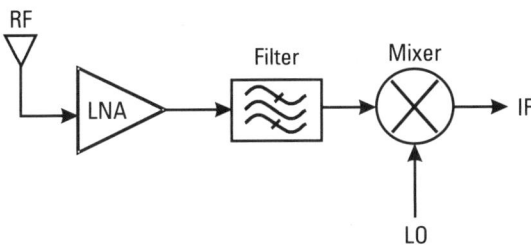

Figure 2.15 A block-level diagram of a superheterodyne receiver front end.

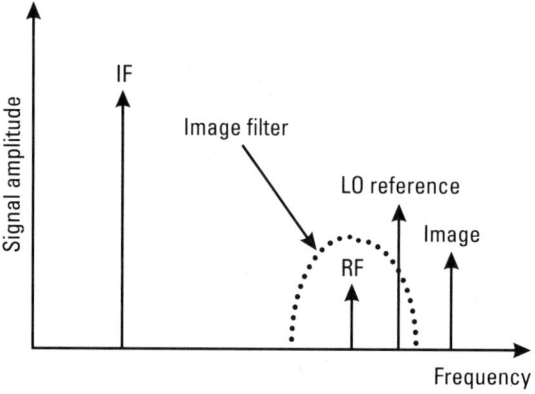

Figure 2.16 Translation of the RF signal to an IF in a superheterodyne receiver.

"on top" of the desired frequency. Thus, before mixing can take place, this unwanted image frequency must be removed. Typically, this is done with a filter that attenuates the image.

Thus, another important specification in a receiver is how much image rejection it has. Image rejection is defined as the ratio of the gain of the desired signal through the receiver G_{sig} to the gain of the image signal through the receiver G_{im}.

$$\text{IR} = 10 \log\left(\frac{G_{sig}}{G_{im}}\right) \quad (2.80)$$

The amount of filtering provided can be calculated by knowing the undesired frequency with respect to the filter center frequency, the filter bandwidth, and filter order. The following equation can be used for this calculation:

$$A_{dB} = \frac{n}{2} \cdot 20 \log\left(\frac{f_{ud} - f_c}{f_{be} - f_c}\right) = \frac{n}{2} \cdot 20 \log\left(2\frac{\Delta f}{f_{BW}}\right) \quad (2.81)$$

where A_{dB} is the attenuation in decibels, n is the filter order (and thus $n/2$ is the effective order on each edge), f_{ud} is the frequency of the undesired signal, f_c is the filter center frequency, f_{be} is the filter band edge, Δf is $f_{ud} - f_c$, and f_{BW} is $2(f_{be} - f_c)$.

Example 2.9 Image Reject Filtering

A system has an RF band from 902 to 928 MHz and a 200-kHz channel bandwidth and channel spacing. The first IF is at 70 MHz. With a 26-MHz image-reject filter, determine the order of filter required to get a worst-case image rejection of better than 50 dB.

Solution
The frequency spectrum is shown in Figure 2.17. At RF, the local oscillator frequency f_{LO} is tuned to be 70 MHz above the desired RF signal so that the desired signal will be mixed down to IF at 70 MHz. Thus, f_{LO} is adjustable between 972 and 998 MHz to allow signals between 902 and 928 MHz to be received. Any signal or noise 70 MHz above f_{LO} will also mix into the IF stage. This is known as the *image frequency*. An image reject filter is required to prevent any image signals from entering the mixer. The worst case will be when the image frequency is closest to the filter frequency. This occurs when the input is at 902 MHz, the LO is at 972 MHz, and the image is 1,042 MHz. The required filter order n can be calculated by solving (2.81) using $f_{BW} = 26$ MHz and $\Delta f = 70 + 44 + 13 = 127$ MHz as follows:

$$n = \frac{2 \cdot A_{dB}}{20 \cdot \log(2\Delta f/f_{BW})} = 5.05$$

Since the order is an even number, a sixth-order filter is used and total attenuation is calculated to be 59.4 dB.

2.5.2 Blockers and Blocker Filtering

Large unwanted signals can block the desired signal. This can happen when the desired signal is small and the undesired signal is large, for example, when the desired signal is far away and the undesired signal is close. If the result is that the receiver is overloaded, the desired signal cannot be received. This situation is known as *blocking*. If the blockers are in the desired frequency band, then filters do not help until the IF stage is reached.

Example 2.10 How Blockers Are Used To Determine Linearity
Consider the typical blocker specifications for a *Global System Mobile* (GSM) receiver shown in Figure 2.18. In the presence of the blockers, the input signal

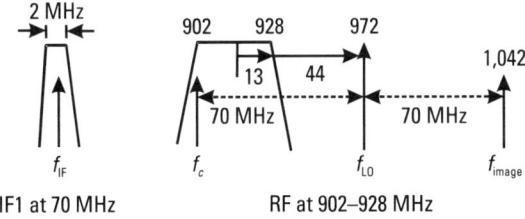

Figure 2.17 Signal spectrum for filter example.

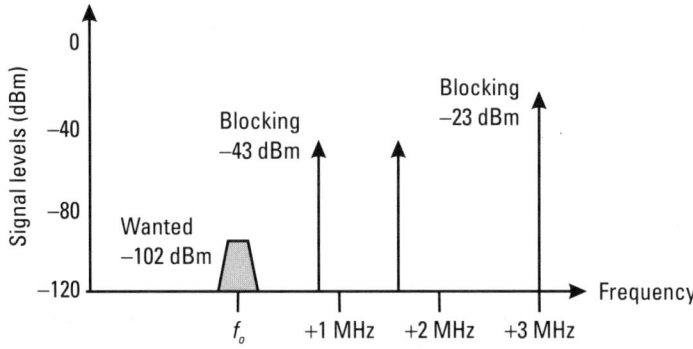

Figure 2.18 GSM minimum detectable signal and blocker levels.

is at −102 dBm and the required signal-to-noise ratio, with some safety margin, is 11 dB. Calculate the required input linearity of the GSM receiver.

Solution
This is an example of the so-called near-far problem that occurs when the desired signal is far away and one or more interfering signals are close by and hence much larger than the wanted signal. So what will be the effect of the blockers? With nonlinearity, third-order intermodulation between the pair of blockers will cause interference directly on top of the signal. The level of this disturbance must be low enough so that the signal can still be detected. The other potential problem is that the large blocker at −23 dBm can cause the amplifier to saturate, rendering the amplifier helpless to respond to the desired signal, which is much smaller. In other words, the receiver has been blocked.

As an estimate, the blocker inputs at −43 dBm will result in third-order intermodulation components (referred to the input) which must be less than −113 dBm, so there is still 11 dB of SNR at the input. Thus, the third-order components (at −113 dBm) are 70 dB below the fundamental components (at −43 dBm). Using (2.57) with P_i at −43 dBm and $[P_1 − P_3] = 70$ dB results in IIP3 of about −8 dBm. Going by this number, the 1-dB compression point is at about −18 dBm at the input. Thus, the single input blocker at −23 dBm is still 5 dB away from the 1-dB compression point. This sounds safe, although there will now be gain through the LNA and the mixer. The blocker will not be filtered until after the mixer, so one must be careful not to saturate any of the components along this path.

The blocking signals can cause problems in a receiver through another mechanism known as *reciprocal mixing*. For a blocker at an offset of Δf from the desired signal, if the oscillator also has a component at the same offset Δf from the carrier, then the blocking signal will be mixed directly to the IF.

Example 2.11 Calculating Maximum Level of Synthesizer Spurs

For the previous GSM specifications, calculate the allowable noise in a synthesizer in the presence of the blocking signals.

Solution Any tone in the synthesizer at 600-kHz offset will mix with the blocker which is at −43 dBm and mix it to the IF stage, where it will interfere with the wanted signal. The blocker can be mixed with noise anywhere in the 200-kHz bandwidth, so a further 53 dB is added to the noise. We note that to be able to detect the wanted signal reliably, as in the previous example, we need the signal to be about 11 dB or so above the mixed-down blocker. Therefore, the mixed-down blocker must be less than −113 dBm. Therefore, the maximum synthesizer noise power at 600-kHz offset is calculated as −113 + 43 − 53 = −123 dB lower than the desired oscillating amplitude measured in a 1-Hz bandwidth. This is an illustration of what is known as *phase noise* and will be discussed in more detail in Chapter 8.

References

[1] Papoulis, A., *Probability, Random Variables, and Stochastic Processes*, New York: McGraw-Hill, 1984.

[2] Sze, S. M., *Physics of Semiconductor Devices*, 2nd ed., New York: John Wiley & Sons, 1981.

[3] Gray, P. R., et al., *Analysis and Design of Analog Integrated Circuits*, 4th ed., New York: John Wiley & Sons, 2001.

[4] Stremler, F. G., *Introduction to Communication Systems*, Reading, MA: Addison-Wesley, 1977.

[5] Jordan, E. C., and K. G. Balmain, *Electromagnetic Waves and Radiating Systems*, 2nd ed., Englewood Cliffs, NJ: Prentice Hall, 1968.

[6] Rappaport, T. S., *Wireless Communications*, Upper Saddle River, NJ: Prentice Hall, 1996.

[7] Proakis, J. G., *Digital Communications*, 3rd ed., New York: McGraw-Hill, 1995.

[8] Gonzalez, G., *Microwave Transistor Amplifiers*, Upper Saddle River, NJ: Prentice Hall, 1997.

[9] Wambacq, P., and W. Sansen, *Distortion Analysis of Analog Integrated Circuits*, Norwell, MA: Kluwer, 1998.

[10] Wambacq, P., et al., "High-Frequency Distortion Analysis of Analog Integrated Circuits," *IEEE Trans. on Circuits and Systems II: Analog and Digital Signal Processing*, Vol. 46, No. 3, March 1999, pp. 335–345.

[11] "Some Notes on Composite Second and Third Order Intermodulation Distortions," *Matrix Technical Notes MTN-108*, Middlesex, NJ: Matrix Test Equipment, http://www.matrixtest.com/Literat/mtn108.htm, accessed Dec. 15, 1998.

[12] "The Relationship of Intercept Points and Composite Distortions," *Matrix Technical Notes MTN-109,* Middlesex, NJ: Matrix Test Equipment, http://www.matrixtest.com/Literat/mtn109.htm, Feb. 18, 1998.

[13] Razavi, B., *RF Microelectronics,* Upper Saddle River, NJ: Prentice Hall, 1998.

[14] Carson, R. S., *Radio Communications Concepts: Analog,* New York: John Wiley & Sons, 1990, Chapter 8.

Selected Bibliography

Fukui, H., *Low Noise Microwave Transistors and Amplifiers,* New York: John Wiley & Sons, 1981.

Larson, L. E., (ed.), *RF and Microwave Circuit Design for Wireless Communications,* Norwood, MA: Artech House, 1997.

Rohde, U. L., J. Whitaker, and A. Bateman, *Communications Receivers: DPS, Software Radios, and Design,* 3rd ed., New York: McGraw-Hill, 2000.

Sklar, B., *Digital Communications: Fundamentals and Applications,* 2nd ed., Englewood Cliffs, NJ: Prentice Hall, 2001.

3

A Brief Review of Technology

3.1 Introduction

At the heart of RF integrated circuits are the transistors used to build them. The basic function of a transistor is to provide gain. Unfortunately, transistors are never ideal, because along with gain comes nonlinearity and noise. The nonlinearity is used to good effect in mixers and in the limiting function in oscillators. Transistors also have a maximum operating frequency beyond which they cannot produce gain.

Metal oxide semiconductor (MOS) transistors and bipolar transistors will be discussed in this chapter. CMOS is the technology of choice in any digital application because of its very low quiescent power dissipation and ease of device isolation. However, traditionally, *MOS field-effect transistors* (MOSFETs) have had inferior speed and noise compared to bipolar transistors. Also, CMOS devices have proved challenging to model for RF circuit simulation, and without good models, RFIC design can be a very frustrating experience. In order to design RFICs, it is necessary to have a good understanding of the high-speed operation of the transistors in the technology that is being used. Thus, in this chapter a basic introduction to some of the more important properties will be provided. For more detail on transistors, the interested reader should consult [1–10].

3.2 Bipolar Transistor Description

Figure 3.1 shows a cross section of a basic npn bipolar transistor. The collector is formed by epitaxial growth in a p− substrate (the n− region). A p region inside the collector region forms the base region; then an n+ emitter region is

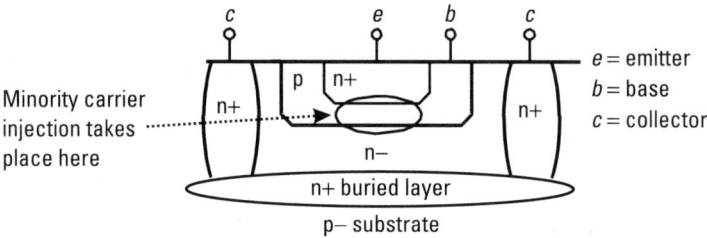

Figure 3.1 Planar bipolar transistor cross-section diagram.

formed inside the base region. The basic transistor action all takes place directly under the emitter in the region shown with an oval. This can be called the *intrinsic transistor*. The intrinsic transistor is connected through the diffusion regions to the external contacts labeled *e*, *b*, and *c*. More details on advanced bipolar structures, such as using SiGe *heterojunction bipolar transistors* (HBTs), and double-poly self-aligned processes can be found in the literature [1, 2]. Note that although Si is the most common substrate for bipolar transistors, it is not the only one; for example, GaAs HBTs are often used in the design of cellular radio power amplifiers and other high-power amplifiers.

Figure 3.2 shows the transistor symbol and biasing sources. When the transistor is being used as an amplifying device, the base-emitter junction is forward biased while the collector-base junction is reverse biased, meaning the collector is at a higher voltage than the base. This bias regime is known as the *forward active region*. Electrons are injected from the emitter into the base region. Because the base region is narrow, most electrons are swept into the collector instead of going to the base contact. This is equivalent to conventional (positive) current from collector to emitter. Some holes are back-injected into the emitter and some electrons recombine in the base, resulting in a small base current that is directly proportional to collector current $i_c = \beta i_b$. Thus, the overall concept is that collector current is controlled by a small base current. The collector current can also be related to the base-emitter voltage in this region of operation by

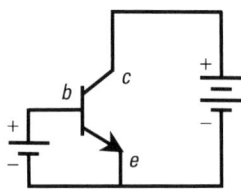

Figure 3.2 Bipolar transistor symbol and bias supplies.

$$I_C = I_S e^{(V_{BE}/v_T)} \tag{3.1}$$

where I_S is a constant known as the *saturation current*, V_{BE} is the dc bias between the base and emitter, and v_T is the thermal voltage given by

$$v_T = \frac{kT}{q} \tag{3.2}$$

where q is the electron charge, T is the temperature in Kelvin, and k is Boltzmann's constant. The thermal voltage is approximately equal to 25 mV at a temperature of 290K, close to room temperature.

Figure 3.3 shows the collector characteristics for a typical bipolar transistor. The transistor has two other regions of operation usually avoided in analog design. When the base-emitter junction is not forward biased, the transistor is cut off. The transistor is in the saturated region if both the base-emitter and collector-emitter junctions are forward biased. In saturation, the base is flooded with minority carriers. This generally leads to a delayed response when the bias conditions change to another region of operation. In saturation, V_{CE} is typically less than a few tenths of a volt. Note that in the active region, the collector current is not constant. There is a slope to the current versus voltage curve, indicating that the collector current will increase with collector-emitter voltage. The slopes of all the lines are such that they will meet at a negative voltage V_A called the *Early voltage*. This voltage can be used to characterize the transistor output impedance.

The intrinsic transistor is connected through the diffusion regions to the external contacts labeled e, b, and c. These connections add series resistance and increase the parasitic capacitance between the regions. The series resistance

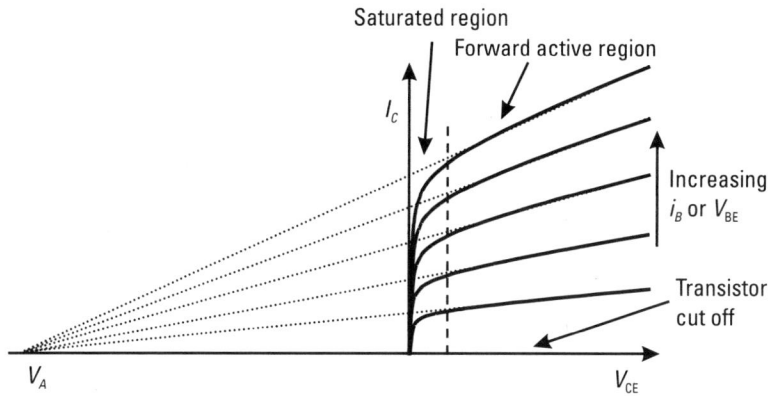

Figure 3.3 Transistor characteristic curves.

in the collector is reduced by the buried layer. The effects of other series resistance are often reduced by the use of multiple contacts, as shown in Figure 3.4.

3.3 β Current Dependence

Figure 3.5 shows the dependence of β on collector current. β drops off at high currents because the electron concentration in the base-collector depletion region becomes comparable to the background dopant ion concentration, leading to a dramatic increase in the effective width of the base. This is called the *Kirk effect* or *base pushout*. As a result, the base resistance is current dependent. Another effect is emitter crowding, which comes about because of the distributed nature of parasitic resistance at the base contact, causing the base-emitter voltage to be higher close to the base contact. This results in the highest current density at the edge of the emitter. In the other extreme, at low currents, β may be reduced due to the excess current resulting from recombination in the emitter-base depletion region.

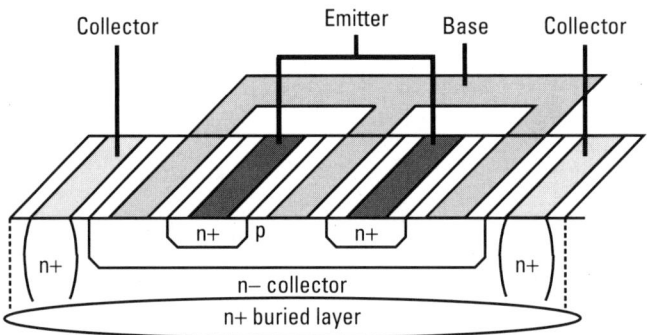

Figure 3.4 Transistor with multiple contacts, shown in three dimensions.

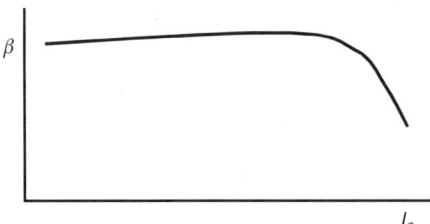

Figure 3.5 Current dependence of β.

3.4 Small-Signal Model

Once the bias voltages and currents are determined for the transistor, it is necessary to determine how it will respond to *alternating current* (ac) signals exciting it. Thus, an ac small-signal model of the transistor is now presented. Figure 3.6 shows a fairly complete small-signal model for the bipolar transistor. The values of the small-signal elements shown, r_π, C_π, C_μ, g_m, and r_o, will depend on the dc bias of the transistor. The intrinsic transistor (shown directly under the emitter region in Figure 3.1) is shown at the center. The series resistances to the base, emitter, and collector are shown respectively by r_b, r_E, and r_c. Also, between each pair of terminals there is some finite capacitance shown as C_{bc}, C_{ce}, and C_{be}. This circuit can be simplified by noting that of the extrinsic resistors, r_b is the largest, and as a result r_E and r_c are often omitted, along with the capacitances C_{bc}, C_{ce}, and C_{be}, as shown in Figure 3.7. Resistor r_E is low due to high doping of the emitter, while r_c is reduced by a heavily

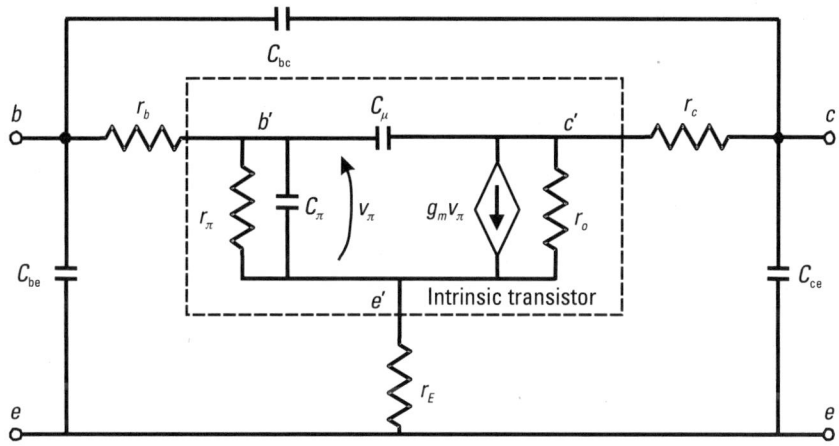

Figure 3.6 Small-signal model for bipolar transistor.

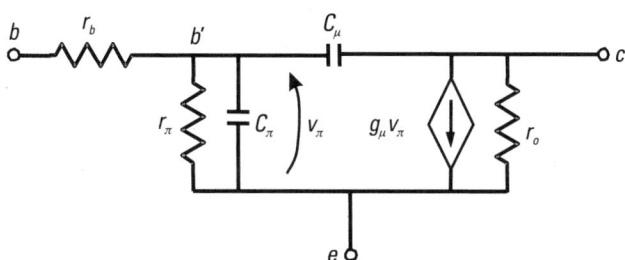

Figure 3.7 Simplified small-signal model for bipolar transistor.

doped buried layer in the collector. The base resistance r_b is the source of several problems. First, it forms an input voltage divider between r_b, r_π, and C_π, which reduces the input signal amplitude and deteriorates high-frequency response. It also directly adds to thermal noise.

3.5 Small-Signal Parameters

Now that the small-signal model has been presented, to help determine appropriate values for model parameters at different operating points, some simple formulas will be presented.

First, the short-circuit current gain β is given by

$$\beta = \underbrace{\frac{i_c}{i_b}}_{\text{small-signal}} = \underbrace{\frac{\Delta I_C}{\Delta I_B}}_{\Delta\text{large-signal}} \tag{3.3}$$

noting that currents can be related by

$$i_c + i_b = i_e \tag{3.4}$$

Transconductance g_m is given by

$$g_m = \frac{i_c}{v_\pi} = \frac{I_C}{v_T} = \frac{I_C q}{kT} \tag{3.5}$$

where I_C is the dc collector current. Note that the small-signal value of g_m in (3.5) is related to the large-signal behavior of (3.1) by differentiation.

At low frequency, where the transistor input impedance is resistive, i_c and i_b can be related by

$$i_c = \beta i_b = g_m v_\pi = g_m i_b r_\pi \tag{3.6}$$

(neglecting current through r_o), which means that

$$\beta = g_m r_\pi \tag{3.7}$$

Also, the output resistance can be determined in terms of the early voltage V_A:

$$r_o = \frac{V_A}{I_C} \tag{3.8}$$

3.6 High-Frequency Effects

There are two typical figures of merit f_T and f_{max} used to describe how fast a transistor will operate. f_T is the frequency at which the short-circuit current gain β is equal to 1. f_{max} is the frequency at which maximum available power gain $G_{A,max}$ is equal to 1.

Referring to Figure 3.7, an expression can be found for the corner frequency f_β, beyond which the current gain β decreases:

$$f_\beta = \frac{1}{2\pi r_\pi (C_\pi + C_\mu)} \approx \frac{1}{2\pi r_\pi C_\pi} \quad (3.9)$$

Since this is a first-order roll-off, f_T is β times higher than f_β.

$$f_T = \beta f_\beta = \frac{g_m}{2\pi (C_\pi + C_\mu)} \approx \frac{g_m}{2\pi C_\pi} = \frac{I_C}{2\pi C_\pi v_T} \quad (3.10)$$

The maximum frequency for which power gain can be achieved is called f_{max}, while $G_{A,max}$ is the maximum achievable gain at a particular frequency. f_{max} and $G_{A,max}$ are measured by conjugately matching source and load to the transistor.

f_{max} can be determined by noting that at f_{max} the impedance of C_π is very low. As a result, r_π can be ignored and the input impedance is approximately equal to r_b (the residual capacitive reactance can be resonated with a series inductor and so can be ignored). Thus, input power is

$$P_{in} = \frac{v_b^2}{r_b} \quad (3.11)$$

where v_b is the input rms voltage on the base. The current source has an output current equal to

$$i_c = g_m v_\pi \quad (3.12)$$

where the magnitude of v_π is given by

$$|v_\pi| = \frac{v_b}{r_b \omega k C_\pi} \quad (3.13)$$

since the current in C_π is much greater than in r_π. Here k is the multiplier due to Miller multiplication of C_μ [3]. This factor is often ignored in the literature, but it will be shown later that at f_{max}, $k = 3/2$, so it should not be ignored. Thus, the output current is

$$i_c = \frac{g_m v_b}{r_b \omega k C_\pi} \quad (3.14)$$

The output impedance is determined by applying an output test voltage v_{cx} and measuring the total output current i_{cx}. The most important component of current comes from the current source $i_{cx} = g_m v_{\pi x}$ where $v_{\pi x}$ is related to v_{cx} by the C_μ, C_π voltage divider described by

$$v_{\pi x} = \frac{v_{cx} C_\mu}{C_\pi + C_\mu} \quad (3.15)$$

Thus, the real part of the output impedance z_o is

$$\Re\{z_o\} = \frac{v_{cx}}{i_{cx}} = \frac{C_\pi + C_\mu}{g_m C_\mu} \approx \frac{C_\pi}{g_m C_\mu} \quad (3.16)$$

Current through C_π will be seen as a reactive part of the output impedance. It turns out that the real and reactive components are roughly equal; however, the reactive component can be ignored, since its effect will be eliminated by output matching. The remaining real component will be loaded with an equal real component, resulting in an output voltage of

$$\frac{v_c}{v_\pi} = -\frac{C_\pi}{2C_\mu} \quad (3.17)$$

Using this result, the Miller multiplication of C_μ results in

$$kC_\pi = C_\pi + \left(1 + \frac{v_c}{v_\pi}\right)C_\mu = C_\pi + \left(1 + \frac{C_\pi}{2C_\mu}\right)C_\mu = C_\pi + C_\mu + \frac{C_\pi}{2} \approx \frac{3}{2} C_\pi \quad (3.18)$$

The output power P_o is

$$P_o = \frac{i_c^2 \Re\{z_o\}}{4} = \frac{g_m v_b^2}{4 r_b^2 \omega^2 k^2 C_\pi C_\mu} \quad (3.19)$$

$$\frac{P_o}{P_i} = \frac{g_m}{4 r_b \omega^2 k^2 C_\pi C_\mu} \quad (3.20)$$

If this is set equal to 1, one can solve for f_{max}:

$$f_{max} = \frac{1}{2\pi}\sqrt{\frac{g_m}{4r_b k^2 C_\pi C_\mu}} = \sqrt{\frac{f_T}{8\pi r_b k^2 C_\mu}} = \frac{1}{4\pi r_{bb}}\sqrt{\frac{\beta}{k^2 C_\pi C_\mu}} \tag{3.21}$$

where r_{bb} is the total base resistance given by $r_{bb} = \sqrt{r_b r_\pi}$. Note that f_{max} can be related to the geometric mean of f_T and the corner frequency defined by r_b and C_μ.

3.6.1 f_T as a Function of Current

The f_T is heavily bias dependent; therefore, only when properly biased at a current of $I_{opt f_T}$ will the transistor have its maximum f_T as shown in Figure 3.8. As seen in (3.10), f_T is dependent on C_π and g_m. The capacitor C_π is often described as being a combination of the base-emitter junction capacitance C_{je} and the diffusion capacitance C_d. The junction capacitance is voltage dependent where the capacitance decreases at higher voltage. The diffusion capacitance is current dependent and increases with increasing current. However, for current levels below the current for peak f_T, g_m increases faster with increasing current; hence f_T is increasing in this region. At high currents, f_T drops due to current crowding and conductivity modulation effects in the base region [1].

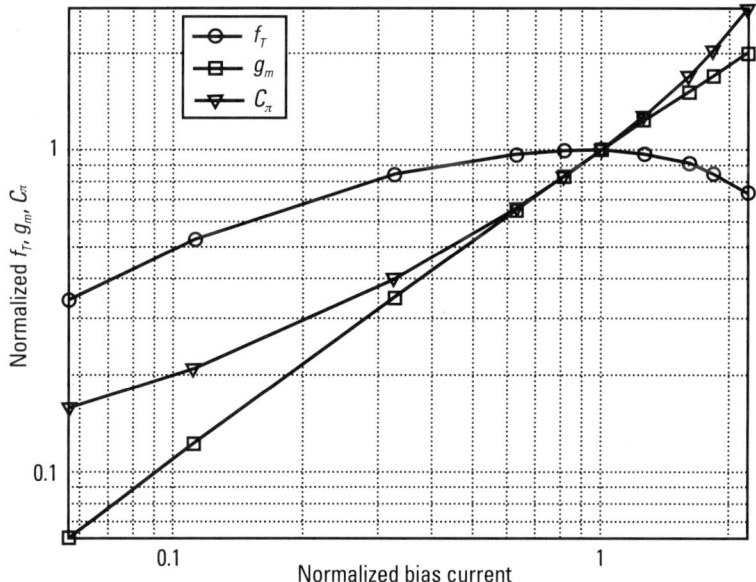

Figure 3.8 Normalized f_T, g_m, and c_π versus bias current.

Note that in many processes, f_T is nearly independent of size for the same current density in the emitter (but always a strong function of current). Some f_T curves that could be for a typical modern 50-GHz SiGe process are shown in Figure 3.9.

Example 3.1 f_T and f_{max} Calculations

From the data in Table 3.1 for a typical 50-GHz bipolar process, calculate z_o, f_T, and f_{max} for the 15x transistor. Use this to verify some of the approximations made in the above derivation for f_{max}.

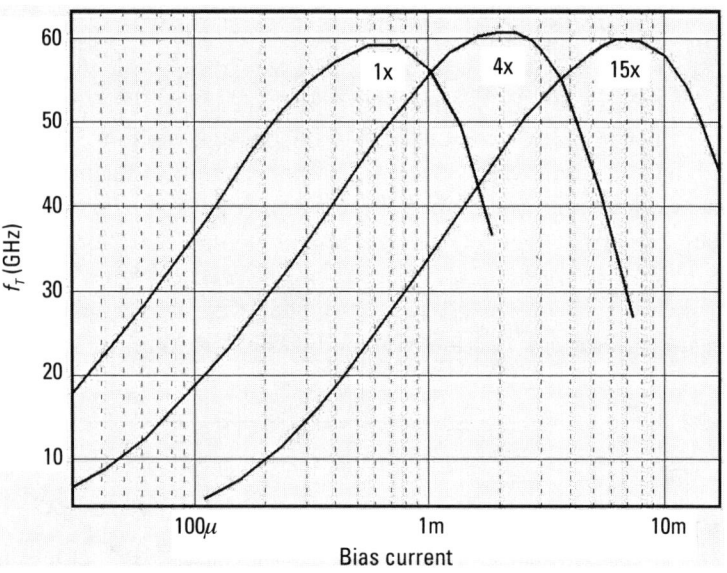

Figure 3.9 f_T as a function of currents for different transistor size relative to a unit transistor size of 1x.

Table 3.1
Example Transistors

Parameter	Transistor Size		
	1x	4x	15x
$I_{opt f_T}$ (mA)	0.55	2.4	7.9
C_π (fF)	50	200	700
C_μ (fF)	2.72	6.96	23.2
r_b (Ω)	65	20.8	5.0

Solution

At 7.9 mA, g_m is equal to 316 mA/V and if $\beta = 100$, then $r_\pi = 316.5\Omega$ and f_T is calculated to be 71.8 GHz. It can be noted that simulation of the complete model resulted in a somewhat reduced value of 60 GHz. At 71.8 GHz, the impedance of C_π is calculated to be $-j3.167\Omega$. Thus, the approximation that this impedance is much less than r_b or r_π is justified. Calculation of f_{max} results in a value of 107 GHz. The real part of the output impedance is calculated as 382Ω. We note that the reactive part of the output impedance will be canceled out by the matching network.

3.7 Noise in Bipolar Transistors

In addition to the thermal noise in resistors, which was discussed in Chapter 2, transistors also have other types of noise. These will be discussed next.

3.7.1 Thermal Noise in Transistor Components

The components in a transistor that have thermal noise are r_b, r_E, and r_c. Given a resistor value R, a noise voltage source must be added to the transistor model of value $4kTR$, as discussed in Chapter 2.

3.7.2 Shot Noise

Shot noise occurs at both the base and the collector and is due to the discrete nature of charge carriers, as they pass a potential barrier, such as a pn junction. That is to say that even though we think about current as a continuous "flow," it is actually made up of many electrons (charge carriers) that move through the conductor. If the electrons encounter a barrier they must cross, then at any given instant, a different number of electrons will cross that barrier even though on average they cross at the rate of the current flow. This random process is called *shot noise* and is usually expressed in amperes per root hertz.

$$i_{bn} = \sqrt{2qI_B} \qquad (3.22)$$

and the collector shot noise is described by

$$i_{cn} = \sqrt{2qI_C} \qquad (3.23)$$

where I_B and I_C are the base and collector bias currents, respectively. The frequency spectrum of shot noise is white.

3.7.3 1/f Noise

This type of noise is also called *flicker noise*, or *excess noise*. The $1/f$ noise is due to variation in the conduction mechanism, for example, fluctuations of surface effects (such as the filling and emptying of traps) and of recombination and generation mechanisms. Typically, the power spectral density of $1/f$ noise is inversely proportional to frequency and is given by the following equation:

$$\overline{i^2_{bf}} = K I_C^m \frac{1}{f^\alpha} \qquad (3.24)$$

where m is between 0.5 and 2, α is about equal to 1, and K is a process constant.

The $1/f$ noise is dominant at low frequencies, as shown in Figure 3.10; however, beyond the corner frequency (shown as 10 kHz), thermal noise dominates. The effect of $1/f$ noise on RF circuits can usually be ignored. An exception is in the design of oscillators, where $1/f$ noise can modulate the oscillator output signal, producing or increasing phase noise. The $1/f$ noise is also important in direct down-conversion receivers, as the output signal is close to dc. Note also that $1/f$ noise is much worse for MOS transistors, where it can be significant up to 1 MHz.

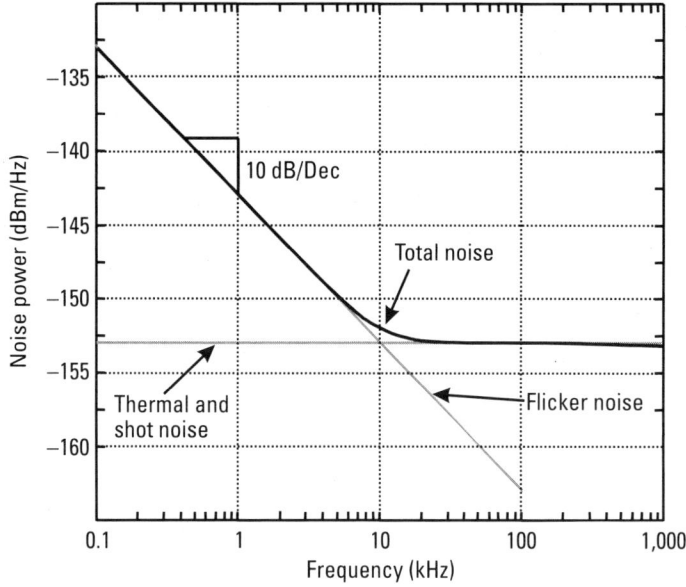

Figure 3.10 Illustration of noise power spectral density.

3.8 Base Shot Noise Discussion

It is interesting that base shot noise can be related to noise in the resistor r_π by noting that the base shot noise current is in parallel with r_π, as shown in Figure 3.11. As shown in the following equation, base shot noise can be related to resistor thermal noise, except that it has a value of $2kTR$ instead of the expected $4kTR$, making use of equations (3.3), (3.5), (3.7), and (3.22).

$$v_{bn} = i_{bn} \cdot r_\pi = \sqrt{2qI_B} \cdot r_\pi = \sqrt{2q\frac{I_C}{\beta}} \cdot r_\pi = \sqrt{2q\frac{I_C}{g_m r_\pi}} \cdot r_\pi$$

$$= \sqrt{2q\frac{I_C}{\frac{I_C q}{kT} r_\pi}} \cdot r_\pi = \sqrt{2kTr_\pi} \qquad (3.25)$$

Thus, base shot noise can be related to thermal noise in the resistor r_π (but is off by a factor of 2). This is sometimes expressed by stating that the diffusion resistance is generating noise *half thermally*. Note that any resistor in thermal equilibrium must generate $\sqrt{4kTR}$ of noise voltage. However, a conducting pn junction is not in thermal equilibrium, and power is being added so it is allowed to break the rules.

3.9 Noise Sources in the Transistor Model

Having discussed the various noise sources in a bipolar transistor, the model for these noise sources can now be added to the transistor model. The noise sources in a bipolar transistor can be shown as in Figure 3.12. These noise sources can also be added to the small-signal model as shown in Figure 3.13.

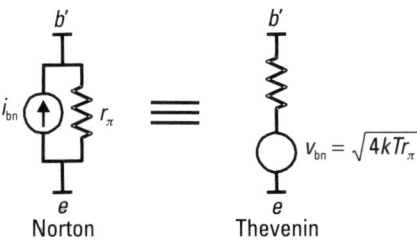

Figure 3.11 Noise model of base shot noise.

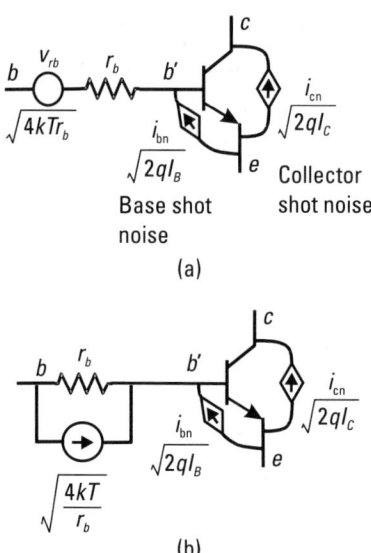

Figure 3.12 Transistor with noise models: (a) base series noise source; and (b) base parallel noise source.

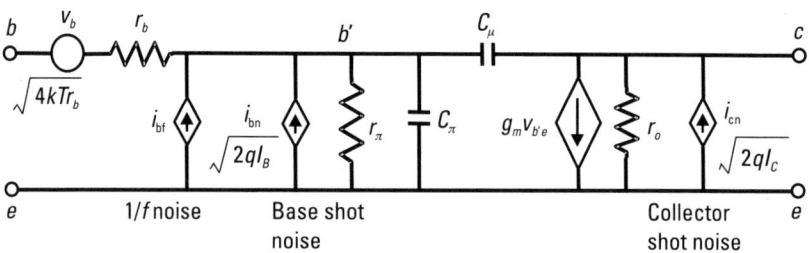

Figure 3.13 Transistor small-signal model with noise.

3.10 Bipolar Transistor Design Considerations

For highest speed, a bias current near peak f_T is suggested. However, it can be noted from Figure 3.8 that the peaks are quite wide. The f_T drops by 10% of its peak value only when current is reduced to half of the optimum value or when it is increased by 50% over its optimum value. Figure 3.8 also shows that junction capacitance is roughly proportional to transistor size, while base resistance is inversely proportional to transistor size.

Thus, a few guidelines are provided as follows:

- Pick lower current to reduce power dissipation with minimal reduction of f_T. For noise consideration, lower current is often used.

- Pick largest transistor size to give lowest base resistance. This will have a direct impact on noise. However, on the down side, large size requires large current for optimal f_T. Another negative impact is that junction capacitances increase with larger transistors.
- Collector shot noise power is proportional to current, but signal power gain is proportional to current squared, so more current can improve noise performance if collector shot noise is dominant.

3.11 CMOS Transistors

Bipolar transistors are preferred for RF circuits due to the higher values of g_m achievable for a given amount of bias current. However, for a complete radio on a chip, it is necessary to use a process that can be used to implement back-end digital or DSP functions. This could be BiCMOS or straight CMOS. BiCMOS would be preferable, since bipolar can then be used for RF, possibly adding *p-channel MOS* (PMOS) transistors for power-control functions. However, for economic reasons or for the need to use a particular CMOS-only process to satisfy the back-end requirements, one may be forced to implement RF circuits in CMOS. For this reason, we give a brief summary of CMOS transistors. Basic PMOS and n-channel MOS (NMOS) transistors are shown in Figure 3.14.

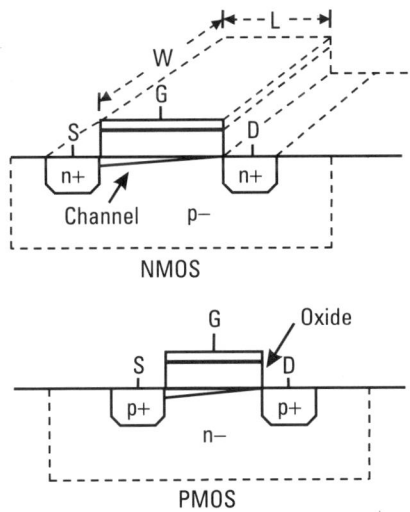

Figure 3.14 CMOS transistors.

3.11.1 NMOS

The drain characteristic curves for an NMOS transistor are similar to the curves for an npn bipolar transistor and are shown in Figure 3.15. When a positive voltage is applied to the gate of the NMOS device, electrons are attracted towards the gate. With sufficient voltage, an n channel is formed under the oxide, allowing electrons to flow between the drain and the source under the control of the gate voltage v_{GS}. Thus, as gate voltage is increased, current increases. For small applied v_{DS} with constant v_{GS}, the current between drain and source is related to the applied v_{DS}. For very low v_{DS}, the relationship is nearly linear. For sufficiently large v_{DS}, the channel becomes restricted at the drain end (pinched off) as shown in Figure 3.14. For larger v_{DS}, current is saturated and remains nearly independent of v_{DS}. This means the output conductance g_o is very low, which is advantageous for high gain in amplifiers.

3.11.2 PMOS

The operation of PMOS is similar to that of NMOS except that negative v_{GS} is applied. This attracts holes, to form a conducting p channel. The characteristic curves for PMOS and NMOS are similar if the absolute value is taken for current and voltage.

3.11.3 CMOS Small-Signal Model Including Noise

As with the bipolar transistor, the noise in a MOSFET can be modeled by placing a noise current source in parallel with the output, as shown in Figure

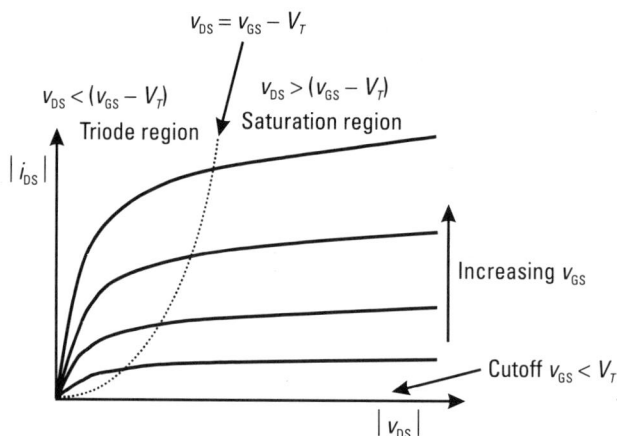

Figure 3.15 CMOS transistor curves.

3.16, representing the thermal channel noise. The input-referred noise is as shown in Figure 3.17.

For analog design, it is often assumed that the input impedance Z_{in} is infinity, in which case the input noise current would be zero. However, at RF, both input-referred noise current and noise voltage are required to account for the actual RF input impedance, which is neither zero nor infinity. Another point to note is that input noise voltage and noise current are correlated sources, since they both have the same origin. This is similar to the case in bipolar transistors with collector shot noise referred to the base.

There are several significant sources of noise not included in the simple model. One example is noise due to gate resistance. We note that gate resistance is also an important factor in determining f_{max}. This gate resistance can be calculated from the dimensions of the gate and the gate resistivity ρ by

$$R_{GATE} = \frac{1}{3}\rho\frac{W}{L} \tag{3.26}$$

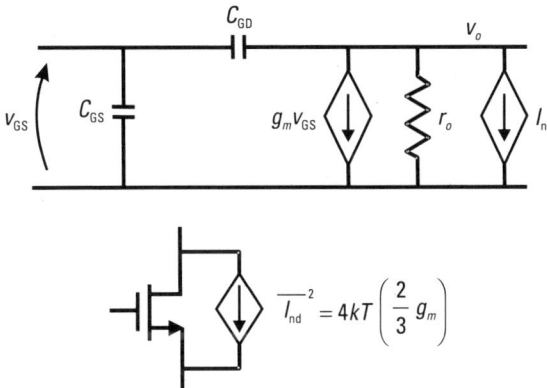

Figure 3.16 CMOS small-signal model with noise.

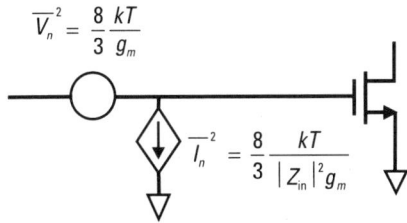

Figure 3.17 Gate-referred noise in NMOS transistor.

for a gate with a contact on one side. Here, ρ is the effective resistivity of the gate poly with typical values between 10 and 20 $\mu\Omega$ cm. We note that the gate poly by itself would have a resistance of $\rho W/Lt$, where t is the effective thickness of the silicided poly gate, with a typical value of 0.1 μm. The factor of 1/3 in (3.26) comes from the fact that the transistor current is flowing under all regions of the gate. The series resistance varies from 0Ω near the contact to $\rho W/Lt$ for the far end of the gate, with an effective value given by (3.26). If the gate is contacted on both sides, the effective resistance drops by a further factor of 4 such that

$$R_{\text{GATE}} = \frac{1}{12} \rho \frac{W}{L} \tag{3.27}$$

This formula still underestimates the noise in the CMOS transistor. Noise modeling in CMOS transistors is still an area of active research.

3.11.4 CMOS Square Law Equations

As with bipolar transistors, some simplistic equations for calculating model parameters will now be shown. In the saturation region of operation, the current can be described by [4]

$$i_{DS} = \frac{\mu C_{ox}}{2} \left(\frac{W}{L}\right) \frac{(v_{GS} - V_T)^2}{1 + \alpha(v_{GS} - V_T)} (1 + \lambda v_{DS}) \tag{3.28}$$

where V_T is the threshold voltage and λ is the output slope factor given by

$$g_o = \frac{di_{DS}}{dv_{DS}} = I\lambda \tag{3.29}$$

and α approximately models the combined mobility degradation and velocity saturation effects given by [4]

$$\alpha = \theta + \frac{\mu_0}{2nv_{\text{sat}}L} \tag{3.30}$$

where θ is the mobility-reduction coefficient and v_{sat} is the saturation velocity. We note that for small values of α or small overdrive voltage ($v_{GS} - V_T$), (3.28) becomes the familiar square law equation, as in (3.31).

$$i_{DS} = \frac{\mu C_{ox}}{2} \left(\frac{W}{L}\right) (v_{GS} - V_T)^2 (1 + \lambda v_{DS}) \tag{3.31}$$

The transconductance is given by the derivative of the current with respect to the input voltage. For the simple square law equation, this becomes

$$g_m = \frac{di_{DS}}{dv_{GS}} = \mu C_{ox}\left(\frac{W}{L}\right)(v_{GS} - V_T)(1 + \lambda v_{DS}) \qquad (3.32)$$

This can also be shown to be equal to (note the λ term has been left out)

$$g_m = \sqrt{2\mu C_{ox}\frac{W}{L}I_{DS}} \qquad (3.33)$$

In the triode region of operation, current is given by

$$i_{DS} = \mu C_{ox}\left(\frac{W}{L}\right)\left(v_{GS} - V_T - \frac{v_{DS}}{2}\right)v_{DS}(1 + \lambda v_{DS}) \qquad (3.34)$$

In practice, for RF design, short-channel devices are used. The equations for these devices are poor; thus, it is necessary to use simulators to find the curves. However, even the more complex models used in the simulators are also typically poor, especially in modeling output conductance and phase shift at high frequency. Thus, measurements are needed to verify any designs or to refine the models.

References

[1] Taur, Y., and T. H. Ning, *Fundamentals of Modern VLSI Devices*, Cambridge, England: Cambridge University Press, 1998.

[2] Plummer, J. D., P. B. Griffin, and M. D. Deal, *Silicon VLSI Technology: Fundamentals, Practice, and Modeling*, Upper Saddle River, NJ: Prentice Hall, 2000.

[3] Sedra, A. S., and K. C. Smith, *Microelectronic Circuits*, 4th ed., New York: Oxford University Press, 1998.

[4] Terrovitis, M. T., and R. G. Meyer, "Intermodulation Distortion in Current-Commutating CMOS Mixers," *IEEE J. of Solid-State Circuits*, Vol. 35, No. 10, Oct. 2000, pp. 1461–1473.

[5] Roulston, D. J., *Bipolar Semiconductor Devices*, New York: McGraw-Hill, 1990.

[6] Streetman, B. G., *Solid-State Electronic Devices*, 3rd ed., Englewood Cliffs, NJ: Prentice Hall, 1990.

[7] Muller, R. S., and T. I. Kamins, *Device Electronics for Integrated Circuits*, New York: John Wiley & Sons, 1986.

[8] Sze, S. M., *High Speed Semiconductor Devices,* New York: John Wiley & Sons, 1990.

[9] Sze, S. M., *Modern Semiconductor Device Physics,* New York: John Wiley & Sons, 1997.

[10] Cooke, H., "Microwave Transistors Theory and Design," *Proc. IEEE,* Vol. 59, Aug. 1971, pp. 1163–1181.

4

Impedance Matching

4.1 Introduction

In RF circuits, we very seldom start with the impedance that we would like. Therefore, we need to develop techniques for transforming an arbitrary impedance into the impedance of choice. For example, consider the RF system shown in Figure 4.1. Here the source and load are 50Ω (a very popular impedance), as are the transmission lines leading up to the IC. For optimum power transfer, prevention of ringing and radiation, and good noise behavior, for example, we need the circuit input and output impedances matched to the system. In general, some matching circuit must almost always be added to the circuit, as shown in Figure 4.2.

Typically, reactive matching circuits are used because they are lossless and because they do not add noise to the circuit. However, using reactive matching components means that the circuit will only be matched over a range of frequencies and not at others. If a broadband match is required, then other techniques may need to be used. An example of matching a transistor amplifier with a capacitive input is shown in Figure 4.3. The series inductance adds an impedance

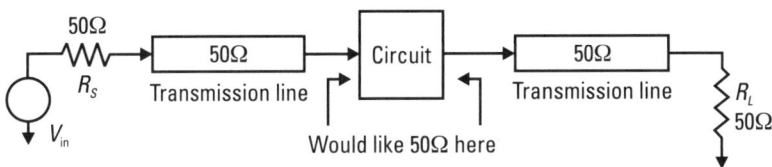

Figure 4.1 Circuit embedded in a 50-Ω system.

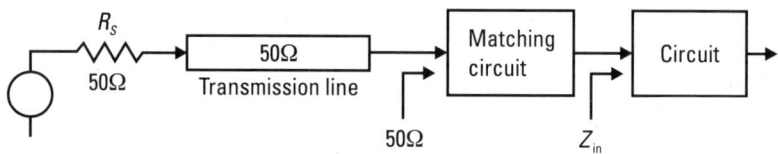

Figure 4.2 Circuit embedded in a 50-Ω system with matching circuit.

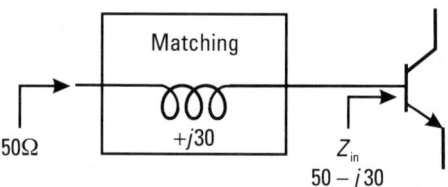

Figure 4.3 Example of a very simple matching network.

of $j\omega L$ to cancel the input capacitive impedance. Note that, in general, when an impedance is complex $(R + jX)$, then to match it, the impedance must be driven from its complex conjugate $(R - jX)$.

A more general matching circuit is required if the real part is not 50Ω. For example, if the real part of Z_{in} is less than 50Ω, then the circuit can be matched using the circuit in Figure 4.4 and described in Example 4.1.

Example 4.1 Matching Using Algebra Techniques

A possible impedance-matching network is shown in Figure 4.4. Use the matching network to match the transistor input impedance $Z_{in} = 40 - j30\Omega$ to $Z_o = 50\Omega$. Perform the matching at 2 GHz.

Solution
We can solve for Z_2 and Y_3, where for convenience we have chosen impedance for series components and admittance for parallel components. An expression for Z_2 is

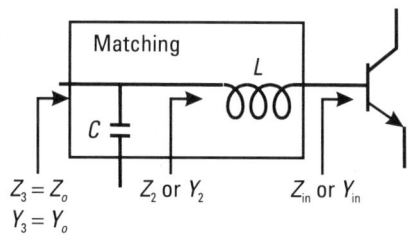

Figure 4.4 A possible impedance-matching network.

$$Z_2 = Z_{in} + j\omega L$$

where $Z_{in} = R_{in} - jX_{in}$. Solving for Y_3 and equating it to the reference admittance Y_o,

$$Y_3 = Y_2 + j\omega C = \frac{1}{Z_o} = Y_o$$

Using the above two equations, to eliminate Y_2 leaving only L and C as unknowns,

$$\frac{1}{Z_{in} + j\omega L} = Y_o - j\omega C$$

Solving the real and imaginary parts of this equation, values for C and L can be found. With some manipulation,

$$\frac{R_{in} - j(\omega L - X_{in})}{R_{in}^2 + (\omega L - X_{in})^2} = Y_o - j\omega C$$

the real part of this equation gives

$$\omega L = X_{in} + \sqrt{\frac{R_{in} - Y_o R_{in}^2}{Y_o}} = 30 + \sqrt{\frac{40 - (0.02)(40)^2}{0.02}} = 50$$

Now using the imaginary half part of the equation,

$$\omega C = \frac{\omega L - X_{in}}{R_{in}^2 + (\omega L - X_{in})^2} = \frac{50 - 30}{40^2 + (50 - 30)^2} = 0.01$$

At 2 GHz it is straightforward to determine that L is equal to 3.98 nH and C is equal to 796 fF. We note that the impedance is matched exactly only at 2 GHz. We also note that this matching network cannot be used to transform all impedances to 50Ω. Other matching circuits will be discussed later.

Although the preceding analysis is very useful for entertaining undergraduates during final exams, in practice there is a more general method for determining a matching network and finding the values. However, first we must review the Smith chart.

4.2 Review of the Smith Chart

The reflection coefficient is a very common figure of merit used to determine how well matched two impedances are. It is related to the ratio of power transmitted to power reflected from the load. A plot of the reflection coefficient is the basis for the Smith chart, which is a very useful way to plot the impedances graphically. The reflection coefficient Γ can be defined in terms of the load impedance Z_L and the characteristic impedance of the system Z_o as follows:

$$\Gamma = \frac{Z_L - Z_o}{Z_L + Z_o} = \frac{z - 1}{z + 1} \quad (4.1)$$

where $z = Z_L/Z_o$ is the normalized impedance. Alternatively, given Γ, one can find Z_L or z as follows:

$$Z_L = Z_o \frac{1 + \Gamma}{1 - \Gamma} \quad (4.2)$$

or

$$z = \frac{1 + \Gamma}{1 - \Gamma} \quad (4.3)$$

For any impedance with a positive real part, it can be shown that

$$0 \leq |\Gamma| \leq 1 \quad (4.4)$$

The reflection coefficient can be plotted on the x-y plane and its value for any impedance will always fall somewhere in the unit circle. Note that for the case where $Z_L = Z_o$, $\Gamma = 0$. This means that the axis of the plot is the point where the load is equal to the characteristic impedance (in other words, perfect matching). Real impedances lie on the real axis from 0Ω at $\Gamma = -1$ to $\infty\Omega$ at $\Gamma = +1$. Purely reactive impedances lie on the unit circle. Thus, impedances can be directly shown normalized to Z_o. Such a plot is called a *Smith chart* and is shown in Figure 4.5. Note that the circular lines on the plot correspond to contours of constant resistance, while the arcing lines correspond to lines of constant reactance. Thus, it is easy to graph any impedance quickly. Table 4.1 shows some impedances that can be used to map out some important points on the Smith chart (it assumes that $Z_o = 50\Omega$).

Just as contours of constant resistance and reactance were plotted on the Smith chart, it is also possible to plot contours of constant conductance and

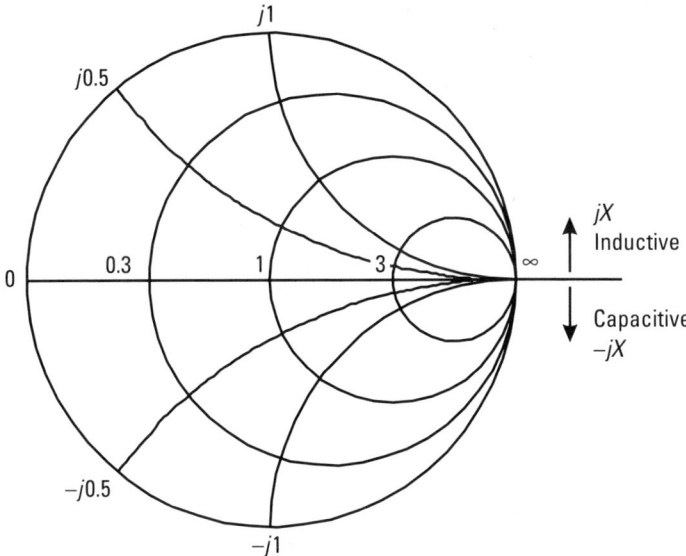

Figure 4.5 A Smith chart.

Table 4.1
Mapping Impedances to Points on the Smith Chart

Z_L	Γ
50	0
0	−1
∞	1
100	0.333
25	−0.333
$j50$	$1\angle 90°$
jX	$1\angle 2\tan^{-1}(X/50)$
$50 - j141.46$	$0.8166\angle -35.26°$

susceptance. Such a chart can also be obtained by rotating the Z chart (impedance Smith chart) by 180°. An admittance Smith chart, or Y chart, is shown in Figure 4.6. This set of admittance curves can be overlaid with the impedance curves to form a ZY Smith chart as shown in Figure 4.7. Then for any impedance Z, the location on the chart can be found, and Y can be read directly or plotted. This will be shown next to be useful in matching, where series or parallel components can be added to move the impedance to the center or to any desired point.

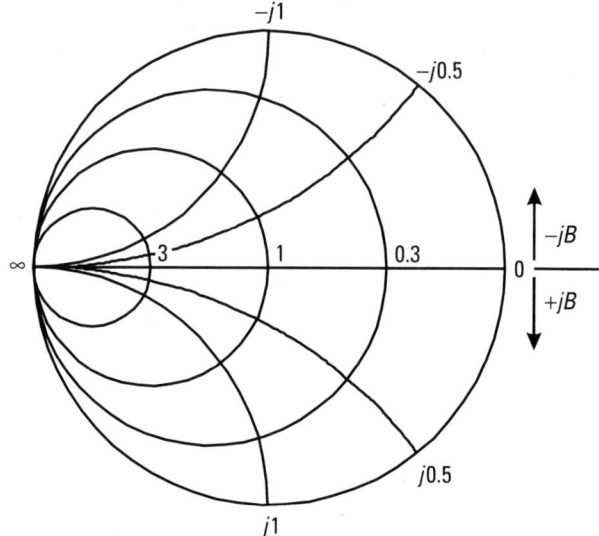

Figure 4.6 An admittance Smith chart or Y Smith chart.

Figure 4.7 A ZY Smith chart.

4.3 Impedance Matching

The input impedance of a circuit can be any value. In order to have the best power transfer into the circuit, it is necessary to match this impedance to the impedance of the source driving the circuit. The output impedance must be similarly matched. It is very common to use reactive components to achieve this impedance transformation, because they do not absorb any power or add noise. Thus, series or parallel inductance or capacitance can be added to the circuit to provide an impedance transformation. Series components will move the impedance along a constant resistance circle on the Smith chart. Parallel components will move the admittance along a constant conductance circle. Table 4.2 summarizes the effect of each component.

With the proper choice of two reactive components, any impedance can be moved to a desired point on the Smith chart. There are eight possible two-component matching networks, also known as *ell* networks, as shown in Figure 4.8. Each will have a region in which a match is possible and a region in which a match is not possible.

In any particular region on the Smith chart, several matching circuits will work and others will not. This is illustrated in Figure 4.9, which shows what matching networks will work in which regions. Since more than one matching network will work in any given region, how does one choose? There are a number of popular reasons for choosing one over another.

1. Sometimes matching components can be used as dc blocks (capacitors) or to provide bias currents (inductors).
2. Some circuits may result in more reasonable component values.
3. Personal preference. Not to be underestimated, sometimes when all paths look equal, you just have to shoot from the hip and pick one.

Table 4.2
Using Lumped Components to Match Circuits

Component Added	Effect	Description of Effect
Series inductor	$z \to z + j\omega L$	Move clockwise along a resistance circle
Series capacitor	$z \to z - j/\omega C$	Smaller capacitance increases impedance $(-j/\omega C)$ to move counterclockwise along a conductance circle
Parallel inductor	$y \to y - j/\omega L$	Smaller inductance increases admittance $(-j/\omega L)$ to move counterclockwise along a conductance circle
Parallel capacitor	$y \to y + j\omega C$	Move clockwise along a conductance circle

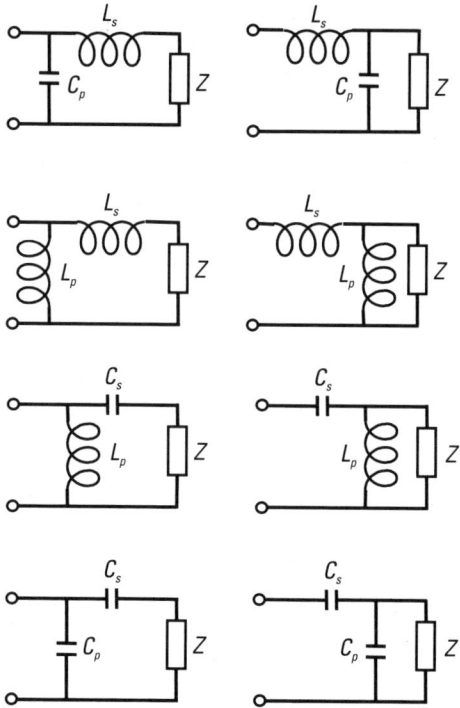

Figure 4.8 The eight possible impedance-matching networks with two reactive components.

4. Stability. Since transistor gain is higher at lower frequencies, there may be a low-frequency stability problem. In such a case, sometimes a highpass network (series capacitor, parallel inductor) at the input may be more stable.

5. Harmonic filtering can be done with a lowpass matching network (series L, parallel C). This may be important, for example, for power amplifiers.

Example 4.2 General Matching Example
Match $Z = 150 - 50j$ to 50Ω using the techniques just developed.

Solution
We first normalize the impedance to 50Ω. Thus, the impedance that we want to match is $3 - 1j$. We plot this on the Smith chart as point A, as shown in Figure 4.10. Now we can see from Figure 4.9 that in this region we have two possible matching networks. We choose arbitrarily to use a parallel capacitor and then a series inductor. Adding a parallel capacitor moves the impedance around a constant conductance circle to point B, which places the impedance

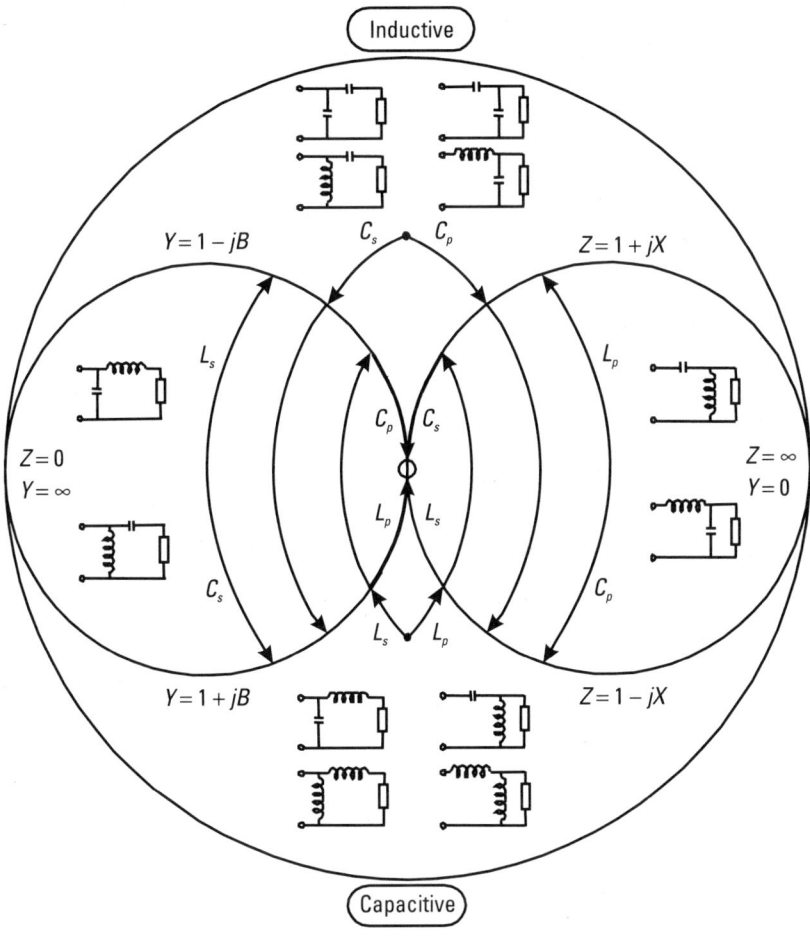

Figure 4.9 Which ell matching networks will work in which regions.

on the 50-Ω resistance circle. Once on the unit circle, a series inductance moves the impedance along a constant resistance circle and moves the impedance to the center at point C. The values can be found by noting that point A is at $Y_A = 1/Z_A = 0.3 + j0.1$ and B is at $Y_B = 0.3 + 0.458j$; therefore, we need a capacitor admittance of $0.348j$. Since $Z_B = 1/Y_B = 1 - 1.528j$, an inductor reactance of $1.528j$ is needed to bring it to the center.

Example 4.3 Illustration of Different Matching Networks

Match a 200-Ω load to a 50-Ω source at 1 GHz with both a lowpass and a highpass matching network, illustrating the filtering properties of the matching network.

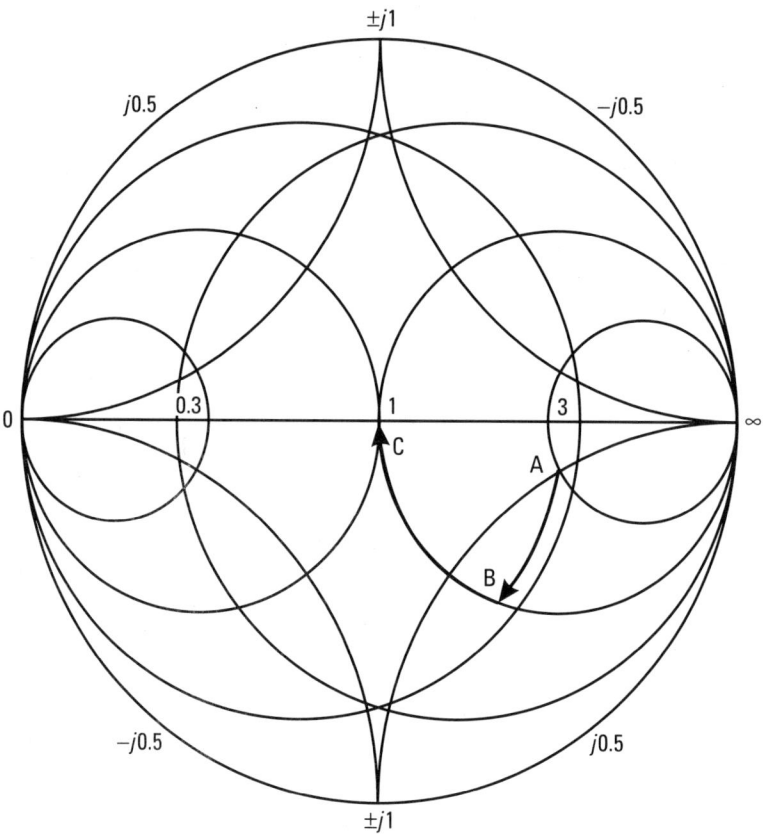

Figure 4.10 Matching process.

Solution

Using the techniques above, two matching circuits as shown in Figure 4.11 are designed. The frequency response can be determined with the results shown in Figure 4.12. It would seem from this diagram that for the lowpass matching network, the signal can be transferred from dc to the −3-dB corner at about 1.53 GHz. However, as seen in the plot of the input impedance in Figure 4.13, the impedance is only matched in a finite band around the center frequency. For the lowpass network, the impedance error is less than 25Ω from 0.78 to 1.57 GHz. It can be noted that if the mismatch between the source and the load is increased, the bandwidth of the matching circuit will be narrower.

For optimal power transfer and minimal noise, impedance should be controlled (although the required impedance for optimal power transfer may not be the value for minimal noise). Also, sources, loads, and connecting cables or transmission lines will be at some specified impedance, typically 50Ω.

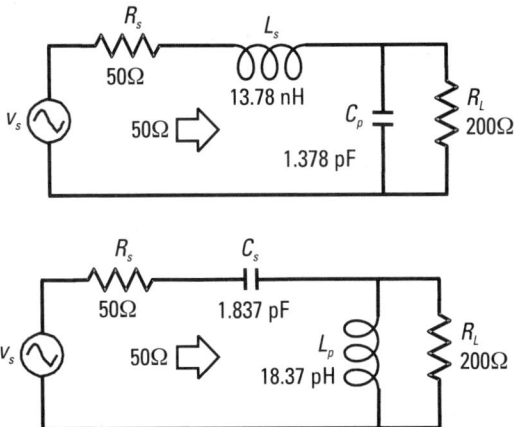

Figure 4.11 Lowpass and highpass matching network for Example 4.4.

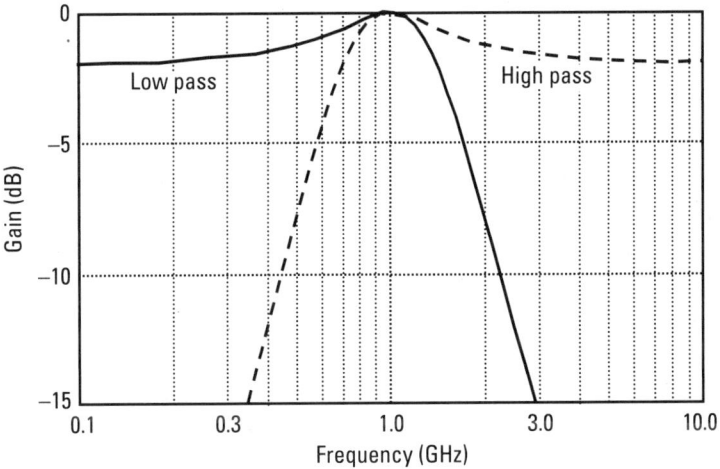

Figure 4.12 Frequency response for lowpass and highpass matching networks.

Example 4.4 The Effect of Matching on Noise
Study the impact of matching on base shot noise. Use the 15x transistor defined in Table 3.1, which is operated at 2 mA at 1 GHz.

Solution
The small-signal model and calculated matching impedances are shown in Figure 4.14. The transistor has an input impedance of 1,250Ω in parallel with 700 fF, which at 1 GHz is equal to $Z_\pi = 40 - j220\Omega$. Using this and the base resistance of 5Ω, the impedance seen by the base shot noise source can be

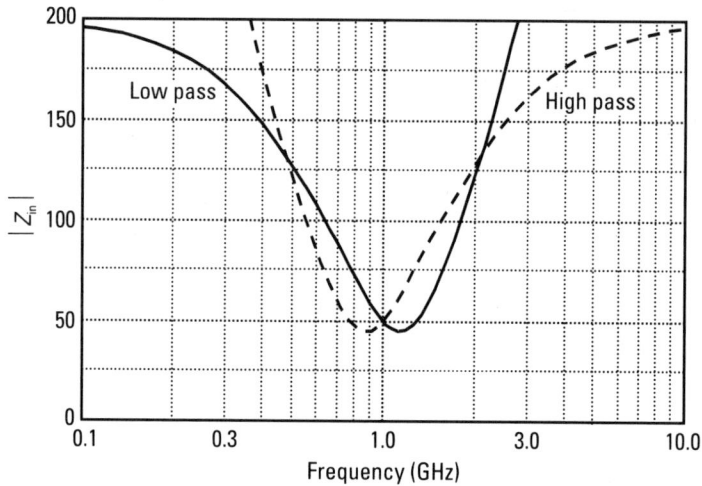

Figure 4.13 Input impedance of lowpass and highpass matching networks.

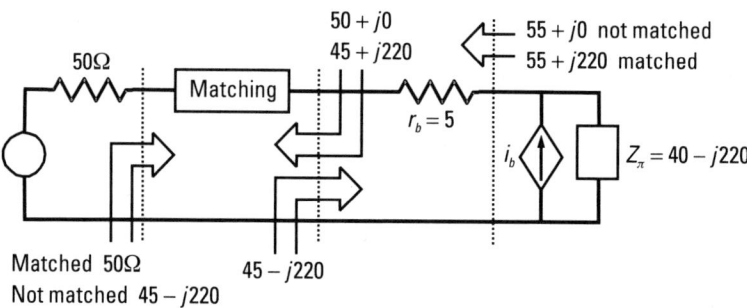

Figure 4.14 Calculation of impact of impedance matching on base shot noise.

determined. Without matching, the input is driven by 50Ω, so the base shot noise sees 55Ω in parallel with $40 - j220\Omega$, which is equal to about $50 - j11.6\Omega$. With matching, the base shot noise sees $50 + j220\Omega$ in parallel with $40 - j220\Omega$ for a net impedance of $560 - j24\Omega$. Thus, with matching, the base shot noise current sees an impedance whose magnitude is about 10 times higher, and thus the impact of the base shot noise is significantly worse with impedance matching.

4.4 Conversions Between Series and Parallel Resistor-Inductor and Resistor-Capacitor Circuits

Series and parallel resistor-capacitor (RC) and resistor-inductor (RL) networks are widely used basic building blocks of matching networks [1, 2]. In this section, conversions between series and parallel forms of these networks will be discussed. All real inductors and capacitors have resistors (either parasitic or intentional) in parallel or series with them. For the purposes of analysis, it is often desirable to replace these elements with equivalent parallel or series resistors, as shown in Figure 4.15.

To convert between a series and parallel RC circuits, we first note that the impedance is

$$Z = R_s + \frac{1}{j\omega C_s} \tag{4.5}$$

Converting to an admittance

$$Y = \frac{j\omega C_s + \omega^2 C_s^2 R_s}{1 + \omega^2 C_s^2 R_s^2} \tag{4.6}$$

Thus, the inverse of the real part of this equation gives R_p:

$$R_p = \frac{1 + \omega^2 C_s^2 R_s^2}{\omega^2 C_s^2 R_s} = R_s(1 + Q^2) \tag{4.7}$$

where Q known as the quality factor is defined as before as $|Z_{Im}|/|Z_{Re}|$ where Z_{Im} is the imaginary part of Z and Z_{Re} is the real part of Z. This definition of Q is convenient for the series network, while the equivalent definition of Q as $|Y_{Im}|/|Y_{Re}|$ is more convenient for a parallel network.

The parallel capacitance is thus

$$C_p = \frac{C_s}{1 + \omega^2 C_s^2 R_s^2} = C_s \left(\frac{Q^2}{1 + Q^2} \right) \tag{4.8}$$

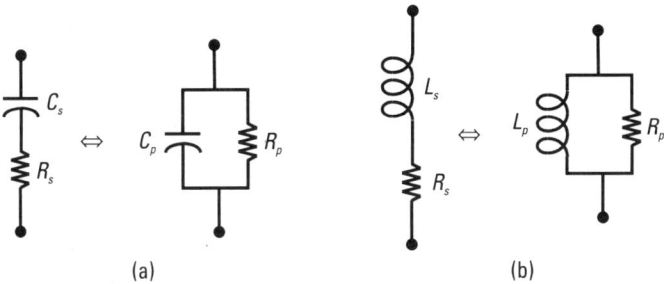

Figure 4.15 Narrowband equivalent models for (a) a capacitor and series resistor, and (b) an inductor and series resistor.

Similarly, for the case of the inductor,

$$R_p = R_s(1 + Q^2) \tag{4.9}$$

$$L_p = L_s \left(\frac{Q^2}{1 + Q^2} \right) \tag{4.10}$$

For large Q, parallel and series L are about the same and similarly parallel and series C are about the same. Also, parallel R is large, while series R is small.

4.5 Tapped Capacitors and Inductors

Another two very common basic circuits that appear often are shown in Figure 4.16. This is the case of two reactive elements with a resistance in parallel with one of the reactive elements. In this case, the two inductors or two capacitors act to transform the resistance into a higher equivalent value in parallel with the equivalent series combination of the two reactances.

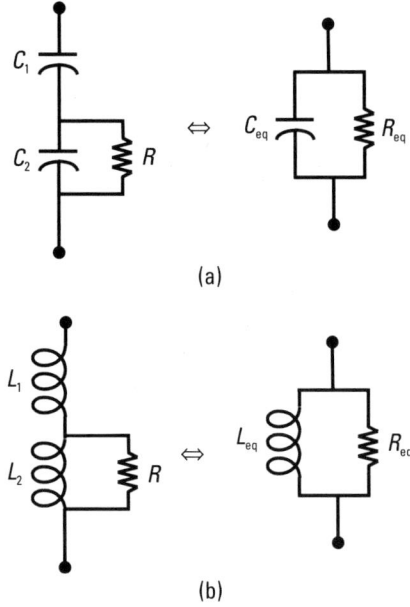

Figure 4.16 Narrowband equivalent models for (a) a tapped capacitor and resistor, and (b) a tapped inductor and resistor.

Much as in the previous section, the analysis of either Figure 4.16(a) or Figure 4.16(b) begins by finding the equivalent impedance of the network. In the case of Figure 4.16(b), the impedance is given by

$$Z_{in} = \frac{j\omega L_1 R + j\omega L_2 R - \omega^2 L_1 L_2}{R + j\omega L_2} \tag{4.11}$$

Equivalently, the admittance can be found:

$$Y_{in} = \frac{j\omega R^2 (L_1 + L_2) - \omega^2 L_2^2 R + j\omega^3 L_1 L_2^2}{-\omega^2 R^2 (L_1 + L_2)^2 - \omega^4 L_1^2 L_2^2} \tag{4.12}$$

Thus, the inverse of the real part of this equation gives R_{eq}:

$$R_{eq} = \frac{-R^2(L_1 + L_2)^2 - \omega^2 L_1^2 L_2^2}{-RL_2^2} = R \left[\frac{(L_1 + L_2)^2 + \frac{L_1^2}{Q_2^2}}{L_2^2} \right] \tag{4.13}$$

where Q_2 is the quality factor of L_2 and R in parallel. As long as Q_2 is large, then a simplification is possible. This is equivalent to stating that the resistance of R is large compared to the impedance of L_2, and the two inductors form a voltage divider.

$$R_{eq} \approx R \left(\frac{L_1 + L_2}{L_2} \right)^2 \tag{4.14}$$

The equivalent inductance of the network can be found as well. Again, the inverse of the imaginary part divided by $j\omega$ is equal to the equivalent inductance:

$$L_{eq} = \frac{\left[R^2(L_1 + L_2)^2 - \omega^2 L_1^2 L_2^2 \right]}{R^2(L_1 + L_2)^2 + \omega^2 L_1 L_2^2} = \frac{(L_1 + L_2)^2 - \frac{L_1^2}{Q_2^2}}{L_1 + L_2 + \frac{L_1}{Q_2^2}} \tag{4.15}$$

Making the same approximation as before, this simplifies to

$$L_{eq} \approx L_1 + L_2 \tag{4.16}$$

which is just the series combination of the two inductors if the resistor is absent.

The same type of analysis can be performed on the network in Figure 4.16(a). In this case,

$$R_{eq} \approx R \left(\frac{C_1 + C_2}{C_1} \right)^2 \tag{4.17}$$

$$C_{eq} \approx \left(\frac{1}{C_1} + \frac{1}{C_2} \right)^{-1} \tag{4.18}$$

4.6 The Concept of Mutual Inductance

Any two coupled inductors that affect each other's magnetic fields and transfer energy back and forth form a transformer. How tightly they are coupled together affects how efficiently they transfer energy back and forth. The amount of coupling between two inductors can be quantified by defining a coupling factor k, which can take on any value between one and zero. Another way to describe the coupling between two inductors is with mutual inductance. For two coupled inductors of value L_p and L_s, coupling factor k and the mutual inductance M as shown in Figure 4.17 are related by

$$k = \frac{M}{\sqrt{L_p L_s}} \tag{4.19}$$

The relationship between voltage and current for two coupled inductors can be written out as follows [3]:

$$V_p = j\omega L_p I_p + j\omega M I_s \tag{4.20}$$
$$V_s = j\omega L_s I_s + j\omega M I_p$$

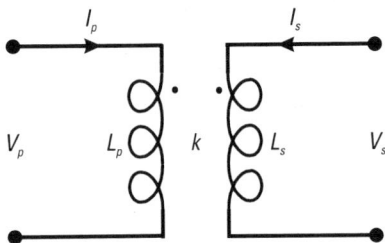

Figure 4.17 A basic transformer structure.

Note that dots in Figure 4.17 are placed such that if current flows in the indicated direction, then fluxes will be added [4]. Equivalently, if I_p is applied and V_s is 0V, current will be induced opposite to I_s to minimize the flux.

For transformers, it is necessary to determine where to place the dots. We illustrate this point in Figure 4.18, where voltages V_1, V_2, and V_3 generate flux through a transformer core. The currents are drawn so that the flux is reinforced. The dots are placed appropriately to agree with Figure 4.17.

An equivalent model for the transformer that uses mutual inductance is shown in Figure 4.19. This model can be shown to be valid if two of the ports are connected together as shown in the figure by writing the equations in terms of I_p and I_s and using the mutual inductance M.

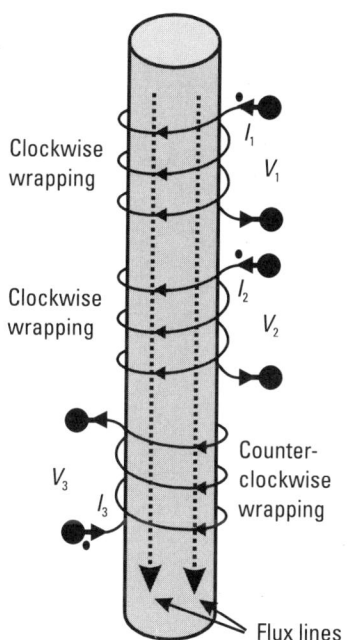

Figure 4.18 Flux lines and determining correct dot placement.

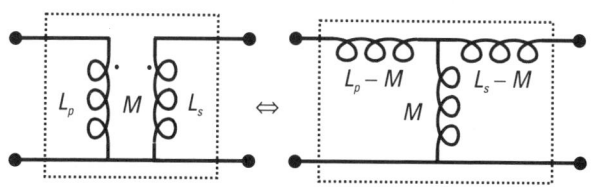

Figure 4.19 Two equivalent models for a transformer.

Example 4.5 Equivalent Impedance of Transformer Networks

Referring to the diagram of Figure 4.20, find the equivalent impedance of each structure, noting the placement of the dots.

Solution

For each structure we apply a test voltage and see what current flows. In the first case, the current flows into the side with the dot of each inductor. In this case, the flux from each structure is added. If we apply a voltage V to the circuit on the left in Figure 4.20, then $V/2$ appears across each inductor. Therefore, for each inductor,

$$\frac{V}{2} = j\omega LI + j\omega MI$$

We can solve for the impedance by

$$Z = \frac{V}{I} = 2j\omega(L + M)$$

Thus, since $Z = j\omega L_{eq}$, we can solve for L_{eq}:

$$L_{eq} = 2L + 2M$$

In the second case for the circuit on the right in Figure 4.20, the dots are placed in such a way that the flux is reduced. We repeat the analysis:

$$\frac{V}{2} = j\omega LI - j\omega MI$$

$$Z = \frac{V}{I} = 2j\omega(L - M)$$

$$L_{eq} = 2L - 2M$$

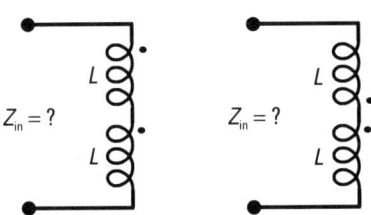

Figure 4.20 Circuits to find the equivalent impedance.

Thus, in the first case the inductance reinforces itself, but in the second case it is decreased.

4.7 Matching Using Transformers

Transformers, as shown in Figure 4.17, can transform one resistance into another resistance depending on the ratio of the inductance of the primary and the secondary. Assuming that the transformer is ideal (that is, the coupling coefficient k is equal to 1, which means that the coupling of magnetic energy is perfect) and lossless, and

$$L_p = NL_s \tag{4.21}$$

then it can be shown from elementary physics that

$$\frac{V_p}{V_s} = \frac{I_s}{I_p} = \sqrt{N} \tag{4.22}$$

Note that here we have defined N as the inductance ratio, but traditionally it is defined as a turns ratio. Since, in an integrated circuit, turns and inductance are not so easily related, this alternative definition is used.

Now if the secondary is loaded with impedance R_s, then the impedance seen in parallel on the primary side R_p will be

$$R_p = \frac{V_p}{I_p} = \frac{V_s\sqrt{N}}{\frac{I_s}{\sqrt{N}}} = \frac{V_s}{I_s}N = R_s N \tag{4.23}$$

Thus, the impedance on the primary and secondary are related by the inductance ratio. Therefore, placing a transformer in a circuit provides the opportunity to transform one impedance into another. However, the above expressions are only valid for an ideal transformer where $k = 1$. Also, if the resistor is placed in series with the transformer rather than in parallel with it, then the resistor and inductor will form a voltage divider, modifying the impedance transformation. In order to prevent the voltage divider from being a problem, the transformer must be tuned or resonated with a capacitor so that it provides an open circuit at a particular frequency at which the match is being performed. Thus, there is a trade-off in a real transformer between near-ideal behavior and bandwidth. Of course, the losses in the winding and substrate cannot be avoided.

4.8 Tuning a Transformer

Unlike the previous case where the transformer was assumed to be ideal, in a real transformer there are losses. Since there is inductance in the primary and secondary, this must be resonated out if the circuit is to be matched to a real impedance. To do a more accurate analysis, we start with the equivalent model for the transformer loaded on the secondary with resistance R_L, as shown in Figure 4.21.

Next, we find the equivalent admittance looking into the primary. Through circuit analysis, it can be shown that

$$Y_{in} = \frac{-R_L \omega^2 (L_s L_p - M^2) - j\omega^3 (L_s L_p - M^2) - j\omega R_L^2 L_p + \omega^2 L_s L_p R_L}{\omega^4 (L_s L_p - M^2)^2 + \omega^2 R_s^2 L_p^2}$$

(4.24)

Taking the imaginary part of this expression, the inductance seen looking into the primary $L_{\text{eff-}p}$ can be found, making use of (4.19) to express the results in terms of the coupling coefficient k:

$$L_{\text{eff-}p} = \frac{\omega^2 L_s^2 L_p (1 - k^2)^2 + R_L^2 L_p}{\omega^2 L_s^2 (1 - k^2) + R_L^2}$$

(4.25)

When $k = 1$ or 0, then the inductance is simply L_p. When k has a value between these two limits, then the inductance will be reduced slightly from this value, depending on circuit values. Thus, a transformer can be made to resonate and have a zero reactive component at a particular frequency using a capacitor on either the primary C_p or secondary C_s:

$$\omega_o = \frac{1}{\sqrt{L_{\text{eff-}p} C_p}} = \frac{1}{\sqrt{L_{\text{eff-}s} C_s}}$$

(4.26)

where $L_{\text{eff-}s}$ is the inductance seen looking into its secondary.

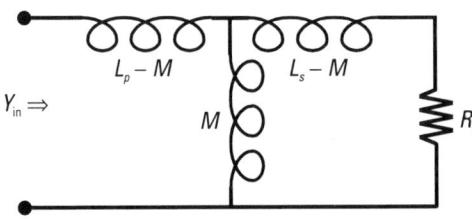

Figure 4.21 Real transformer used to transform one resistance into another.

The exact resistance transformation can also be extracted and is given by

$$R_{\text{eff}} = \frac{R_L^2 L_p - \omega^2 L_s^2 L_p (1-k^2)^2}{R_L L_s k^2} \quad (4.27)$$

Note again that if $k = 1$, then $R_{\text{eff}} = R_L \cdot N$ and goes to infinity as k goes to zero.

4.9 The Bandwidth of an Impedance Transformation Network

Using the theory already developed, it is possible to make most matching networks into equivalent parallel or series inductance, resistance, capacitance (LRC) circuits, such as the one shown in Figure 4.22. The transfer function for this circuit is determined by its impedance, which is given by

$$\frac{V_{\text{out}}(s)}{I_{\text{in}}(s)} = \frac{1}{C}\left(\frac{s}{s^2 + \frac{s}{RC} + \frac{1}{LC}}\right) \quad (4.28)$$

In general, this second-order transfer function has the form

$$A(s) = \frac{A_o s}{s^2 + s\text{BW} + \omega_o^2} \quad (4.29)$$

where

$$\text{BW} = \frac{1}{RC} \quad (4.30)$$

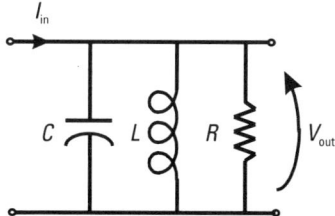

Figure 4.22 An inductor, capacitor (LC) resonator with resistive loss.

and

$$\omega_o = \sqrt{\frac{1}{LC}} \qquad (4.31)$$

This is an example of a damped second-order system with poles in the left-hand half plane, as shown in Figure 4.23 (provided that R is positive and finite). This system will have a frequency response that is centered on a given resonance frequency ω_o and will fall off on either side of this frequency, as shown in Figure 4.24. The distance on either side of the resonance frequency where the transfer function falls in amplitude by 3 dB is usually defined as the circuit bandwidth. This is the frequency at which the gain of the transfer function is down by 3 dB relative to the gain at the center frequency.

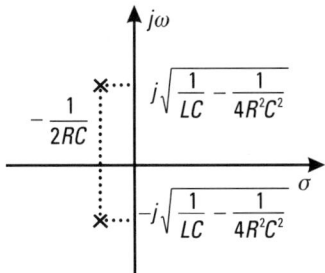

Figure 4.23 Pole plot of an undamped LC resonator.

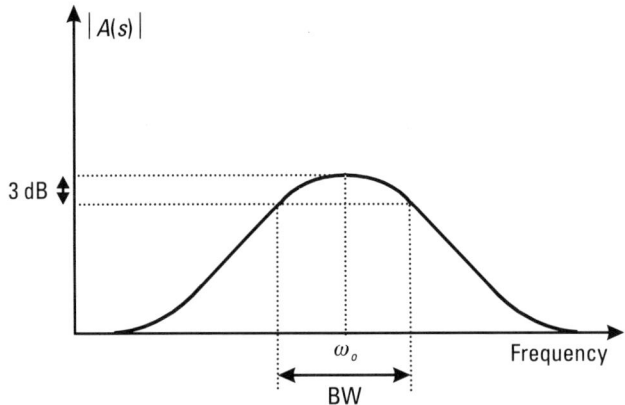

Figure 4.24 Plot of a general second-order bandpass transfer function.

4.10 Quality Factor of an LC Resonator

The Q (quality factor) of an LC resonator is another figure of merit used. It is defined as

$$Q = 2\pi \left(\frac{E_{\text{stored/cycle}}}{E_{\text{lost/cycle}}} \right) \quad (4.32)$$

This can be used as a starting point to define Q in terms of circuit parameters.

We first note that all the loss must occur in the resistor, because it is the only element present capable of dissipating any energy and the energy dissipated per cycle is

$$E_{\text{lost/cycle}} = \int_0^T \frac{V_{\text{osc}}^2 \sin^2(\omega_{\text{osc}} t)}{R} dt = \frac{1}{2} V_{\text{osc}}^2 \frac{T}{R} \quad (4.33)$$

Energy is also stored each cycle in the capacitor and the Q is therefore given by

$$E_{\text{stored/cycle}} = \frac{1}{2} CV_{\text{osc}}^2 \Rightarrow Q = 2\pi \frac{CR}{T} = CR\omega_{\text{osc}} = R\sqrt{\frac{C}{L}} \quad (4.34)$$

Another definition of Q that is particularly useful is [5]

$$Q = \frac{\omega_o}{2} \left| \frac{d\phi}{d\omega} \right| \quad (4.35)$$

where ϕ is the phase of the resonator and $d\phi/d\omega$ is the rate of change of the phase transfer function with respect to frequency. This can be shown to give the same value in terms of circuit parameters as (4.32).

The Q of a resonator can also be related to its center frequency and bandwidth, noting that

$$Q = R\sqrt{\frac{C}{L}} = \frac{RC}{\sqrt{LC}} = \frac{\omega_o}{\text{BW}} \quad (4.36)$$

Example 4.6 Matching a Transistor Input with a Transformer

A circuit has an input that is made up of a 1-pF capacitor in parallel with a 200-Ω resistor. Use a transformer with a coupling factor of 0.8 to match it to a source resistance of 50Ω. The matching circuit must have a bandwidth of 200 MHz and the circuit is to operate at 2 GHz.

Solution
The matching circuit will look much like that shown in Figure 4.25. We will use the secondary of the transformer as a resonant circuit so that there will be no reactance at 2 GHz. We first add capacitance in parallel with the input capacitance so that the circuit will have the correct bandwidth: Using (4.30),

$$C_{total} = \frac{1}{RBW} = \frac{1}{(100\Omega)(2\pi \cdot 200 \text{ MHz})} = 7.96 \text{ pF}$$

Note that the secondary "sees" 100Ω total, due to 200Ω from the load and 200Ω from the source resistance. This means that C_{extra} must be 6.96 pF. Now, to resonate at 2 GHz, this means that the secondary of the transformer must have an inductance of

$$L_s = \frac{1}{\omega_o^2 C_s} = \frac{1}{(2\pi \cdot 2 \text{ GHz})^2 (7.96 \text{ pF})} = 0.8 \text{ nH}$$

Now we must set the inductance ratio to turn 200Ω into 50Ω:

$$L_p = \frac{R_{eff} R_L L_s k^2}{R_L^2 - \omega_o^2 L_s^2 (1 - k^2)^2}$$

$$= \frac{50\Omega \cdot 200\Omega \cdot 0.8 \text{ nH} \cdot (0.8)^2}{(200\Omega)^2 - (2\pi \cdot 2 \text{ GHz})^2 (0.8 \text{ nH})^2 (1 - 0.8^2)^2}$$

$$= 0.13 \text{ nH}$$

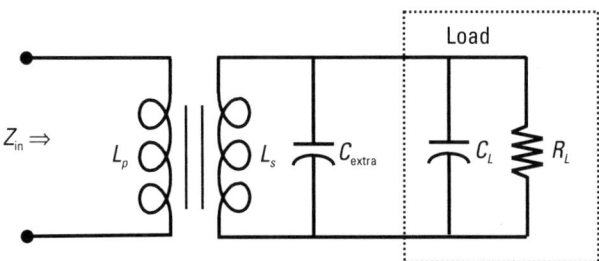

Figure 4.25 Transformer matching network used to match the input of a transistor.

Example 4.7 Matching Using a Two-Stage Ell Network

Match 200Ω to 50Ω at 1 GHz using an ell matching network. Do it first in one step, then do it in two steps matching it first to 100Ω. Compare the bandwidth of the two matching networks.

Solution

Figure 4.26 illustrates matching done in "one step" (with movement from a to b to c) versus matching done in "two steps" (with movement from a to d to c). One-step matching was previously shown in Example 4.3 and Figure 4.11. Two-step matching calculations are also straightforward, with an ell network converting from 200Ω to 100Ω, and then another ell network converting from 100Ω to 50Ω. The resulting network is shown in Figure 4.27.

A comparison of frequency response shown in Figure 4.28 clearly shows the bandwidth broadening effect of matching in two steps. To quantify the effect, the magnitude of the input impedance is shown in Figure 4.29.

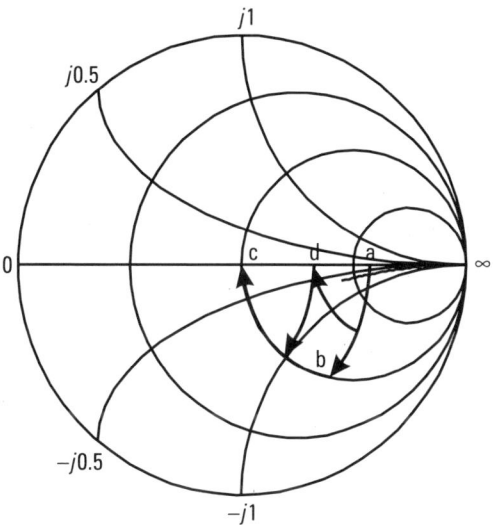

Figure 4.26 Smith chart illustration of one-step versus two-step matching.

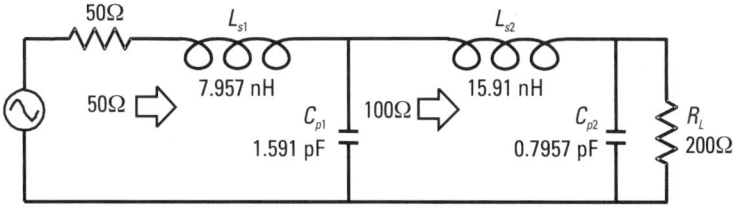

Figure 4.27 Circuit for two-step matching.

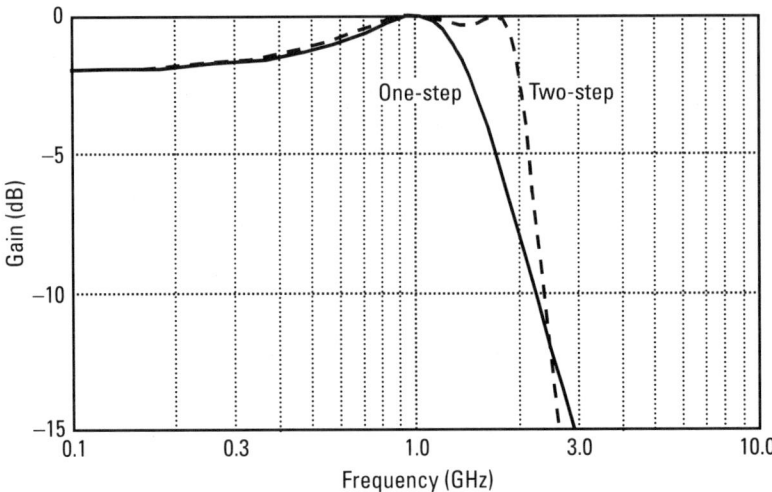

Figure 4.28 Frequency response for one-step and two-step matching.

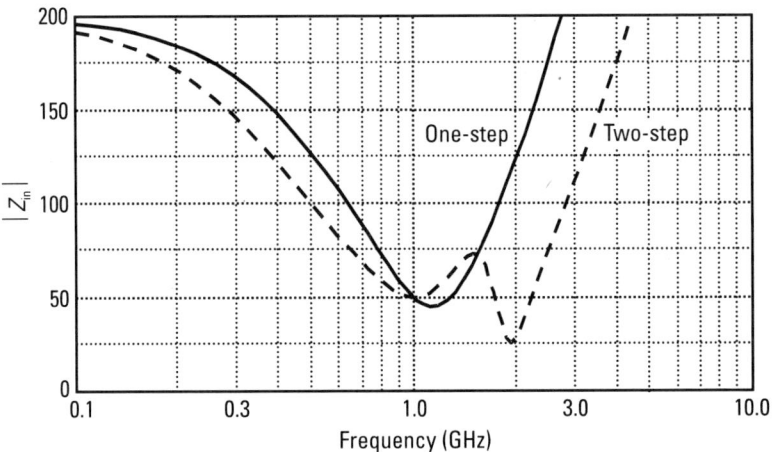

Figure 4.29 Input impedance for one-step and two-step matching.

4.11 Transmission Lines

When designing circuits on chip, transmission line effects can often be ignored, but at chip boundaries they are very important. Transmission lines have effects that must be considered at these interfaces in order to match the input or output of an RFIC. As already discussed, transmission lines have a characteristic impedance, and when they are loaded with an impedance different from this characteristic impedance, they cause the impedance looking into the transmission line to change with distance. If the transmission line, such as that shown in Figure 4.30, is considered lossless, then the input impedance at any distance d from the load is given by [6]

$$Z_{in}(d) = Z_o \left[\frac{Z_L + jZ_o \tan\left(\frac{2\pi}{\lambda}d\right)}{Z_o + jZ_L \tan\left(\frac{2\pi}{\lambda}d\right)} \right] \quad (4.37)$$

where λ is one wavelength of an electromagnetic wave at the frequency of interest in the transmission line. A brief review of how to calculate λ will be given in Chapter 5. Thus, the impedance looking into the transmission line is periodic with distance. It can be shown from (4.37) that for each distance λ traveled down the transmission line, the impedance makes two clockwise rotations about the center of the Smith chart.

Transmission lines can also be used to synthesize reactive impedances. Note that if Z_L is either an open or short circuit, then by making the transmission line an appropriate length, any purely reactive impedance can be realized. These types of transmission lines are usually referred to as *open-circuit* and *short-circuit stubs*.

4.12 S, Y, and Z Parameters

S, Y, and Z parameters (scattering, admittance, and impedance parameters, respectively) are widely used in the analysis of RF circuits. For RF measurements, for example, with a network analyzer, S parameters are typically used. These may be later converted to Y or Z parameters in order to perform certain analyses. In this section, S parameters and conversions to other parameters will be described.

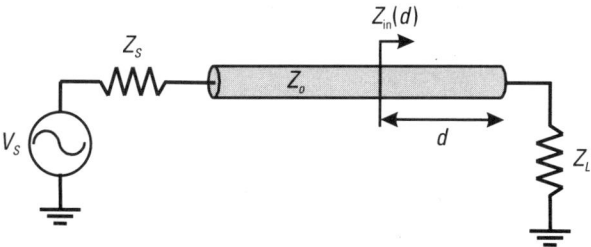

Figure 4.30 Impedance seen moving down a transmission line.

S parameters are a way of calculating a two-port network in terms of incident and reflected (or scattered) power. Referring to Figure 4.31, assuming port 1 is the input, a_1 is the input wave, b_1 is the reflected wave, and b_2 is the transmitted wave. We note that if a transmission line is terminated in its characteristic impedance, then the load absorbs all incident power traveling along the transmission line and there is no reflection.

The S parameters can be used to describe the relationship between these waves as follows:

$$b_1 = S_{11}a_1 + S_{12}a_2 \quad (4.38)$$

$$b_2 = S_{22}a_2 + S_{21}a_1 \quad (4.39)$$

This can also be written in a matrix as

$$\begin{bmatrix} b_1 \\ b_2 \end{bmatrix} = \begin{bmatrix} S_{11} & S_{12} \\ S_{21} & S_{22} \end{bmatrix} \begin{bmatrix} a_1 \\ a_2 \end{bmatrix} = [b] = [S][a] \quad (4.40)$$

Thus, S parameters are reflection or transmission coefficients and are usually normalized to a particular impedance. S_{11}, S_{22}, S_{21}, and S_{12} will now each be defined.

$$S_{11} = \frac{b_1}{a_1}\bigg|_{a_2=0} \quad (4.41)$$

S_{11} is the input reflection coefficient measured with the output terminated with Z_o. This means the output is matched and all power is transmitted into the load; thus a_2 is zero.

$$S_{21} = \frac{b_2}{a_1}\bigg|_{a_2=0} \quad (4.42)$$

S_{21}, the forward transmission coefficient, is also measured with the output terminated with Z_o. S_{21} is equivalent to gain.

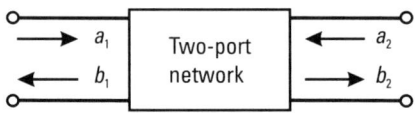

Figure 4.31 General two-port system with incident and reflected waves.

$$S_{22} = \frac{b_2}{a_2}\bigg|_{a_1=0} \qquad (4.43)$$

S_{22} is the output reflection coefficient measured by applying a source at the output and with the input terminated with Z_o.

$$S_{12} = \frac{b_1}{a_2}\bigg|_{a_1=0} \qquad (4.44)$$

S_{12} is the reverse transmission coefficient measured with the input terminated with Z_o.

In addition to S parameters, there are many other parameter sets that can be used to characterize a two-port network. Since engineers are used to thinking in terms of voltages and currents, another popular set of parameters are the Z and Y parameters shown in Figure 4.32.

$$\begin{bmatrix} v_1 \\ v_2 \end{bmatrix} = \begin{bmatrix} Z_{11} & Z_{12} \\ Z_{21} & Z_{22} \end{bmatrix} \begin{bmatrix} i_1 \\ i_2 \end{bmatrix} \qquad (4.45)$$

$$\begin{bmatrix} i_1 \\ i_2 \end{bmatrix} = \begin{bmatrix} Y_{11} & Y_{12} \\ Y_{21} & Y_{22} \end{bmatrix} \begin{bmatrix} v_1 \\ v_2 \end{bmatrix} \qquad (4.46)$$

It is also useful to be able to translate from one set of these parameters to the other. These relationships are well known and are summarized in Table 4.3.

Microwave transistors or amplifiers are often completely (and exclusively) characterized with S parameters. For radio frequency integrated circuits, detailed transistor models are typically used that allow the designer (with the help of the simulator) to design circuits. The models and simulators can be used to find S parameters, which can be used with the well-known microwave techniques to find, for example, maximum gain, optimal noise figure, and stability. However, the simulators, which use the models to generate the S parameters, can be used directly to find maximum gain, optimal noise figure, and stability

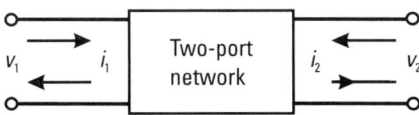

Figure 4.32 General two-port system with input and output currents and voltages.

Table 4.3
Relationships Between Different Parameter Sets

S	S	Z	Y
S_{11}	S_{11}	$\dfrac{(Z_{11}-Z_o)(Z_{22}+Z_o)-Z_{12}Z_{21}}{(Z_{11}+Z_o)(Z_{22}+Z_o)-Z_{12}Z_{21}}$	$\dfrac{(Y_o-Y_{11})(Y_{22}+Y_o)+Y_{12}Y_{21}}{(Y_{11}+Y_o)(Y_{22}+Y_o)-Y_{12}Y_{21}}$
S_{12}	S_{12}	$\dfrac{2Z_{12}Z_o}{(Z_{11}+Z_o)(Z_{22}+Z_o)-Z_{12}Z_{21}}$	$\dfrac{-2Y_{12}Y_o}{(Y_{11}+Y_o)(Y_{22}+Y_o)-Y_{12}Y_{21}}$
S_{21}	S_{21}	$\dfrac{2Z_{21}Z_o}{(Z_{11}+Z_o)(Z_{22}+Z_o)-Z_{12}Z_{21}}$	$\dfrac{-2Y_{21}Y_o}{(Y_{11}+Y_o)(Y_{22}+Y_o)-Y_{12}Y_{21}}$
S_{22}	S_{22}	$\dfrac{(Z_{11}+Z_o)(Z_{22}-Z_o)-Z_{12}Z_{21}}{(Z_{11}+Z_o)(Z_{22}+Z_o)-Z_{12}Z_{21}}$	$\dfrac{(Y_o+Y_{11})(Y_o-Y_{22})+Y_{12}Y_{21}}{(Y_{11}+Y_o)(Y_{22}+Y_o)-Y_{12}Y_{21}}$
Z_{11}	$Z_o\dfrac{(1+S_{11})(1-S_{22})+S_{12}S_{21}}{(1-S_{11})(1-S_{22})-S_{12}S_{21}}$	Z_{11}	$\dfrac{Y_{22}}{Y_{11}Y_{22}-Y_{12}Y_{21}}$
Z_{12}	$Z_o\dfrac{2S_{12}}{(1-S_{11})(1-S_{22})-S_{12}S_{21}}$	Z_{12}	$\dfrac{-Y_{12}}{Y_{11}Y_{22}-Y_{12}Y_{21}}$
Z_{21}	$Z_o\dfrac{2S_{21}}{(1-S_{11})(1-S_{22})-S_{12}S_{21}}$	Z_{21}	$\dfrac{-Y_{21}}{Y_{11}Y_{22}-Y_{12}Y_{21}}$
Z_{22}	$Z_o\dfrac{(1+S_{22})(1-S_{11})+S_{12}S_{21}}{(1-S_{11})(1-S_{22})-S_{12}S_{21}}$	Z_{22}	$\dfrac{Y_{11}}{Y_{11}Y_{22}-Y_{12}Y_{21}}$
Y_{11}	$Y_o\dfrac{(1+S_{22})(1-S_{11})+S_{12}S_{21}}{(1+S_{11})(1+S_{22})-S_{12}S_{21}}$	$\dfrac{Z_{22}}{Z_{11}Z_{22}-Z_{12}Z_{21}}$	Y_{11}
Y_{12}	$Y_o\dfrac{-2S_{12}}{(1+S_{11})(1+S_{22})-S_{12}S_{21}}$	$\dfrac{-Z_{12}}{Z_{11}Z_{22}-Z_{12}Z_{21}}$	Y_{12}
Y_{21}	$Y_o\dfrac{-2S_{21}}{(1+S_{11})(1+S_{22})-S_{12}S_{21}}$	$\dfrac{-Z_{21}}{Z_{11}Z_{22}-Z_{12}Z_{21}}$	Y_{21}
Y_{22}	$Y_o\dfrac{(1-S_{22})(1+S_{11})+S_{12}S_{21}}{(1+S_{11})(1+S_{22})-S_{12}S_{21}}$	$\dfrac{Z_{11}}{Z_{11}Z_{22}-Z_{12}Z_{21}}$	Y_{22}

without the need to generate a list of S parameters. However, it is worthwhile to be familiar with these design techniques, since they can give insight into circuit design, which can be of much more value than simply knowing the location of the "simulate" button.

References

[1] Krauss, H. L., C. W. Bostian, and F. H. Raab, *Solid State Radio Engineering*, New York: John Wiley & Sons, 1980.

[2] Smith, J. R., *Modern Communication Circuits*, 2nd ed., New York: McGraw-Hill, 1998.

[3] Irwin, J. D., *Basic Engineering Circuit Analysis*, New York: Macmillan Publishing Company, 1993.

[4] Sadiku, M. N. O., *Elements of Electromagnetics*, 2nd ed., Fort Worth, TX: Sanders College Publishing, 1994.

[5] Razavi, B., "A Study of Phase Noise in CMOS Oscillators," *IEEE J. Solid-State Circuits*, Vol. 31, March 1996, pp. 331–343.

[6] Pozar, D. M., *Microwave Engineering*, 2nd ed., New York: John Wiley & Sons, 1998.

5

The Use and Design of Passive Circuit Elements in IC Technologies

5.1 Introduction

In this chapter, passive circuit elements will be discussed. First, metallization and back-end processing (away from the silicon) in integrated circuits will be described. This is the starting point for many of the passive components. Then design, modeling, and use of passive components will be discussed. These components are interconnect lines, inductors, capacitors, transmission lines, and transformers. Finally, there will be a discussion of the impact of packaging.

Passive circuit elements such as inductors and capacitors are necessary components in RF circuits, but these components often limit performance, so it is worthwhile to study their design and use. For example, inductors have many applications in RF circuits, as summarized in Table 5.1. An important property of the inductor is that it can simultaneously provide low impedance to dc while providing finite ac impedance. In matching circuits or tuned loads, this allows active circuits to be biased at the supply voltage for maximum linearity. However, inductors are lossy, resulting in increased noise when used in an LNA or oscillator. When used in a power amplifier, losses in inductors can result in decreased efficiency. Also, substrate coupling is a serious concern because of the typically large physical dimensions of the inductor.

5.2 The Technology Back End and Metallization in IC Technologies

After all the front-end processing is complete, the active devices are connected using metal (the back end), which is deposited above the transistors as shown

Table 5.1
Applications and Benefits of Inductors

Circuit	Application	Benefit
LNA	Input match, degeneration	Simultaneous power and noise matching, improved linearity
	Tuned load	Biasing for best linearity, filtering, less problems with parasitic capacitance now part of resonant circuit.
Mixer	Degeneration	Increased linearity, reduced noise
Oscillator	Resonator	Sets oscillating frequency, high Q circuit results in reduced power requirement, lower phase noise
Power amplifier	Matching, loads	Maximize voltage swings, higher efficiency due to swing (inductor losses reduce the efficiency)

in Figure 5.1. The metals must be placed in an insulating layer of silicon dioxide (SiO_2) to prevent different layers of metal from shorting with each other. Most processes have several layers of metal in their back end. These metal layers can also be used to build capacitors, inductors, and even resistors.

The bottom metal is usually tungsten, which is highly resistive. However, unlike aluminum, gold, or copper, this metal has the property that it will not

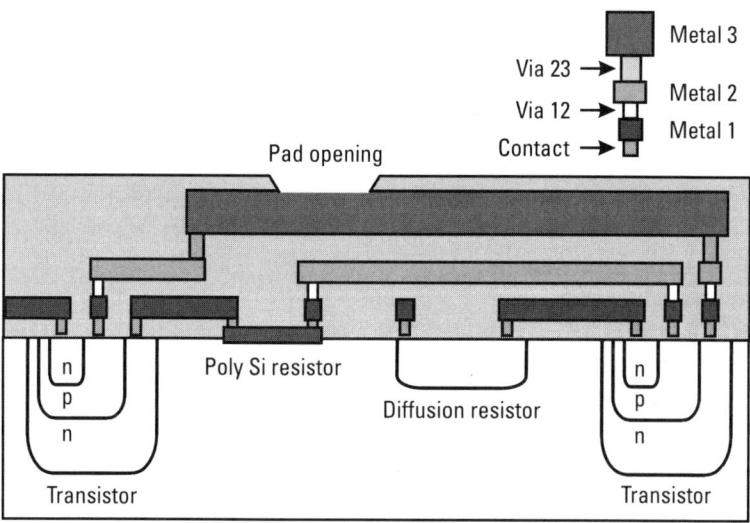

Figure 5.1 Cross section of a typical bipolar back-end process.

diffuse into the silicon. When metals such as copper diffuse into silicon, they cause junctions to leak, seriously impairing the performance of transistors. A contact layer is used to connect this tungsten layer to the active circuitry in the silicon. Higher levels of metal can be connected to adjacent layers using conductive plugs that are commonly called *vias*. Whereas metal can be made in almost any shape desired by the designer, the vias are typically limited to a standard square size. However, it is possible to use arrays of vias to reduce the resistance.

Higher metal layers are often made out of aluminum, as it is much less resistive than tungsten. In some modern processes, copper, which has even lower resistance than aluminum, may be available. The top level of metal will often be made much thicker than the lower levels to provide a low resistance routing option. However, the lithography for this layer may be much coarser than that of underling layers. Thus, the top layers can accommodate a lower density of routing lines.

5.3 Sheet Resistance and the Skin Effect

All conductive materials can be characterized by their resistivity ρ or their conductivity σ. These two quantities are related by

$$\rho = \frac{1}{\sigma} \qquad (5.1)$$

Resistivity is expressed as ohm-meters (Ωm). Knowing the geometry of a metal and its resistivity is enough to estimate the resistance between any two points connected by the metal. As an example, consider the conductor shown in Figure 5.2. To find the resistance along its length, divide the resistivity of the metal by the cross-sectional area and multiply by the length.

$$R = \frac{\rho L}{Wt} \qquad (5.2)$$

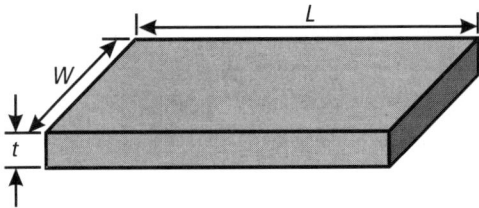

Figure 5.2 Rectangular conductor with current flowing in the direction of L.

Often in IC technologies, sheet resistance is used instead of resistivity. Sheet resistance is given by

$$\rho_s = \frac{\rho}{t} = R\left(\frac{W}{L}\right) \qquad (5.3)$$

Given the sheet resistance, typically expressed as ohms per square ($\Omega/\square$), to find resistance, multiply by the number of squares between input and output. That is to say, for every distance traveled down the conductor equal to its width W, a square has been defined. If the conductor has a length equal to twice its width, then it is two squares long.

As the metal gets thicker, the resistance of the line decreases. However, the lithography of the process becomes harder to control. Thick metal lines close to one another also suffer from capacitance between the two adjacent side walls. At high frequencies, another effect comes into play as well. EM waves suffer attenuation as they enter a conductor, so as the frequency approaches the gigahertz range, the distance that the waves can penetrate becomes comparable to the size of the metal line. The result is that the current becomes concentrated around the outside of the conductor with very little flowing in the center. The depth at which the magnitude of the EM wave is decreased to 36.8% (e^{-1}) of its intensity at the surface is called the *skin depth* of the metal. The skin depth is given by

$$\delta = \sqrt{\frac{\rho}{\pi f \mu}} \qquad (5.4)$$

where f is the frequency and μ is the permeability of the metal. Table 5.2 shows the skin depth of some common metals over the frequency band of interest.

Table 5.2
Skin Depth of Various Metals at Various Frequencies

Metal	$\rho\,(\mu\Omega\cdot\text{cm})$	500 MHz	1 GHz	2 GHz	5 GHz	10 GHz
Gold	2.44	3.5 μm	2.5 μm	1.8 μm	1.1 μm	0.79 μm
Tungsten	5.49	5.3 μm	3.7 μm	2.6 μm	1.7 μm	1.2 μm
Aluminum	2.62	3.6 μm	2.6 μm	1.8 μm	1.2 μm	0.82 μm
Copper	1.72	3.0 μm	2.1 μm	1.5 μm	0.93 μm	0.66 μm
Silver	1.62	2.9 μm	2.0 μm	1.4 μm	0.91 μm	0.64 μm
Nickel	6.90	5.9 μm	4.2 μm	3.0 μm	1.9 μm	1.3 μm

Since most of the applications lie in the 900-MHz to 5-GHz band, it is easy to see that making lines much thicker than about 4 μm will lead to diminishing returns. Going any thicker will yield little advantage at the frequencies of interest, because the center of the conductor will form a *dead zone*, where little current will flow anyway.

Example 5.1 Effect of Skin Depth on Resistance

A rectangular aluminum line has a width of 20 μm, a thickness of 3 μm, and a length of 100 μm. Compute the resistance of the line at dc and at 5 GHz assuming that all the current flows in an area one skin depth from the surface. Assume that aluminum has a resistivity of 3 $\mu\Omega \cdot$ cm. Note that there are more complex equations that describe the resistance due to skin effects, especially for circular conductors [1]; however, the simple estimate used here will illustrate the nature of the skin effect.

Solution

The dc resistance is given by

$$R = \frac{\rho L}{Wt} = \frac{3\ \mu\Omega\text{cm} \cdot 100\ \mu\text{m}}{20\ \mu\text{m} \cdot 3\ \mu\text{m}} = 50\ \text{m}\Omega$$

The skin depth at 5 GHz of aluminum is

$$\delta = \sqrt{\frac{\rho}{\pi f \mu}} = \sqrt{\frac{3\ \mu\Omega \cdot \text{cm}}{\pi \cdot 5\ \text{GHz} \cdot 4\pi \times 10^{-7}\ \frac{N}{A^2}}} = 1.23\ \mu\text{m}$$

We now need to modify the original calculation and divide by the useful cross-sectional area rather than the actual cross-sectional area.

$$R = \frac{\rho L}{Wt - (W - 2\delta)(t - 2\delta)}$$

$$= \frac{3\ \mu\Omega \cdot \text{cm} \cdot 100\ \mu\text{m}}{20\ \mu\text{m} \cdot 2\ \mu\text{m} - 17.5\ \mu\text{m} \cdot 0.54\ \mu\text{m}}$$

$$= 98.2\ \text{m}\Omega$$

This is almost a 100% increase. Thus, while we may be able to count on process engineers to give us thicker metal, this may not solve all our problems.

5.4 Parasitic Capacitance

Metal lines, as well as having resistance associated with them, also have capacitance. Since the metal in an IC technology is embedded in an insulator over a conducting substrate, the metal trace and the substrate form a parallel-plate capacitor. The parasitic capacitance of a metal line can be approximated by

$$C = \frac{\epsilon_0 \epsilon_r A}{h} \tag{5.5}$$

where A is the area of the trace and h is the distance to the substrate.

Since metal lines in ICs can often be quite narrow, the fringing capacitance can be important, as the electric fields cannot be approximated as being perpendicular to the conductor, as shown in Figure 5.3.

For a long line, the capacitance per unit length, taking into account fringing capacitance, can be determined from [2]

$$C = \epsilon_0 \epsilon_r \left[\frac{W}{h} + 0.77 + 1.06 \left(\frac{W}{h} \right)^{1/4} + 1.06 \left(\frac{t}{h} \right)^{1/2} \right] \tag{5.6}$$

We note that the terms in the square brackets are unitless; the final capacitance has the same units as ϵ_0 (ϵ_0 is 8.85×10^{-12} F/m, ϵ_r for SiO$_2$ is 3.9). The first term accounts for the bottom-plate capacitance, while the other three terms account for fringing capacitance. As will be seen in Example 5.2, wider lines will be less affected by fringing capacitance.

Note there is also capacitance between lines vertically and horizontally. A rough estimate of capacitance would be obtained by using the parallel-plate capacitance formula; however, this omits the fringing capacitance, so would be an underestimate. So what is the effect of such capacitance? For one, it can lead to crosstalk between parallel lines, or between lines that cross over. For parallel lines, crosstalk can be reduced by further separation, or by placing a ground line between the two signal-carrying lines.

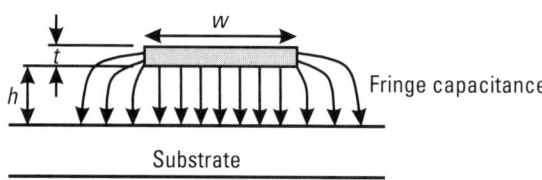

Figure 5.3 Electric field lines showing the effect of fringing capacitance.

Example 5.2 Calculation of Capacitance

Calculate bottom plate capacitance and fringing capacitance for a 1 poly, 4 metal process with distances to substrate and conductor thickness as given in the first two rows of Table 5.3. Calculate for metal widths of 1 μm and 50 μm.

Solution

Bottom plate capacitance can be estimated from (5.5), which is equivalent to the first term in (5.6). Total capacitance can be calculated from (5.6) and the difference attributed to fringing capacitance. Results are shown for the 1- and 50-μm lines in Table 5.3. It can be seen that bottom plate capacitance is a very poor estimate of total capacitance for a 1-μm line. When calculated for a 50-μm width, the bottom plate capacitance and the total capacitance are in much closer agreement. This example clearly shows the inaccuracies inherent in a simple calculation of capacitance. Obviously, it is essential that layout tools have the ability to determine parasitic capacitance accurately.

5.5 Parasitic Inductance

In addition to capacitance to the substrate, metal lines in ICs have inductance. The current flowing in the line will generate magnetic field lines as shown in Figure 5.4. Note that the Xs indicate current flow into the page.

For a flat trace of width w and a distance h above a ground plane, an estimate for inductance in nanohenry per millimeter is [3]

$$L \approx \frac{1.6}{K_f} \cdot \frac{h}{w} \qquad (5.7)$$

Table 5.3
Capacitance for a Line with Width of 1 μm and 50 μm

	Poly	Metal 1	Metal 2	Metal 3	Metal 4
Height above substrate h (μm)	0.4	1.0	2.5	4.0	5.0
Conductor thickness t (μm)	0.4	0.4	0.5	0.6	0.8
Bottom-plate capacitance (aF/μm^2)	86.3	34.5	13.8	8.6	6.9
Total capacitance (aF/μm^2) (1-μm line)	195.5	120.8	85.8	75.2	72.6
Total capacitance (aF/μm^2) (50-μm line)	90.0	37.5	16.2	10.8	9.0

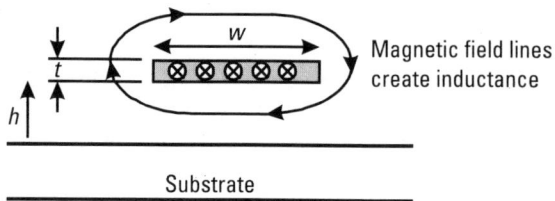

Figure 5.4 Magnetic field lines around an IC line carrying current.

Here K_f is the fringe factor, which can be approximated as

$$K_f \approx 0.72 \cdot \frac{h}{w} + 1 \tag{5.8}$$

Example 5.3 Calculation of Inductance
Calculate the inductance per unit length for traces with a h/w of 0.5, 1, and 2.

Solution
Application of (5.7) and (5.8) shows that for h/w is 0.5, 1, and 2, and the resultant L is 0.59, 0.93, and 1.31 nH/mm. A typical rule of thumb is that bond wires have an inductance of 1 nH/mm. This rule of thumb can also be seen to apply approximately to a metal line on chip.

5.6 Current Handling in Metal Lines

As one can imagine, there is a finite amount of current that can be forced down an IC interconnect before it fails. However, even if the line refrains from exploding, this does not necessarily mean that the current is acceptable for long-term reliability. The main mechanism for loss of reliability is metal migration. Metal migration is related to the level of dc current, and this information is used to specify current limits in an IC. To explain metal migration, consider that normally the diffusion process is random, but with dc current, metal atoms are bombarded more from one side than from the other. This causes the movement of metal atoms, which is referred to as *metal migration*. Sufficient movement in the metal can result in gaps or open circuits appearing in metal and subsequent circuit failure. Any defects or grain boundaries can make the problems worse.

The maximum allowable current in a metal line also depends on the material. For example, aluminum, though lower in resistance, is worse than tungsten for metal migration, due to its much lower melting temperature. Thus, even though there is less energy dissipated per unit length in aluminum, it is less able to handle that energy dissipation.

For 1-μm-thick aluminum, a typical value for maximum current would be 1 mA of dc current for every micrometer of metal width. Similarly, a 2-μm-thick aluminum line would typically be able to carry 2 mA of dc current per micrometer of metal width. The ac current component of the current can be larger (a typical factor of 4 is often used). We note that other metals like copper and gold are somewhat lower in resistance than aluminum; however, due to better metal migration properties, they can handle more current than aluminum.

Example 5.4 Calculating Maximum Line Current

If a line carries no dc current, but has a peak ac current of 500 mA, a 1-μm-thick metal line would need to be about 500 mA/4 mA/μm = 125 μm wide. However, if the dc current is 500 mA, and the peak ac current is also 500 mA (i.e., 500 ± 500 mA), then the 500-μm-wide line required to pass the dc current is no longer quite wide enough. To cope with the additional ac current, another 125 μm is required for a total width of 625 μm.

Current limitations have implications for inductors as well (these will be considered next). Integrated inductors are typically 10 or 20 μm wide and can therefore handle only 10 to 40 mA of dc current, and up to 160 mA of ac current. This obviously limits the ability to do on-chip tuning or matching for power amplifiers or other circuits with high bias current requirements.

5.7 Poly Resistors and Diffusion Resistors

Poly resistors are made out of conductive polycrystalline silicon that is directly on top of the silicon front end. Essentially, this layer acts like a resistive metal line. Typically, these layers have a resistivity in the 10 $\Omega/\square$ range.

Diffusion resistors are made by doping a layer of silicon to give it the desired resistivity, typically 1 k$\Omega/\square$ or more, and can be made with either p doping or n doping as shown in Figure 5.5. If n doping is used, then the structure can be quite simple, because the edge of the doping region will form a pn junction with the substrate. Since this junction can never be forward biased, current will not flow into the substrate. If, however, p doping is used, then it must be placed in an n well to provide isolation from the substrate.

5.8 Metal-Insulator-Metal Capacitors and Poly Capacitors

We have already discussed that metal lines have parasitic capacitance associated with them. However, since it is generally desirable to make capacitance between metal layers as small as possible, they make poor deliberate capacitors. In order to improve this and conserve chip area, when capacitance between two metal

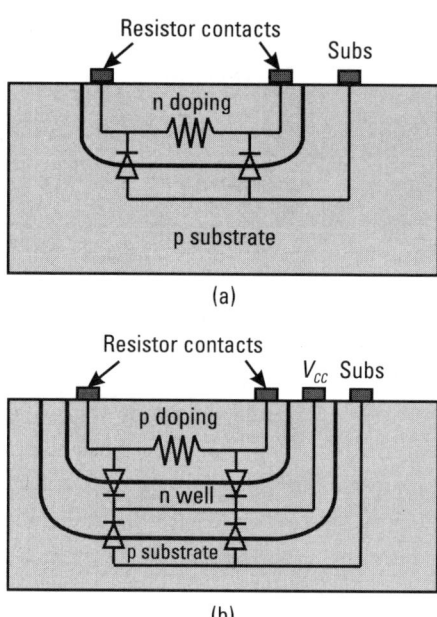

Figure 5.5 Diffusion resistors: (a) diffusion resistor without well isolation (n doping); and (b) diffusion resistor with well isolation (p doping).

lines is deliberate, the oxide between the two lines is thinned to increase the capacitance per unit area. This type of capacitor is called a *metal-insulator-metal capacitor* (MIM cap). More capacitance per unit area saves chip space. The capacitance between any two parallel-plate capacitors is given by (5.5), as discussed previously. Since this expression holds for a wide range of applied voltages, these types of capacitors are extremely linear. However, if there is too much buildup of charge between the plates, they can actually break down and conduct. This is of particular concern during the processing of wafers; thus there are often rules (called *antenna rules*) governing how much metal can be connected directly to the capacitors.

Capacitors can also be made from two layers of poly silicon separated by a layer of dielectric. However, since poly silicon is closer to the substrate, they will therefore have more bottom-plate capacitance. A simple model for an integrated capacitor is shown in Figure 5.6.

5.9 Applications of On-Chip Spiral Inductors and Transformers

The use of the inductor is illustrated in Figure 5.7, in which three inductors are shown in a circuit that is connected to a supply of value V_{CC}. A similar

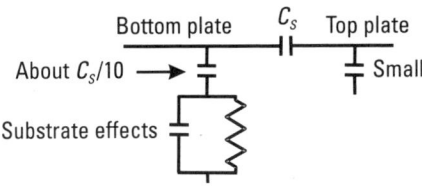

Figure 5.6 Model for an integrated capacitor.

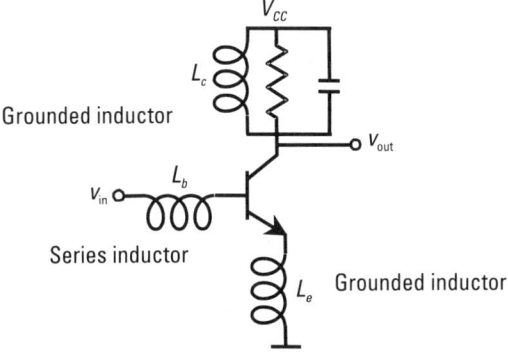

Figure 5.7 Application of inductors and capacitors.

circuit that employs a transformer is shown in Figure 5.8. These two circuits are examples of LNAs, which will be discussed in detail in Chapter 6. The first job of the inductor is to resonate with any parasitic capacitance, potentially allowing higher frequency operation. A side effect (often wanted) is that such resonance results in filtering. Inductors L_b and L_e form the input match and

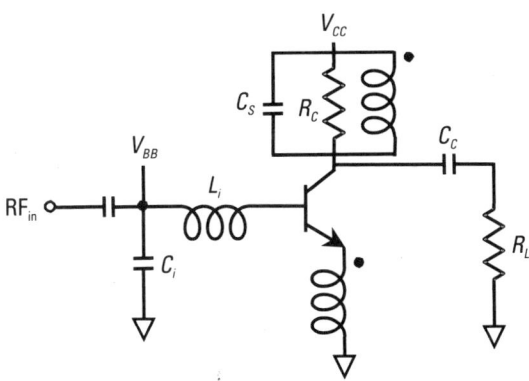

Figure 5.8 Application of transformer.

degeneration, while L_c forms a tuned load. As a load or as emitter degeneration, one side of the inductor sees ac ground. This allows increased output swing, since there is ideally no dc voltage drop across the inductor. Similarly, the input series inductor has no dc drop across it. The disadvantage is that, being in series, it has parasitic capacitance from both sides to the substrate. As a result, a signal can be injected into the substrate, with implications for noise and matching.

In summary, the following advantages of using inductors are seen:

1. It provides bias current with no significant dc drop, which improves linearity.
2. The emitter degeneration increases linearity without an increase in noise.
3. Parasitic capacitance is resonated out.
4. Inductive degeneration can lead to simultaneous noise and power matching.

The transformer-based circuit as shown in Figure 5.8 and described by [4] has similar advantages to those described above for inductor-based tuned circuits.

One difference is that the gain is determined (partially at least) by the transformer turns ratio, thus removing or minimizing dependence on transistor parameters. This has advantages since, unlike the transistor, the transformer has high linearity and low noise.

On the negative side, fully integrated transformers are lossy and more difficult to model, and as a result, they have not been widely used.

5.10 Design of Inductors and Transformers

Of all the passive structures used in RF circuits, high-quality inductors and transformers or baluns are the most difficult to realize monolithically. In silicon, they suffer from the presence of lossy substrates and high-resistivity metal. However, over the past few years much research has been done in efforts to improve fabrication methods for building inductors, as well as modeling, so that better geometries could be used in their fabrication.

When inductors are made in silicon technology with aluminum interconnects, they suffer from the presence of relatively high-resistance interconnect structures and lossy substrates, typically limiting the Q to about 5 at around 2 GHz. This causes many high-speed RF components, such as *voltage-controlled oscillators* (VCOs) or power amplifiers using on-chip inductors, to have limited performance compared to designs using off-chip components. The use of off-chip components adds complexity and cost to the design of these circuits, which has led to intense research aimed at improving the performance of on-chip inductors [5–19].

Traditionally, due to limitations in modeling and simulation tools, inductors were made as square spirals, as shown in Figure 5.9. The wrapping of the metal lines allows the flux from each turn to be added, thus increasing the inductance per unit length of the structure and providing a compact way of achieving useful values of inductance. Square inductors, however, have less than optimum performance due to the 90° bends present in the layout which add to the resistance of the structures. A better structure is shown in Figure 5.10 [7, 18]. Since this inductor is made circular, it has less series resistance. This geometry is more symmetric than traditional inductors (its *S* parameters look the same from either side). Thus, it can be used in differential circuits without needing two inductors to get good symmetry. Also, bias can be applied through the axis of symmetry of this structure if needed in a differential application (i.e., it is a virtual ground point).

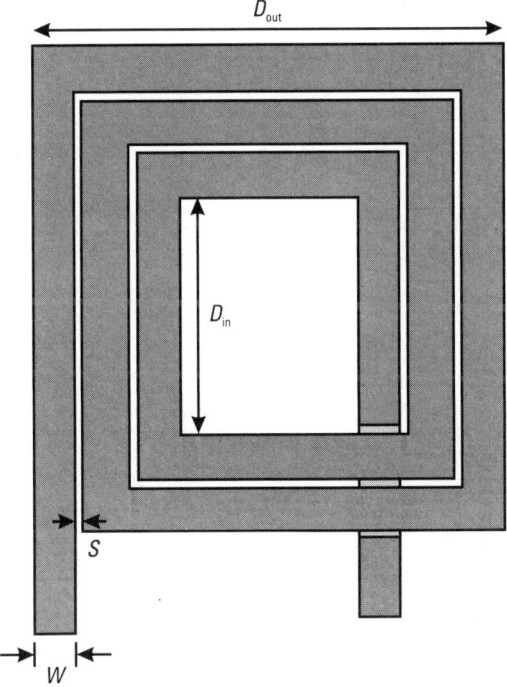

Figure 5.9 A conventional single-ended inductor layout.

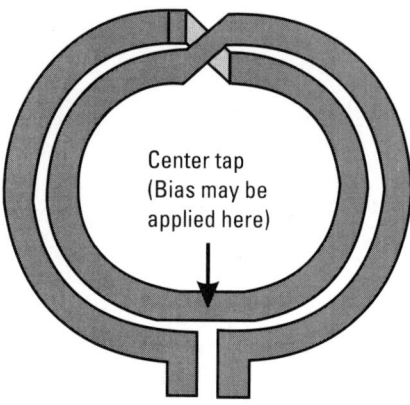

Figure 5.10 A circular differential inductor layout.

5.11 Some Basic Lumped Models for Inductors

When describing on-chip inductors, it is useful to build an equivalent model for the structure. Figure 5.11 shows capacitance between lines, capacitance through the oxide, the inductance of the traces, series resistance, and substrate effects. These effects are translated into the circuit model shown in Figure 5.12, which shows a number of nonideal components. R_s models the series resistance of the metal lines used to form the inductor. Note that the value of R_s will increase at higher frequencies due to the skin effect. C_{oxide} models the capacitance from the lines to the substrate. This is essentially a parallel-plate capacitor formed between the inductor metal and the substrate. C_{sub} and R_{sub} model the losses due to magnetic effects, capacitance, and the conductance of the substrate. They are proportional to the area of the metal in the inductor, and their exact

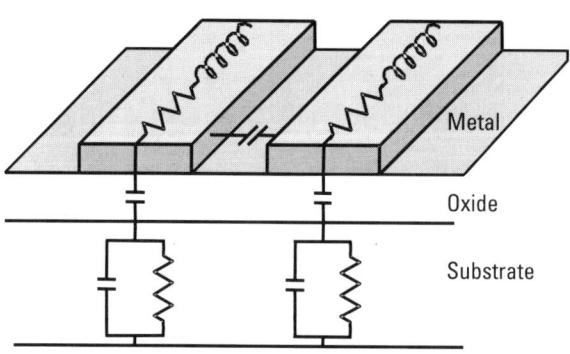

Figure 5.11 Elements used to build an inductor model.

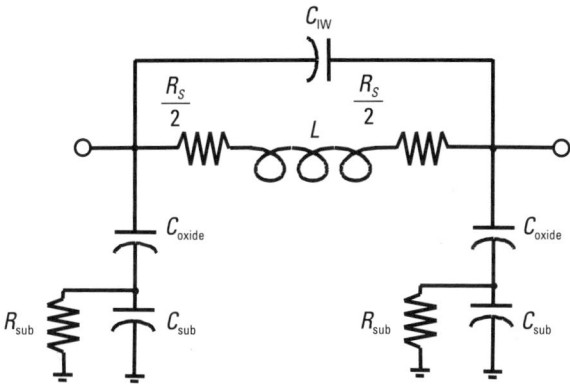

Figure 5.12 Basic π model for a regular inductor.

value depends on the properties of the substrate in question. C_{IW} models the inter-winding capacitance between the traces. This is another parallel-plate capacitor formed by adjacent metal lines. Note that in a regular inductor both sides are not symmetric, partly due to the added capacitance on one side of the structure caused by the underpass. The underpass connects the metal at the center of the planar coil with metal at the periphery.

The model for the symmetric or so-called differential inductor is shown in Figure 5.13 [18]. Here the model is broken into two parts with a pin at the axis of symmetry where a bias can be applied if desired. Note also that since the two halves of the spiral are interleaved, there is magnetic coupling between both halves of the device. This is modeled by the coupling coefficient k.

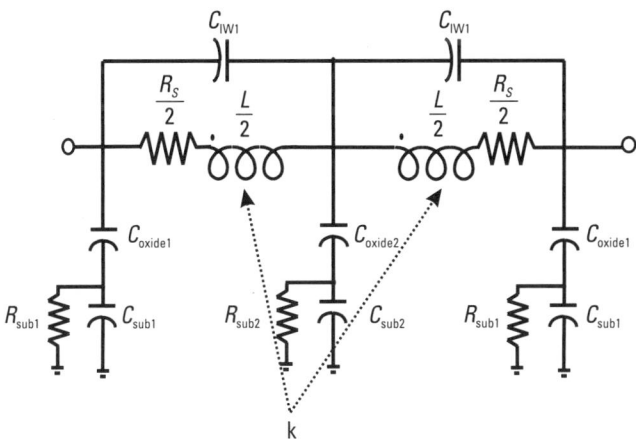

Figure 5.13 Basic model for a differential inductor.

5.12 Calculating the Inductance of Spirals

Recently, some formulas for calculating on-chip inductors have been proposed for both square and octagonal geometries [5]. The following simple expressions can be used:

$$L = 2.34\mu_o \frac{n^2 d_{avg}}{1 + 2.75\psi} \quad (5.9)$$

for square inductors, where n is the number of turns and d_{avg} is given by (see Figure 5.9)

$$d_{avg} = \frac{1}{2}(D_{out} + D_{in}) \quad (5.10)$$

and ψ is given by

$$\psi = \frac{(D_{out} - D_{in})}{(D_{out} + D_{in})} \quad (5.11)$$

and for octagonal inductors,

$$L = 2.25\mu_o \frac{n^2 d_{avg}}{1 + 3.55\psi} \quad (5.12)$$

The formulas can be quite accurate and their use will be demonstrated in Example 5.5. However, often it is easier to use simulators like ASITIC [14] or three-dimensional EM solvers. Since the substrate complicates matters, an EM simulator is often the only option for very complicated geometries. These can be quite slow, which makes them cumbersome to use as a design tool, but the speed is improving as computer power grows.

5.13 Self-Resonance of Inductors

At low frequencies, the inductance of an integrated inductor is relatively constant. However, as the frequency increases, the impedance of the parasitic capacitance elements starts to become significant. At some frequency, the admittance of the parasitic elements will cancel that of the inductor and the inductor will self-resonate. At this point, the reactive part of the admittance will be zero. The inductance is nearly constant at frequencies much lower than the self-resonance frequency; however, as the self-resonance frequency is approached, the inductance rises and then abruptly falls to zero. Beyond the self-resonant frequency, the parasitic capacitance will dominate and the inductor will look capacitive. Thus, the inductor has a finite bandwidth over which it can be used. For reliable operation, it is necessary to stay well below the self-resonance frequency. Since parasitic capacitance increases in proportion to the size of the inductor, the self-resonant frequency decreases as the size of the inductor increases. Thus, the size of on-chip inductors that can be built is severely limited.

5.14 The Quality Factor of an Inductor

The quality factor, or Q, of a passive circuit element can be defined as

$$Q = \frac{|\text{Im}(Z_{\text{ind}})|}{|\text{Re}(Z_{\text{ind}})|} \quad (5.13)$$

where Z_{ind} is the impedance of the inductor. This is not necessarily the most fundamental definition of Q, but it is a good way to characterize the structure. A good way to think about this is that Q is a measure of the ratio of the desired quantity (inductive reactance) to the undesired quantity (resistance). Obviously, the higher-Q device is more ideal.

The Q of an on-chip inductor is affected by many things. At low frequencies, the Q tends to increase with frequency, because the losses are relatively constant (mostly due to metal resistance R_s), while the imaginary part of the impedance is increasing linearly with frequency. However, as the frequency increases, currents start to flow in the substrate through capacitive and, to a lesser degree, magnetic coupling. This loss of energy into the substrate causes an effective increase in the resistance. In addition, the skin effect starts to raise the resistance of the metal traces at higher frequencies. Thus, most integrated inductors have Qs that rise at low frequencies and then have some peak beyond which the losses make the resistance rise faster than the imaginary part of the impedance, and the Q starts to fall off again. Thus, it is easy to see the need for proper optimization to ensure that the inductor has peak performance at the frequency of interest.

Example 5.5 Calculating Model Values for the Inductor

Given a square inductor with the dimensions shown in Figure 5.14, determine a model for the structure including all model values. The inductor is made out of 3-μm-thick aluminum metal. The inductor is suspended over 5 μm of oxide above a substrate. The underpass is 1-μm aluminum and is 3 μm above the substrate. Assume the vias are lossless.

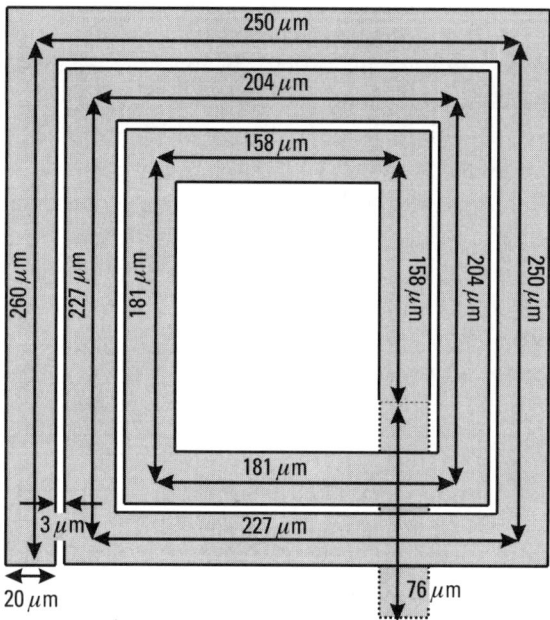

Figure 5.14 Inductor with dimensions.

Solution

We can start by estimating the inductance of the structure by using the formulas in Section 5.11 and referring to Figure 5.9.

$$d_{avg} = \frac{1}{2}(D_{out} + D_{in}) = \frac{1}{2}(270 \ \mu m + 161 \ \mu m) = 215.5 \ \mu m$$

$$\psi = \frac{(D_{out} - D_{in})}{(D_{out} + D_{in})} = \frac{(270 \ \mu m - 161 \ \mu m)}{(270 \ \mu m + 161 \ \mu m)} = 0.253$$

$$L = 2.34 \mu_o \frac{n^2 d_{avg}}{1 + 2.75\psi}$$

$$= 2.34 \cdot 4\pi \times 10^{-7} \frac{N}{A^2} \frac{3^2 \cdot 215.5 \ \mu m}{1 + 2.75 \cdot 0.253}$$

$$= 3.36 \ nH$$

Next, let us estimate the oxide capacitance. First, the total length of the inductor metal is 2.3 mm. Thus, the total capacitance through the oxide is

$$C_{\text{oxide}} = \frac{\epsilon_o \epsilon_r A}{h}$$

$$= \frac{8.85 \times 10^{-12} \frac{C^2}{N \cdot m^2} \cdot 3.9 \cdot 2.3 \text{ mm} \cdot 20 \ \mu\text{m}}{5 \ \mu\text{m}}$$

$$= 317.6 \text{ fF}$$

The underpass must be taken into account here as well:

$$C_{\text{underpass}} = \frac{\epsilon_o \epsilon_r A}{h}$$

$$= \frac{8.85 \times 10^{-12} \frac{C^2}{N \cdot m^2} \cdot 3.9 \cdot 76 \ \mu\text{m} \cdot 20 \ \mu\text{m}}{3 \ \mu\text{m}}$$

$$= 17.4 \text{ fF}$$

Now we must consider the interwinding capacitance:

$$C_{\text{IW}} = \frac{\epsilon_o \epsilon_r A}{d} = \frac{8.85 \times 10^{-12} \frac{C^2}{N \cdot m^2} \cdot 11.9 \cdot 2.3 \text{ mm} \cdot 3 \ \mu\text{m}}{3 \ \mu\text{m}} = 241 \text{ fF}$$

The dc resistance of the line can be calculated from

$$R_{\text{dc}} = \frac{\rho L}{Wt} = \frac{3 \ \mu\Omega \cdot \text{cm} \cdot 2.3 \text{ mm}}{20 \ \mu\text{m} \cdot 3 \ \mu\text{m}} = 1.15 \Omega$$

The skin effect will begin to become important when the thickness of the metal is two skin depths. This will happen at a frequency of

$$f = \frac{\rho}{\pi \mu \delta^2} = \frac{3 \ \mu\Omega \cdot \text{cm}}{\pi \cdot 4\pi \times 10^{-7} \cdot (1.5 \ \mu\text{m})^2} = 3.38 \text{ GHz}$$

Let us ignore the resistance in the underpass. Thus, above 3.38 GHz the resistance of the line will be a function of frequency:

$$R_{\text{ac}}(f) = \frac{\rho L}{Wt - \left(W - 2\sqrt{\frac{\rho}{\pi f \mu}}\right)\left(H - 2\sqrt{\frac{\rho}{\pi f \mu}}\right)}$$

The other thing that must be considered is the substrate. This is an issue for which we really do need a simulator. However, as mentioned above, the capacitance and resistance will be a function of the area. This also means that once several structures in a given technology have been measured, it may be possible to predict these values for future structures. For this example, assume that reasonable values for the fitting parameters have been determined. Thus, R_{sub} and C_{sub} could be something like 870Ω and 115 fF. The complete model with values is shown in Figure 5.15.

Example 5.6 Determining Inductance, Q, and Self-Resonant Frequency

Take the model just created for the inductor in the previous example and compute the equivalent inductance and Q versus frequency. Also, find the self-resonance frequency. Assume that the side of the inductor with the underpass is grounded.

Solution

The equivalent circuit in this case is as shown in Figure 5.16.

This is just an elementary impedance network, so we will skip the details of the analysis and give the results. The inductance is computed by taking the

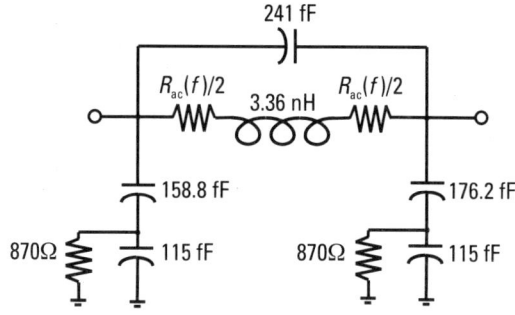

Figure 5.15 Inductor π model with numbers.

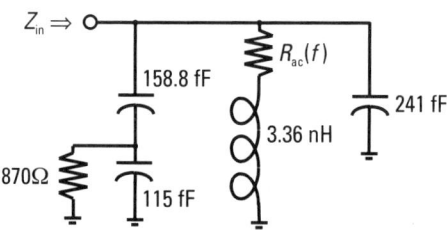

Figure 5.16 Inductor model with one side grounded.

imaginary part of Z_{in} and dividing by $2\pi f$. The Q is computed as in (5.13) and the results are shown in Figure 5.17.

This example shows a Q of almost 20, but in reality, due to higher substrate losses and line resistance leading up to the inductor, the Q will be somewhat lower than shown here, although in most respects this example has shown very realistic results.

5.15 Characterization of an Inductor

Once some inductors have been built and measured, S-parameter data will then be available for these structures. It is then necessary to take these numbers and convert them, for example, into inductance, Q, and self-resonance frequency.

The definitions of Q have already been given in (5.13) and ωL is equal to the imaginary part of the impedance. These definitions seem like simple ones, but the impedance still needs to be defined. Traditionally, we have assumed that one port of the inductor is grounded. In such a case, we can define the impedance seen from port 1 to ground.

Starting with the Z-parameter matrix (which can be easily derived from S-parameter data):

$$\begin{bmatrix} V_1 \\ V_2 \end{bmatrix} = \begin{bmatrix} Z_{11} & Z_{12} \\ Z_{21} & Z_{22} \end{bmatrix} \begin{bmatrix} I_1 \\ I_2 \end{bmatrix} \qquad (5.14)$$

Since the second port is grounded, $V_2 = 0$. Thus, two equations result:

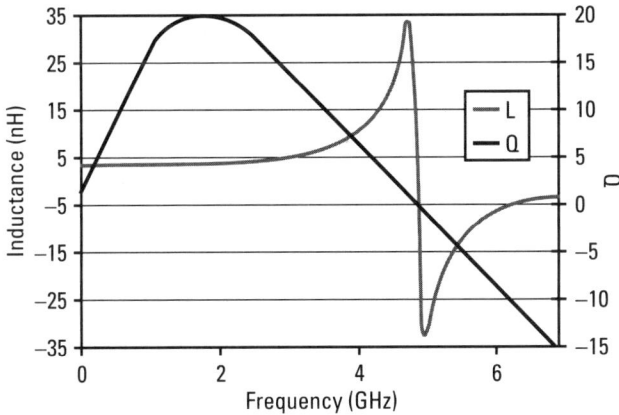

Figure 5.17 Inductor plot of L and Q versus frequency.

$$V_1 = Z_{11}I_1 + Z_{12}I_2 \qquad (5.15)$$

$$0 = Z_{21}I_1 + Z_{22}I_2$$

The second equation can be solved for I_2:

$$I_2 = -\frac{Z_{21}I_1}{Z_{22}} \qquad (5.16)$$

Thus, I_2 can now be removed from the first equation, and solving for $Z_{in} = V_1/I_1$:

$$Z_{port1} = Z_{11} - \frac{Z_{12}Z_{21}}{Z_{22}} \qquad (5.17)$$

Equivalently, if we look from port 2 to ground, the impedance becomes

$$Z_{port2} = Z_{22} - \frac{Z_{12}Z_{21}}{Z_{11}} \qquad (5.18)$$

Note that, referring to Figure 5.12, this effectively grounds out both C_1 and R_1 or C_2 and R_2. Thus, the Q will not necessarily be the same looking from both ports. In fact, the Q will be marginally higher in the case of a regular structure looking from the side with no underpass, as there will be less loss. Also note that the side with no underpass will have a higher self-resonance frequency.

Often designers want to use inductors in a differential configuration. This means that both ends of the inductor are connected to active points in the circuit and neither side is connected to ground. In this case, we can define the impedance seen between the two ports:

Starting again with the Z parameters, the voltage difference applied across the structure is now

$$V_1 - V_2 = Z_{11}I_1 + Z_{12}I_2 - Z_{21}I_1 - Z_{22}I_2 \qquad (5.19)$$

$$V_1 - V_2 = I_1(Z_{11} - Z_{21}) - I_2(Z_{22} - Z_{12}) \qquad (5.20)$$

Because the structure is symmetric, we make the assumption that $I_1 = -I_2$. Thus,

$$Z_{diff} = \frac{V_1 - V_2}{I_1} = Z_{11} + Z_{22} - Z_{12} - Z_{21} \qquad (5.21)$$

In this case, the substrate capacitance and resistance from both halves of the inductor are in series. When the inductor is excited in this mode, it "sees" less loss and will give a higher Q. Thus, the differential Q is usually higher than the single-ended Q. The self-resonance of the inductor in this mode will also be higher than the self-resonance frequency looking from either side to ground. Also, the frequency at which the differential Q peaks is usually higher than for the single-ended excitation. Care must be taken, therefore, when optimizing an inductor for a given frequency, to keep in mind its intended configuration in the circuit.

Important note: Every inductor has a differential Q and a single-ended Q regardless of its layout. Which Q should be used in analysis depends on how the inductor is used in a circuit.

5.16 Some Notes About the Proper Use of Inductors

Designers are very hesitant to place a nonsymmetric regular inductor across a differential circuit. Instead, two regular inductors are usually used. In this case, the center of the two inductors is effectively ac grounded and the effective Q for the two inductors is equal to their individual single-ended Qs. To illustrate this point, take a simplified model of an inductor with only substrate loss, as shown in Figure 5.18. In this case, the single-ended Q is given by

$$Q_{SE} = \frac{R}{\omega L} \tag{5.22}$$

and the differential Q is given by

$$Q_{diff} = \frac{2R}{\omega L} \tag{5.23}$$

Now if two inductors are placed in series as shown in Figure 5.19, the differential Q of the overall structure is given by

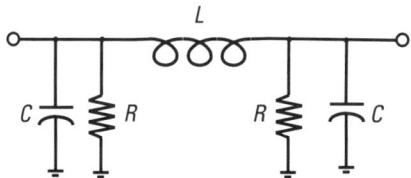

Figure 5.18 Simplified inductor model with only substrate loss.

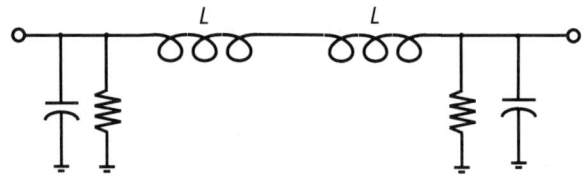

Figure 5.19 Two simplified inductors connected in series with only substrate loss.

$$Q_{\text{diff2}} = \frac{2R}{2\omega L} = \frac{R}{\omega L} = Q_{\text{SE}} \quad (5.24)$$

Thus, the differential Q of the two inductors is equal to the single-ended Q of one of the inductors. The advantage of using symmetric structures should be obvious. Note that, here, substrate losses have been assumed to dominate. If the series resistance is dominant in the structure, then using this configuration will be less advantageous. However, due to the mutual coupling of the structure, this configuration would still be preferred, as it makes more efficient use of chip area.

If using a differential inductor in the same circuit, the designer would probably use only one structure. In this case, the effective Q of the circuit will be equal to the differential Q of the inductor. Note that if a regular inductor were used in its place, the circuit would see its differential Q as well.

When using a regular inductor with one side connected to ground, the side with the underpass should be the side that is grounded, as this will result in a higher Q and a higher self-resonance frequency.

Example 5.7 Single-Ended Versus Differential Q

Take the inductor of Example 5.5 and compute the single-ended Q from both ports as well as the differential Q. Also, compare the self-resonant frequency under these three conditions.

Solution

As before, equivalent circuits can be made from the model in the previous example (see Figure 5.20).

It is a matter of elementary circuit analysis to compute the input impedance of these three networks. The inductance of all three is shown in Figure 5.21. Note that in the case where the underpass is not grounded, the circuit has the lowest self-resonant frequency, while the differential configuration leads to the highest self-resonant frequency.

The Q of these three networks is plotted in Figure 5.22. Note here that at low frequencies where substrate effects are less important, the Qs are all equal, but as the frequencies increase, the case where the circuit is driven

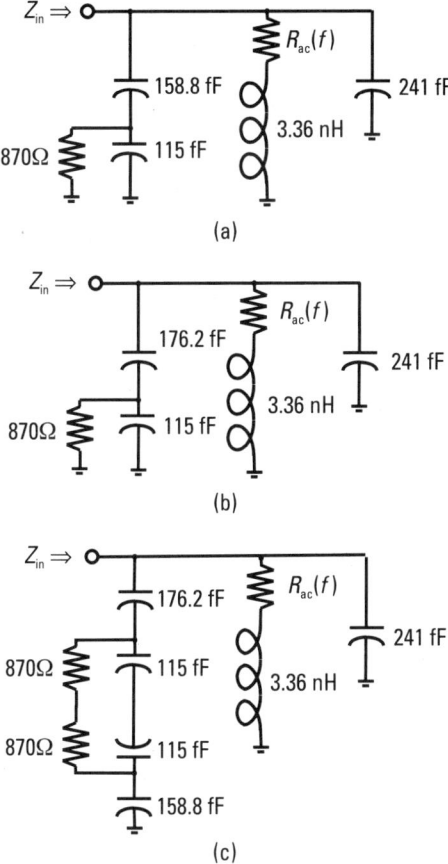

Figure 5.20 Equivalent circuits for the case where (a) the side with the underpass is grounded; (b) the underpass is not grounded; and (c) the circuit is driven differentially.

differentially is clearly better. In addition, in this case, the Q keeps rising to a higher frequency and higher overall value.

5.17 Layout of Spiral Inductors

The goal of any inductor layout is to design a spiral inductor of specified inductance, with Q optimized for best performance at the frequency of interest. In order to achieve this, careful layout of the structure is required. The resistance of the metal lines causes the inductor to have a high series resistance, limiting its performance at low frequencies, while the proximity of the substrate causes substrate loss, raising the effective resistance at higher frequencies. Large coupling

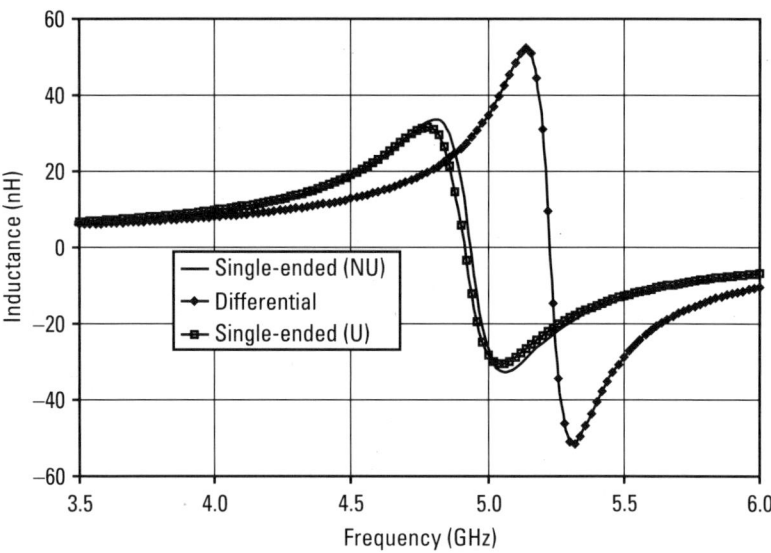

Figure 5.21 The inductance plotted versus frequency for the three modes of operation.

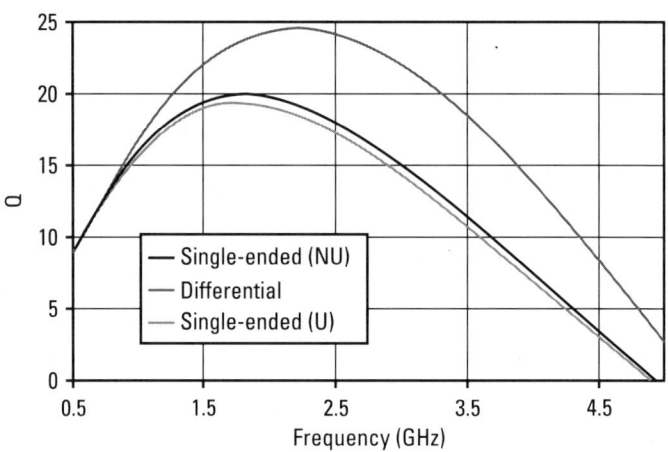

Figure 5.22 The Q plotted versus frequency for the three modes of operation.

between the inductor and the substrate also causes the structures to have low self-resonance frequencies. As a result, there are limitations on the size of the device that can be built.

Traditionally, on-chip inductors have been square as shown in Figure 5.9. This is because these have been easier to model than geometries that are more complicated. A square geometry is by no means optimal, however. The presence of the 90° bends adds unnecessary resistance to the structure, and as the structure is made circular, the performance will improve.

Some guidelines for optimum layout will now be provided. These rules are based on considerations of the effect of geometry on the equivalent model shown in Figure 5.12.

1. *Line spacing:* At low frequencies (2 GHz or less), keep the line spacing as tight as possible. At higher frequencies, due to coupling between turns, larger spacing may be desirable.

2. *Line width:* Increasing metal width will reduce the inductance (fewer turns in a given area as well as less inductance per unit length) and will decrease the series resistance of the lines at low frequencies. Large inductance area means bigger capacitance, which means lower self-resonance, and more coupling of current into the substrate. Therefore, as W goes up, inductance comes down and the frequency of Q_{peak} gets lower (and vice versa). Line widths for typical 1- to 5-nH inductors in the 2- to 5-GHz range would be expected to be from 10 to 25 μm.

3. *Area:* Bigger area means that more current is present in the substrate, so high-frequency losses tend to be increased. Bigger area (for the same line width) means longer spirals, which means more inductance. Therefore, as the area goes up, inductance goes up, and the frequency of Q_{peak} gets lower (and vice versa).

4. *Number of turns:* This is typically a third degree of freedom. It is usually best to pick fewer rather than more turns, provided that the inductor does not get to be huge. Huge is, of course, a relative term, and it is ultimately up to the designers to decide how much space they are willing to devote to the inductor layout. Inner turns add less to the inductance but more resistance, so it is best to keep the inductor hollow. By changing the area and line width, the peak frequency and inductance can be fine-tuned.

5.18 Isolating the Inductor

Inductors tend to be extremely large structures, and as such they tend to couple signals into the substrate; therefore, isolation must be provided. Typically, a ring of substrate contacts is added around each inductor. These substrate contacts are usually placed at a distance of about five line widths away from the inductor. The presence of a patterned (slotted) ground shield, discussed in the next section, may also help in isolating the inductor from the substrate.

5.19 The Use of Slotted Ground Shields and Inductors

In an inductor, currents flow into the substrate through capacitive coupling and are induced into the substrate through magnetic coupling. Current flowing in the substrate causes additional loss. Of the two, generally capacitive coupling is the more dominant loss mechanism. One method to reduce substrate loss is to place a ground plane above the substrate, preventing currents from entering the substrate [12]. However, with a ground plane, magnetically generated currents will be increased, reducing the inductance. One way to get around this problem is to pattern the ground plane such that magnetically generated currents are blocked from flowing. An example of a patterned ground shield designed for a square inductor is shown in Figure 5.23. Slots are cut into the plane perpendicular to the direction of magnetic current flow. The ground shield has the disadvantage of increasing capacitance to the inductor, causing its self-resonant frequency to drop significantly. For best performance, the ground shield should be placed far away from the inductor, but remain above the substrate. In a typical bipolar process, the polysilicon layer is a good choice.

The model for the ground-shielded inductor compared to the standard inductor model is shown in Figure 5.24. For the ground-shielded inductor, the lossy substrate capacitance has been removed, leaving only the lossless oxide capacitance and the parasitic resistance of the shield. As a result, the inductor will have a higher Q.

5.20 Basic Transformer Layouts in IC Technologies

Transformers in silicon are as yet not very common. They are more complicated than inductors and therefore harder to model in many cases. Transformers or

Figure 5.23 Patterned ground shield for a square spiral inductor (inductor not shown).

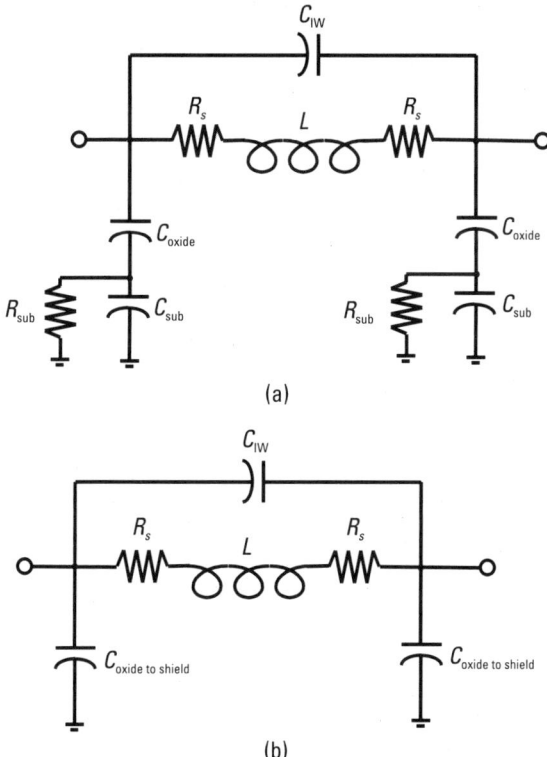

Figure 5.24 Comparison of the basic π model for (a) regular inductor, and (b) shielded inductor.

baluns consist of two interwound spirals that are magnetically coupled. A sample layout of a basic structure is shown in Figure 5.25. In this figure, two spirals are interwound in a 3:3 turns ratio structure. The structure can be characterized by a primary and secondary inductance and a mutual inductance or coupling factor, which describes how efficiently energy can be transferred from one spiral to the other. A symmetric structure with a turns ratio of 2:1 is shown in Figure 5.26.

A simplified transformer model is shown in Figure 5.27. This is modeled as two inductors, but with the addition of coupling coefficient k between them, and interwinding capacitance C_{IW} from input to output.

Example 5.8 Placing the Dots

Place the dots on the transformer shown in Figure 5.26.

Solution

We will start by assuming that a current is flowing in the primary winding, as shown in Figure 5.28(a). This will cause a flux to flow into the page at the

Figure 5.25 Sample layout of two interwound inductors forming a transformer.

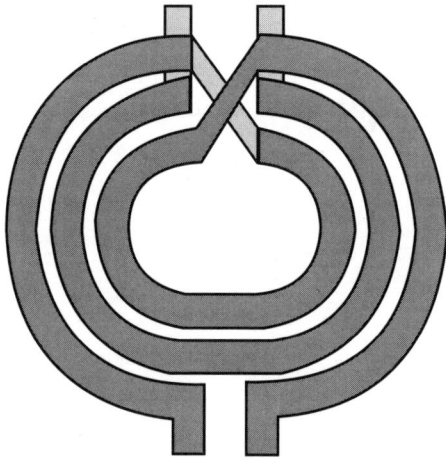

Figure 5.26 Sample layout of a circular 2:1 turns ratio transformer.

center of the winding, as shown in Figure 5.28(b). Thus, in order for a current in the secondary to reinforce the flux, it must flow in the direction shown in Figure 5.28(c). Therefore, the dots go next to the ports where the current flows out of the transformer, as shown in Figure 5.28(d).

5.21 Multilevel Inductors

Inductors can also be made using more than one level of metal. Especially in modern processes, which can have as many as five or more metal layers, it can be advantageous to do so. There are two common ways to build multilevel inductors. The first is simply to strap two or more layers of metal together with

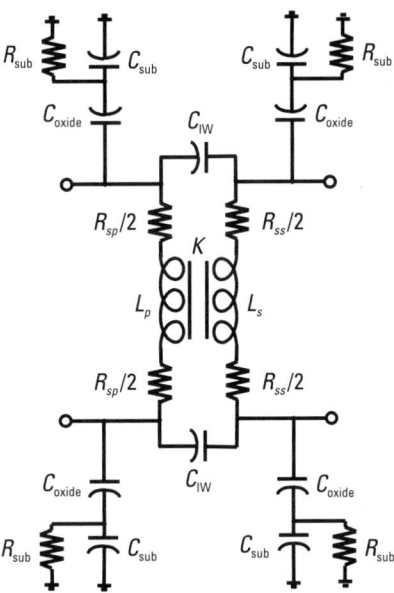

Figure 5.27 Basic model of transformer.

vias to decrease the effective series resistance. This will increase the Q, but at the expense of increased capacitance to the substrate and a resultant decrease in self-resonant frequency. This technique is of benefit for small inductors for which the substrate loss is not dominant and that are at low enough frequency, safely away from the self-resonant frequency.

The second method is to connect two or more layers in series. This results in increased inductance for the same area or allows the same inductance to be realized in a smaller area. A drawing of a two-level inductor is shown in Figure 5.29. Note that the fluxes through the two windings will reinforce one another and the total inductance of the structure will be $L_{\text{top}} + L_{\text{bottom}} + 2M$ in this case. If perfect coupling is assumed and the inductors are of equal size, then this gives $4L$. In general, this is a factor of n^2 more inductance, where n is the number of levels. Thus, this is a way to get larger inductance without using as much chip area.

To determine the capacitance associated with the inductor, we consider the top and bottom spirals as two plates of a capacitor with total distributed capacitance C_1 [17]. In addition, we consider the bottom spiral and the substrate to form a distributed capacitance C_2. Now the total equivalent capacitance of the structure can be approximated. First, note that if a voltage V_1 is applied across the terminals of the inductor, the voltage across C_1 will go from V_1 at the terminals down to zero at the via. Similarly, the voltage across C_2 will go from zero at the terminal (assuming this point is grounded) to $V_1/2$ at the via.

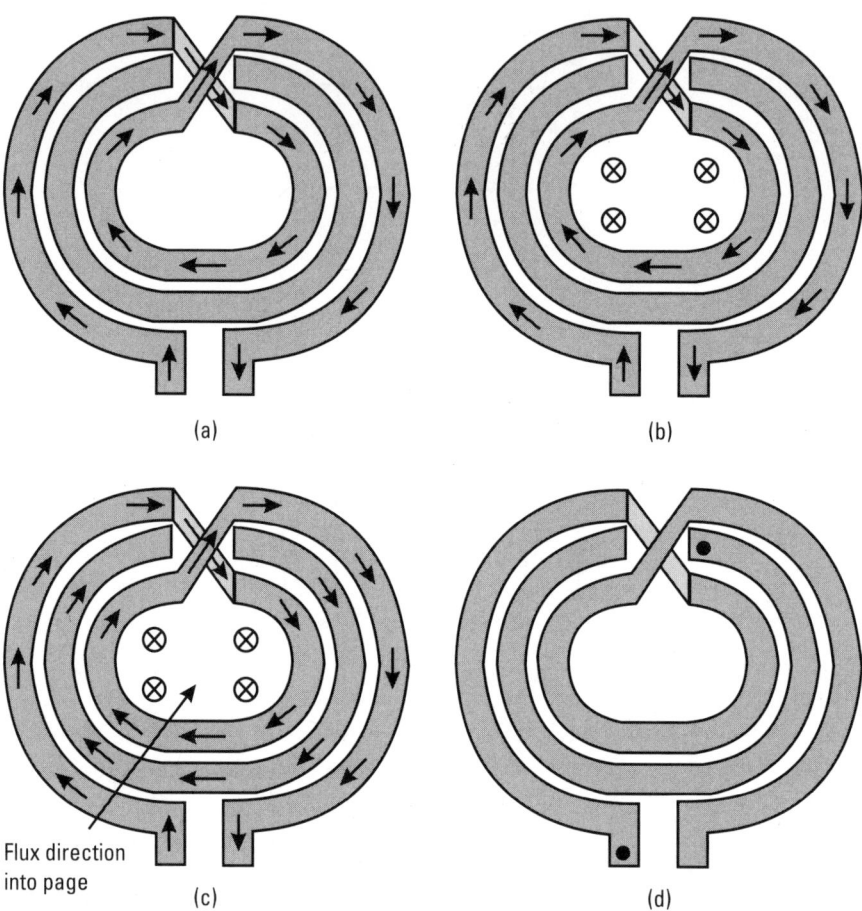

Figure 5.28 Determining dot placement: (a) arbitrary current flow; (b) direction of flux; (c) secondary current flow that adds to the flux; and (d) dot placement.

Thus, the total energy stored in C_1 is

$$E_{C_1} = \frac{1}{2} C_1 V_1^2 \int_0^1 (1-x)^2 \, dx = \frac{1}{2} \frac{C_1}{3} V_1^2 \qquad (5.25)$$

where x is a dummy variable representing the normalized length of the spiral. Note that $V_1(1-x)$ is an approximation to the voltage across C_1 at any point along the spiral. Thus, the equivalent capacitance of C_1 is

$$C_{eq1} = \frac{C_1}{3} \qquad (5.26)$$

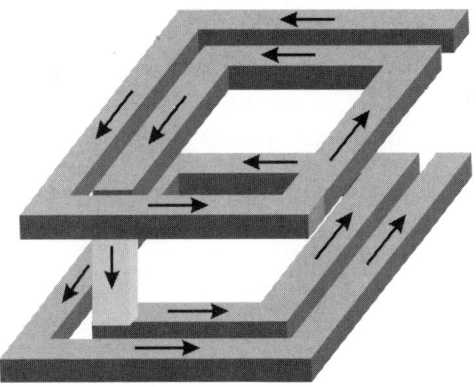

Figure 5.29 Three-dimensional drawing of a multilevel inductor.

The total capacitance can be found in C_2 in much the same way.

$$E_{C_2} = \frac{1}{2} C_2 \left(\frac{V_1}{2}\right)^2 \int_0^1 x^2 \, dx = \frac{1}{2} \frac{C_2}{3 \cdot 2^2} V_1^2 \tag{5.27}$$

Note that $V_1/2 \cdot x$ approximates the voltage across C_2 at any point along the spiral.

$$C_{eq2} = \frac{C_1}{3 \cdot 2^2} \tag{5.28}$$

Thus, the total capacitance is

$$C_{eq} = \frac{C_1}{3} + \frac{C_2}{12} \tag{5.29}$$

Note that the capacitor C_2 is of less importance than C_1. Thus, it would be advantageous to space the two spirals far apart even if this means there is more substrate capacitance (C_2). Note also that C_1 will have a low loss associated with it, since it is not dissipating energy in the substrate.

5.22 Characterizing Transformers for Use in ICs

Traditionally, transformers are characterized by their S parameters. While correct, this gives little directly applicable information about how the transformer will behave in an application when loaded with impedances other than 50Ω. It would be more useful to extract an inductance and Q for both windings and plot the coupling (k factor) or mutual inductance for the structure instead. These properties have the advantage that they do not depend on the system reference impedance.

In the following narrowband model, all the losses are grouped into a primary and secondary resistance as shown in Figure 5.30.

The model parameters can be found from the Z parameters starting with

$$\left.\frac{V_1}{I_1}\right|_{I_2=0} = Z_{11} = R_p + j\omega(L_p - M) + j\omega M = R_p + j\omega L_p \quad (5.30)$$

Similarly,

$$Z_{22} = R_s + j\omega L_s \quad (5.31)$$

Thus, the inductance of the primary and secondary and the primary and secondary Q can be defined as

$$L_s = \frac{\text{Im}(Z_{22})}{j\omega} \quad (5.32)$$

$$L_p = \frac{\text{Im}(Z_{11})}{j\omega} \quad (5.33)$$

$$Q_s = \frac{\text{Im}(Z_{22})}{\text{Re}(Z_{22})} \quad (5.34)$$

$$Q_p = \frac{\text{Im}(Z_{11})}{\text{Re}(Z_{11})} \quad (5.35)$$

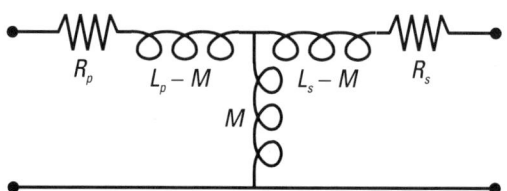

Figure 5.30 Narrowband equivalent model for a transformer.

The mutual inductance can also be extracted as

$$\left.\frac{V_1}{I_2}\right|_{I_1=0} = Z_{12} = j\omega M \qquad (5.36)$$

Therefore,

$$M = \frac{Z_{12}}{j\omega} = \frac{Z_{21}}{j\omega} \qquad (5.37)$$

5.23 On-Chip Transmission Lines

Any on-chip interconnect can be modeled as a transmission line. Transmission line effects on chip can often be ignored if lines are significantly shorter than a quarter wavelength at the frequency of interest. Thus, transmission line effects are often ignored for frequencies between 0 and 5 GHz. However, as higher frequency applications become popular, these effects will become more important.

One of the simplest ways to build a transmission line is by placing a conductor near a ground plane separated by an insulator, as shown in Figure 5.31. Another way to build a transmission line is called a *coplanar waveguide*, as shown in Figure 5.32. Note that in this case a ground plane is not needed, although it will be present in an IC.

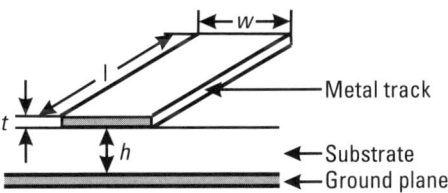

Figure 5.31 Microstrip transmission line.

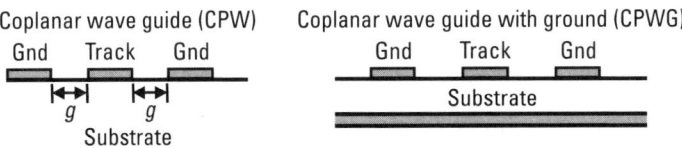

Figure 5.32 Coplanar waveguide transmission lines.

The effect of on-chip transmission lines is to cause phase shift and possibly some loss. Since dimensions in an IC are typically much less than those on the *printed circuit board* (PCB), this is often ignored. The magnitude of these effects can be estimated with a simulator (for example, Agilent's LineCalc) and included as transmission lines in the simulator if it turns out to be important. As a quick estimate for delay, consider that in free space (vacuum) a 1-GHz signal has a wavelength of 30 cm. However, oxide has $\epsilon_{ox} = 3.9$, which slows the speed of propagation by $\sqrt{3.9} = 1.975 \approx 2$. Thus, the wavelength is 15.19 cm and the resultant phase shift is about 2.37°/mm/GHz. The on-chip line can be designed to be a nearly lossless transmission line by including a shield metal underneath. Without such a shield metal, the substrate forms the ground plane and there will be losses. If the substrate were much further away from the conductors in silicon technology, coplanar waveguides would be possible, but with only a few micrometers of oxide, the substrate cannot be avoided except with a shield. On silicon, a 50Ω line is about twice as wide as dielectric thickness. Some quick simulations show that for a 4-μm dielectric, widths of 3, 6, and 12 μm result in about 72Ω, 54Ω, and 36Ω, respectively. Characteristic impedance can be calculated by the formula

$$Z_o = \sqrt{\frac{L}{C}} \tag{5.38}$$

where L and C are the per-unit inductance and capacitance, respectively. We note that if a length of transmission line is necessary on-chip, it may be advantageous to design it with a characteristic impedance higher than 50Ω, as this will result in less current necessary in the circuits matched to it.

5.23.1 Effect of Transmission Line

Matching an amplifier must include the effect of the transmission line up to the matching components. This transmission line causes phase shift (seen as rotation around the center of the Smith chart). If this effect is not considered, matching components can be completely incorrect. As an example, consider an RF circuit on a printed circuit board with off-chip impedance matching. At 5 GHz, with a dielectric constant of 4, a quarter wavelength is about 7.5 mm. This could easily be the distance to the matching components, in which case the circuit impedance has been rotated halfway around the Smith chart, and impedance matching calculated without taking this transmission line into account would result in completely incorrect matching. For example, if a parallel capacitor is needed directly at the RF circuit, at a quarter wavelength distant a series inductor will be needed.

A number of tools are available that can do calculations of transmission lines, and simulators can directly include transmission line models to show the effect of these lines.

5.23.2 Transmission Line Examples

At RF frequencies, any track on a printed circuit board behaves as a transmission line, such as a *microstrip line* (MLIN) (Figure 5.31), a *coplanar waveguide* (CPWG) (Figure 5.32), or a *coplanar waveguide with ground* (CPWG). Differential lines are often designed as coupled *microstrip lines* (MCLINs) (Figure 5.33) or they can become coupled simply because they are close together, for example, at the pins of an integrated circuit. For these lines, differential and common mode impedance can be defined (in microwave terms, these are described as odd-order and even-order impedance, respectively). On an integrated circuit, all lines are transmission lines, even though it may be possible to ignore transmission line effects for short lines. The quality of such transmission lines may suffer due to lossy ground plane (the substrate) or because of poor connection between coplanar ground and substrate ground.

For example, a 400-μm by 3-μm line on-chip, with oxide thickness of 4 μm is simulated to have a characteristic impedance of about 72Ω. The capacitance is estimated by

$$C = \frac{3.9 \cdot 8.85 \times 10^{-12} \text{ F/m} \cdot 3 \times 10^{-6} \cdot 400 \times 10^{-6}}{4 \times 10^{-6}} = 10.35 \text{ fF}$$

Because of fringing and edge effects, the capacitance is probably more like 20 fF, with the result that the inductance L is about 0.104 nH. Note that for a wider line, such as 6 μm, the capacitance is estimated to be about 30 fF, the characteristic impedance is closer to 50Ω, and inductance is about 0.087 nH.

In making use of a simulator to determine transmission line parameters, one needs to specify the substrate thickness, line widths and gaps, dielectric constant, loss tangent, metal conductor conductivity, and thickness. Most simu-

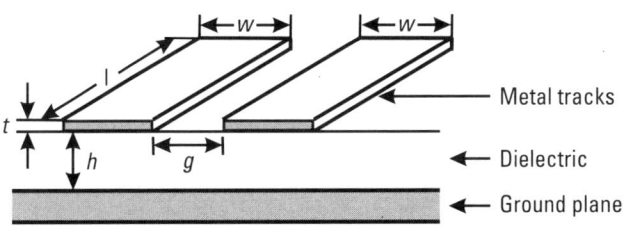

Figure 5.33 Coupled microstrip lines.

lators need dimensions specified in mils (thousandths of an inch), where a mil is equal to 25.4 μm. A typical substrate thickness for a double-sided printed circuit board is 40 to 64 mils. Multilayered boards can have effective layers that are 10 mils or even less. Surface material is often copper with a thickness typically specified by weight; for example, half-ounce copper translates to 0.7 mil. In simulators, the conductivity ρ is typically specified relative to the conductivity of gold. Thus, using Table 5.2, $\rho_{Au} = 1.42\ \rho_{Cu}$, or $\rho_{Cu} = 0.70\ \rho_{Au}$. Table 5.4 shows parameters for a variety of materials, including on-chip material (SiO_2, Si, GaAs) printed circuit board material (FR4, 5880, 6010), and some traditional substrate material for microwave, for example, ceramic.

Example 5.9 Calculation of Transmission Lines

Using a simulator, determine line impedance at 1.9 GHz versus dielectric thickness for microstrip lines, coupled microstrip lines, and coplanar waveguide with a ground plane. Use FR4 material with a dielectric constant of 4.3, and 0.7-mil copper with a line width of 20 mils and a 20-mil gap or space between the lines.

Solution

Calculations were done and the results are shown in Figure 5.34. It can be seen that 50Ω is realized with a dielectric thickness of about 11 mils for the microstrip line and the coplanar waveguide and about 14 mils for the coupled microstrip lines. Thus, the height is just over half of the line width. It can also be seen that a microstrip line and a coplanar waveguide with ground have very similar behavior until the dielectric height is comparable to the gap dimension.

Example 5.10 Transmission Lines

Using a simulator, determine line impedance at 1.9 GHz versus line width, gap, and space for microstrip lines, coupled microstrip lines, and coplanar waveguide with a ground plane. Use material with a dielectric constant of 2.2 and height of 15 mils, and 0.7-mil copper.

Table 5.4
Properties of Various Materials

Material	Loss Tangent	Permittivity	Material	Loss Tangent	Permittivity
SiO_2	0.004–0.04	3.9	Al_2O_3 (ceramic)	0.0001	9.8
Si	0.015	11.9	Sapphire	0.0001	9.4; 1.6
GaAs	0.002	12.9	Quartz	0.0001	3.78
FR4	0.022	4.3	6010	0.002	10.2
5880	0.001	2.20			

The Use and Design of Passive Circuit Elements in IC Technologies 133

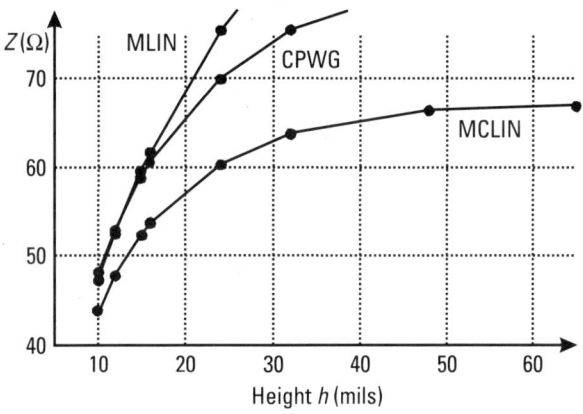

Figure 5.34 Impedance versus dielectric thickness for FR4 with line width of 20 mils at 1.9 GHz.

Solution

Calculations were done, and the results for characteristic impedance versus track width are shown in Figure 5.35. Figure 5.36 shows the track width versus gap or space dimension to result in $Z = 50\Omega$.

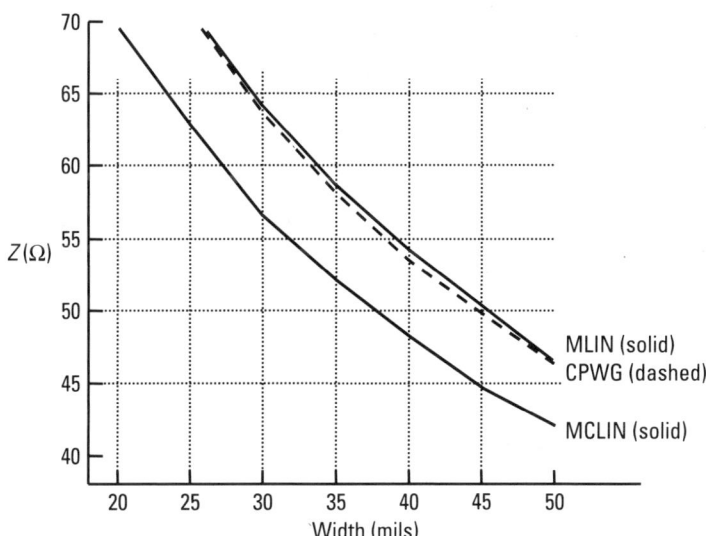

Figure 5.35 Impedance versus track width with dielectric thickness of 15 mils, gap or space of 20 mils, and dielectric constant of 2.2 at 1.9 GHz.

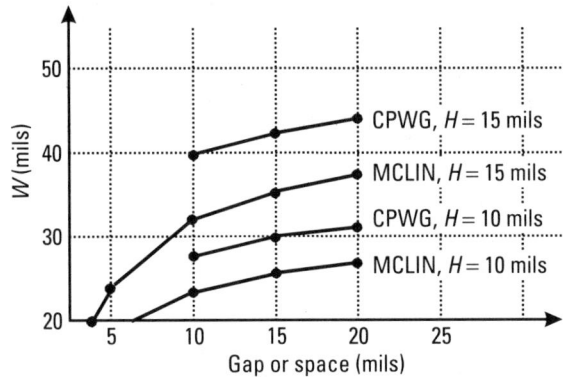

Figure 5.36 Track width versus gap or space to result in $Z = 50\Omega$ for coupled microstrip lines and coplanar waveguide.

5.24 High-Frequency Measurement of On-Chip Passives and Some Common De-Embedding Techniques

So far, we have considered inductors, transformers, and their Z parameters. In this section, we will discuss how to obtain those Z parameters from measurements. A typical set of test structures for measuring an inductor in a pad frame is shown in Figure 5.37. High-frequency ground-signal-ground probes will be landed on these pads so that the S parameters of the structure can be measured. However, while measuring the inductor, the pads themselves will also be measured, and

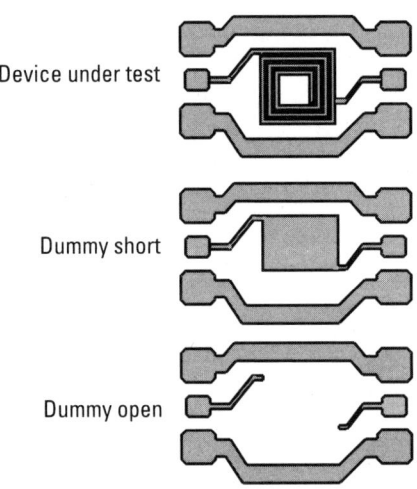

Figure 5.37 Example of high-frequency structures used for measuring on-chip passives.

therefore two additional de-embedding structures will be required. Once the S parameters have been measured for all three structures, a simple calculation can be performed to remove the unwanted parasitics.

The dummy open and dummy short are used to account for parallel and series parasitic effects, respectively. The first step is to measure the three structures, the device as Y_{DUT}, the dummy open as $Y_{dummy-open}$, and the dummy short as $Y_{dummy-short}$. Then the parallel parasitic effects represented by $Y_{dummy-open}$ are removed, as shown by (5.39), leaving the partially corrected device admittance as Y'_{DUT} and the corrected value for the dummy short as $Y'_{dummy-short}$.

$$Y'_{DUT} = Y_{DUT} - Y_{dummy-open} \tag{5.39}$$

$$Y'_{dummy-short} = Y_{dummy-short} - Y_{dummy-open}$$

The final step is to subtract the series parasitics by making use of the dummy short. Once this is done, this leaves only Z_{device}, the device itself as shown in (5.40).

$$Z_{device} = Z'_{DUT} - Z'_{dummy-short} \tag{5.40}$$

where Z'_{DUT} is equal to $1/Y'_{DUT}$ and $Z'_{dummy-short}$ is equal to $1/Y'_{dummy-short}$.

5.25 Packaging

With any IC, there comes a moment of truth, a point where the IC designer is forced to admit that the design must be packaged so that it can be sold and the designers can justify their salaries. Typically, the wafer is cut up into dice with each die containing one copy of the IC. The die is then placed inside a plastic package, and the pads on the die are connected to the leads in the package with wire bonds (metal wires), as shown in Figure 5.38. The package is then sealed and can be soldered to a board.

Once the signals from the chip reach the package leads, they are entering a low-loss 50-Ω environment and life is good. The main trouble is the impedance and coupling of the bond wires, which form inductors and transformers. For a wire of radius r a distance h above a ground plane, an estimate for inductance is [3]

$$L \approx 0.2 \ln \frac{2h}{r} \tag{5.41}$$

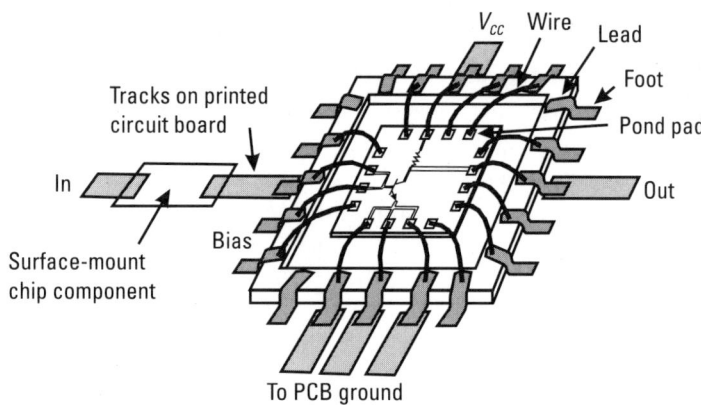

Figure 5.38 An IC in a package (delidded).

where L is expressed as nanohenry per millimeter. For typical bond wires, this results in about 1 nH/mm.

For two round wires separated by d and a distance h above a ground plane, the mutual inductance is estimated by

$$M \approx 0.1 \ln\left[1 + \left(\frac{2h}{d}\right)^2\right] \quad (5.42)$$

where M is expressed as nanohenry per millimeter. As an example, for a pair of bond wires separated by 150 μm (a typical spacing of bond pads on a chip), 1 mm from the ground plane, their mutual inductance would be 0.52 nH/mm, which is a huge number. If the height is dropped to 150 μm above the ground plane, then the mutual inductance of 0.16 nH/mm is still quite significant. Note that parallel bond wires are sometimes used deliberately in an attempt to reduce the inductance. For two inductors in parallel, each of value L_s, one expects the effective inductance to be $L_s/2$. However, with a mutual inductance of M, the effective inductance is $(L_s + M)/2$. Thus, with the example above, two bond wires in parallel would be expected to have 0.5 nH/mm. However, because of the mutual inductance of 0.5 nH/mm, the result is 0.75 nH/mm. Some solutions are to place the bond wires perpendicular to each other or to place ground wires between the active bond wires (obviously of little use if we were trying to reduce the inductance of the ground connection). Another interesting solution is to couple differential signals where the current is flowing in opposite directions. This could apply to a differential circuit or to power and ground. In such a case, the effective inductance is $(L_s - M)/2$, which, in the above example, results in 0.25 nH/mm.

Figure 5.39 is an example of an approximate model for a 32-pin, 5-mm (3-mm die attach area) *thin quad flat pack* (TQFP) package.

At 900 MHz, the impedances would be as shown in Figure 5.40.

The series inductor is dominant at 900 MHz. At the input and output, there is often a matching inductance, so it can simply be reduced to account for the package inductance. At the power supply, the inductance is in series with the load resistor, so gain is increased and the phase is shifted. Thus, if the intended load impedance is 50Ω, the new load impedance is 50 + j17, or in radial terms, 52.8∠18.8°. The most important effect of the package occurs at the ground pad, which is on the emitter of the common-emitter amplifier. This inductance adds emitter degeneration, which can be beneficial in that it can improve linearity and can cause the amplifier input impedance to be less capacitive and thus easier to match. A harmful effect is that the gain is reduced. Also, with higher impedance to external ground, noise injected into this node can be injected into other circuits due to common on-chip ground connections.

Usually it is beneficial to keep ground and substrate impedance low, for example, by using a number of bond pads in parallel, as shown in Figure 5.41.

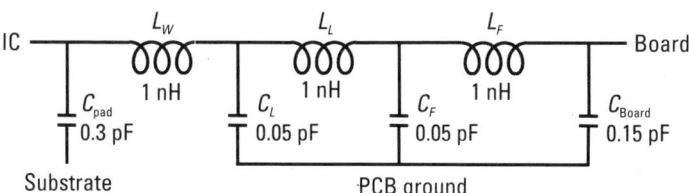

Figure 5.39 Approximate package model.

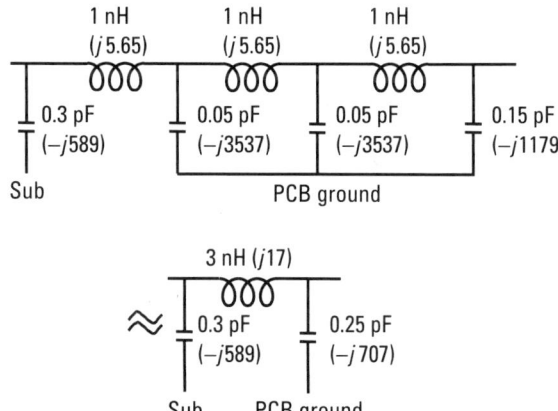

Figure 5.40 Impedances for the approximate package model of Figure 5.39 at 900 MHz.

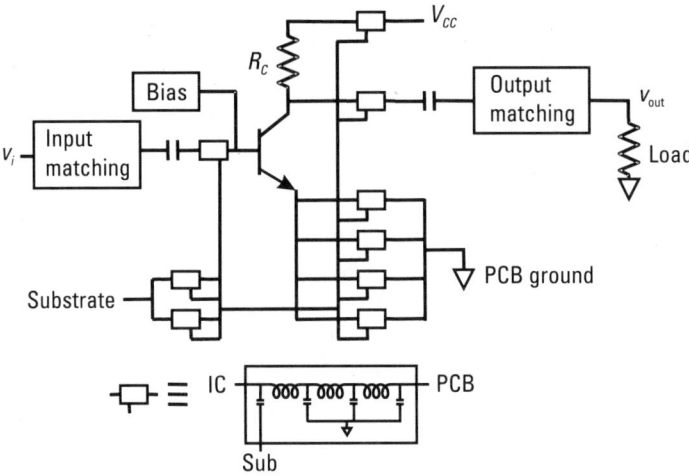

Figure 5.41 Simple amplifier with bond pad models shown.

For example, with four parallel bond pads, the impedance is $j4.2\Omega$. However, it must be noted that with n bond wires in parallel and close together, mutual inductance between them can increase the inductance so that inductance is not decreased by a factor of n, but by something less.

The input source and the load are referenced to the PCB ground. Multiple pads are required for the on-chip ground to have low impedance to PCB ground. Here, the emitter is at on-chip ground. The bond pads have capacitance to substrate, as do any on-chip elements, as previously shown in Figure 5.6, including capacitors, inductors, transistors, and tracks. The substrate has substrate resistance, and substrate contacts are placed all over the chip and are here shown connected to several bond pads. The pads are then connected through bond wires and the package to the PCB, where they could be connected to PCB ground. While bringing the substrate connection out to the printed circuit board is common for mixed-signal designs, for RF circuits, the substrate is usually connected to the on-chip ground.

The lead and foot of the package are over the PCB and so have capacitance to the PCB ground. Note: PCBs usually have a ground plane except where there are tracks.

5.25.1 Other Packaging Techniques

Other packages are available that have lower parasitics. Examples are flip-chip and chip-on-board.

When using flip-chip packaging, a solder ball (or other conducting material) is placed on a board with a matching pattern, and the circuit is connected as shown in Figure 5.42. This results in low inductance (a few tenths of a nanohenry) and very little extra capacitance. One disadvantage is that the pads must be further apart; however, the patterning of the PCB is still very fine, requiring a specialized process. Once flipped and attached, it is also not possible to probe the chip.

When using chip-on-board packaging, the chip is mounted directly on the board and bond wires run directly to the board, eliminating the package, as shown in Figure 5.43. The PCB may be recessed so the top of the chip is level with the board. This may require a special surface on the PCB (gold, for example) to allow bonding to the PCB.

Packaging has an important role in the removal of heat from the circuit, which is especially important for power amplifiers. Thermal conduction can be through contact. For example, the die may touch the metal backing. Thermal conduction can also be through metal connections to bond pads, wires to package, or directly to PCB for chip-on-board. In the case of the flip-chip, thermal conduction is through solder bumps to the printed circuit board.

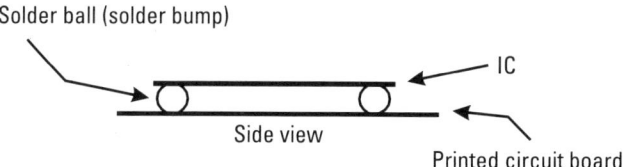

Figure 5.42 Flip-chip packaging.

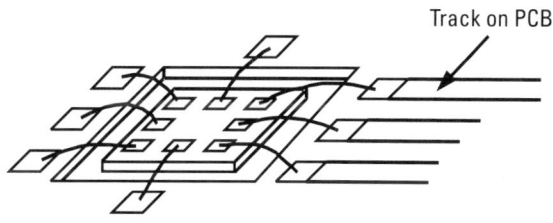

Figure 5.43 Chip-on-board packaging.

References

[1] Abrie, P. L. D., *RF and Microwave Amplifiers and Oscillators,* Norwood, MA: Artech House, 2000.

[2] Barke, E., "Line-to-Ground Capacitance Calculations for VLSI: A Comparison," *IEEE Trans. on Computer-Aided-Design,* Vol. 7, Feb. 1988, pp. 195–298.

[3] Verghese, N. K., T. J. Schmerbeck, and D. J. Allstot, *Simulation Techniques and Solutions for Mixed-Signal Coupling in Integrated Circuits,* Norwell, MA: Kluwer, 1995.

[4] Long, J. R., "A Narrowband Radio Receiver Front-End for Portable Communications Applications," Ph.D. dissertation, Carleton University, 1996.

[5] Mohan, S. S., et al., "Simple Accurate Expressions for Planar Spiral Inductances," *IEEE J. Solid-State Circuits,* Vol. 34, Oct. 1999, pp. 1419–1424.

[6] Razavi, B., *Design of Analog CMOS Integrated Circuits,* New York: McGraw-Hill, 2000, Chapters 17 and 18.

[7] Danesh, M., et al., "A Q-Factor Enhancement Technique for MMIC Inductors," *Proc. RFIC Symposium,* 1998, pp. 183–186.

[8] Cheung, D. T. S., J. R. Long, and R. A. Hadaway, "Monolithic Transformers for Silicon RFIC Design," *Proc. BCTM,* Sept. 1998, pp. 105–108.

[9] Long, J. R., and M. A. Copeland, "The Modeling, Characterization, and Design of Monolithic Inductors for Silicon RF IC's," *IEEE J. Solid-State Circuits,* Vol. 32, March 1997, pp. 357–369.

[10] Edelstein, D. C., and J. N. Burghartz, "Spiral and Solenoidal Inductor Structures on Silicon Using Cu-Damascene Interconnects," *Proc. IITC,* 1998, pp. 18–20.

[11] Hisamoto, D., et al., "Suspended SOI Structure for Advanced 0.1-μm CMOS RF Devices," *IEEE Trans. on Electron Devices,* Vol. 45, May 1998, pp. 1039–1046.

[12] Yue, C. P., and S. S. Wong, "On-Chip Spiral Inductors with Patterned Ground Shields for Si-Based RF IC's," *IEEE J. Solid-State Circuits,* Vol. 33, May 1998, pp. 743–752.

[13] Craninckx, J., and M. S. J. Steyaert, "A 1.8-GHz Low-Phase-Noise CMOS VCO Using Optimized Hollow Spiral Inductors," *IEEE J. Solid-State Circuits,* Vol. 32, May 1997, pp. 736–744.

[14] Niknejad, A. M., and R. G. Meyer, "Analysis, Design, and Optimization of Spiral Inductors and Transformers for Si RF IC's," *IEEE J. Solid-State Circuits,* Vol. 33, Oct. 1998, pp. 1470–1481.

[15] Rogers, J. W. M., J. A. Macedo, and C. Plett, "A Completely Integrated Receiver Front-End with Monolithic Image Reject Filter and VCO," *IEEE RFIC Symposium,* June 2000, pp. 143–146.

[16] Rogers, J. W. M., et al., "Post-Processed Cu Inductors with Application to a Completely Integrated 2-GHz VCO," *IEEE Trans. on Electron Devices,* Vol. 48, June 2001, pp. 1284–1287.

[17] Zolfaghari, A., A. Chan, and B. Razavi, "Stacked Inductors and Transformers in CMOS Technology," *IEEE J. Solid-State Circuits,* Vol. 36, April 2001, pp. 620–628.

[18] Niknejad, A. M., J. L. Tham, and R. G. Meyer, "Fully-Integrated Low Phase Noise Bipolar Differential VCOs at 2.9 and 4.4 GHz," *Proc. European Solid-State Circuits Conference,* 1999, pp. 198–201.

[19] van Wijnen, P. J., "On the Characterization and Optimization of High-Speed Silicon Bipolar Transistors," Beaverton, OR: Cascade Microtech, 1995.

6

LNA Design

6.1 Introduction and Basic Amplifiers

The LNA is the first block in most receiver front ends. Its job is to amplify the signal while introducing a minimum amount of noise to the signal.

Gain can be provided by a single transistor. Since a transistor has three terminals, one terminal should be ac grounded, one serves as the input, and one is the output. There are three possibilities, as shown in Figure 6.1. Each one of the basic amplifiers has many common uses and each is particularly suited to some tasks and not to others. The common-emitter amplifier is most often used as a driver for an LNA. The common-collector, with high input impedance and low output impedance, makes an excellent buffer between stages or before the output driver. The common-base is often used as a cascode in combination with the common-emitter to form an LNA stage with gain to high frequency, as will be shown. The loads shown in the diagrams can be made either with resistors for broadband operation, or with tuned resonators for narrow-band operation. In this chapter, LNAs with resistors will be discussed first, followed by a discussion of narrowband LNAs. Also, refinements such as feedback can be added to the amplifiers to augment their performance.

6.1.1 Common-Emitter Amplifier (Driver)

To start the analysis of the common-emitter amplifier, we replace the transistor with its small-signal model, as shown in Figure 6.2. Z_L represents some arbitrary load that the amplifier is driving.

At low frequency, the voltage gain of the amplifier can be given by

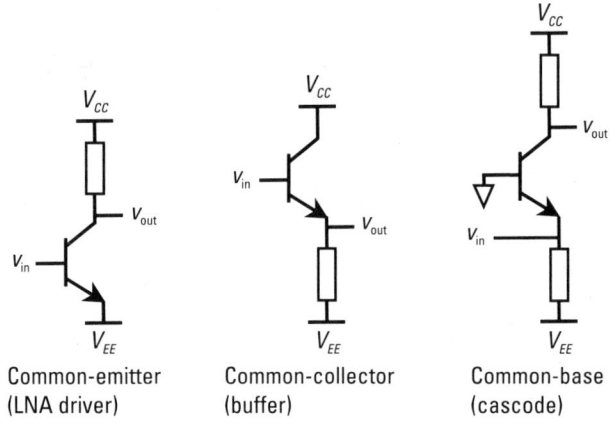

Figure 6.1 Simple transistor amplifiers.

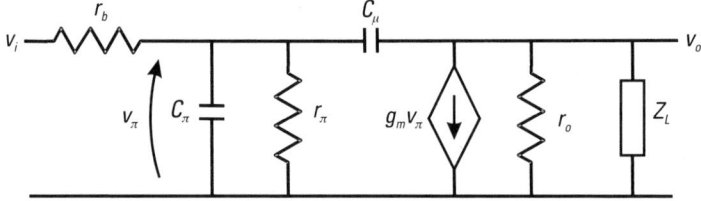

Figure 6.2 The common-emitter amplifier with transistor replaced with its small-signal model.

$$A_{vo} = \frac{v_o}{v_i} = -\frac{r_\pi}{r_b + r_\pi} g_m Z_L \approx \frac{Z_L}{r_e} \quad (6.1)$$

where r_e is the small-signal base-emitter diode resistance as seen from the emitter. Note that $r_\pi = \beta r_e$ and $g_m = 1/r_e$. For low frequencies, the parasitic capacitances have been ignored and r_b has been assumed to be low compared to r_π.

The input impedance of the circuit at low frequencies is given by

$$Z_{in} = r_b + r_\pi \quad (6.2)$$

However, at RF, C_π will provide a low impedance across r_π, and C_μ will provide a feedback (and feedforward) path. The frequency at which the low-frequency gain is no longer valid can be estimated by using Miller's Theorem to replace C_μ with two capacitors C_A and C_B, as illustrated in Figure 6.3, where C_A and C_B are

$$C_A = C_\mu \left(1 - \frac{v_o}{v_\pi}\right) = C_\mu (1 + g_m Z_L) \approx C_\mu g_m Z_L \quad (6.3)$$

LNA Design

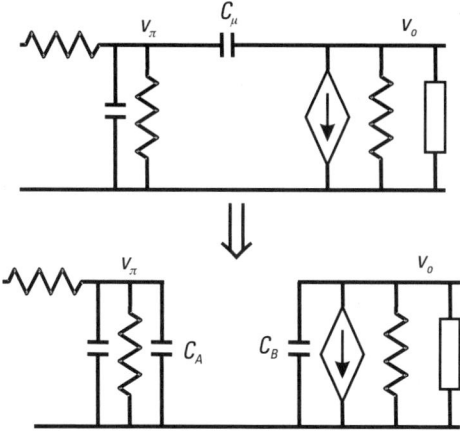

Figure 6.3 C_μ is replaced with two equivalent capacitors C_A and C_B in the common-emitter amplifier.

$$C_B = C_\mu \left(1 - \frac{v_\pi}{v_o}\right) = C_\mu \left(1 + \frac{1}{g_m Z_L}\right) \approx C_\mu \qquad (6.4)$$

There are now two equivalent capacitors in the circuit: one consisting of $C_A + C_\pi$ and the other consisting of C_B. This means that there are now two RC time constants or two poles in the system. The dominant pole is usually the one formed by C_A and C_π. The pole occurs at

$$f_{P1} = \frac{1}{2\pi \cdot \left[r_\pi \| (r_b + R_S)\right]\left[C_\pi + C_A\right]} \qquad (6.5)$$

where R_S is the resistance of the source driving the amplifier. We note that as the load impedance decreases, the capacitance C_A is reduced and the dominant pole frequency is increased.

When calculating f_T, the input is driven with a current source and the output is loaded with a short circuit. This removes the Miller multiplication, and the two capacitors C_π and C_μ are simply connected in parallel. Under these conditions, as explained in Chapter 3, the frequency f_β, where the current gain is reduced by 3 dB, is given by

$$f_\beta = \frac{1}{2\pi \cdot r_\pi (C_\pi + C_\mu)} \qquad (6.6)$$

The unity current gain frequency can be found by noting that with a first-order roll-off, the ratio of f_T to f_β is equal to the low-frequency current gain β. The resulting expression for f_T is

$$f_T = \frac{g_m}{2\pi \cdot (C_\pi + C_\mu)} \qquad (6.7)$$

It is also useful to note that above the pole frequency, we could ignore r_π and just use C_π in the transistor model with little error. This simplified model is shown in Figure 6.4.

Example 6.1 Calculation of Pole Frequency

A 15x transistor, as described in Chapter 3, has the following bias conditions and properties: $I_C = 5$ mA, $r_\pi = 500\Omega$, $\beta = 100$, $C_\pi = 700$ fF, $C_\mu = 23.2$ fF, $g_m = I_C/v_T = 200$ mA/V, and $r_b = 5\Omega$. If $Z_L = 100\Omega$ and $R_S = 50\Omega$. Find the frequency f_{P1} at which the gain drops by 3 dB from its dc value.

Solution

Since $g_m Z_L = 20$,

$$C_A \approx C_\mu g_m Z_L = 23.2 \text{ fF} \cdot 20 = 464 \text{ fF}$$

Thus, the pole is at a frequency of

$$f_{P1} = \frac{1}{2\pi\left[500 \| (5 + 50)\right]\left[700f + 464f\right]} = 2.76 \text{ GHz}$$

Example 6.2 Calculation of Unity Gain Frequency

For the transistor in Example 6.1, compute f_β and f_T.

Solution

Using (6.6) and (6.7), the result is

$$f_\beta = \frac{1}{2\pi r_\pi (C_\pi + C_\mu)} = \frac{1}{2\pi \cdot 500\Omega (700 \text{ fF} + 23.2 \text{ fF})} = 440 \text{ MHz}$$

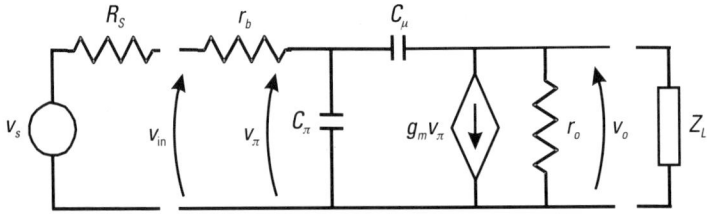

Figure 6.4 Simplified small-signal model for the transistor in the common-emitter amplifier above the dominant pole frequency.

$$f_T = \frac{1}{2\pi(C_\pi + C_\mu)} = \frac{1}{2\pi \cdot (700 \text{ fF} + 23.2 \text{ fF})} = 44 \text{ GHz}$$

Knowing the pole frequency, we can estimate the gain at higher frequencies, assuming that there are no other poles present, with

$$A_v(f) = \frac{A_{vo}}{1 + j\dfrac{f}{f_{P1}}} \qquad (6.8)$$

Example 6.3 Calculation of Gain of Single-Pole Amplifier

For the above example, for $A_{vo} = 20$ with $f_{P1} = 2.76$ GHz, calculate the gain at 5.6 GHz.

Solution

With $f_{P1} = 2.76$ GHz, at 5.6 GHz, the gain can be calculated to be 8.84, or 18.9 dB. This is down by about 7 dB from the low-frequency gain.

The exact expression for v_o/v_s (after about one page of algebra) is

$$\frac{v_o}{v_s} = \frac{s - \dfrac{g_m}{C_\mu}}{C_\pi R_S' \left[s^2 + s\left(\dfrac{1}{C_\pi R_{S\pi}} + \dfrac{1}{C_{\mu\pi} Z_L'} + \dfrac{g_m}{C_\pi}\right) + \left(\dfrac{1}{C_\pi R_{S\pi} C_\mu Z_L'}\right) \right]} \qquad (6.9)$$

where: $R_S' = R_S + r_b$, $R_{S\pi} = R_S' \| r_\pi$, $C_{\mu\pi} = C_\mu$ in series with C_π, and $Z_L' = Z_L \| r_o$.

As expected, this equation features a zero in the right-half plane and real, well-separated poles, similar to that of a pole-splitting operational amplifier [1].

Example 6.4 Calculation of Exact Poles and Zeros

Calculate poles and zeros for the transistor amplifier as in the previous example.

Solution

Results for the previous example: 15x npn, 5 mA, $Z_L = 100\Omega$, $C_\mu = 23.2$ fF, $C_\pi = 700$ fF, $R_S = 50\Omega$, $r_b = 5\Omega$. Using (6.9), the results are that the poles are at 2.66 and 118.3 GHz; the zero is at 1,384 GHz. Thus, the exact equation has been used to verify the original assumptions that the two poles are well separated, that the dominant pole is approximately at the frequency given by the previous equations, and the second pole and feedforward zero in this expression are well above the frequency of interest.

6.1.2 Simplified Expressions for Widely Separated Poles

If a system can be described by a second-order transfer function given by

$$\frac{v_o}{v_i} = \frac{A(s-z)}{s^2 + sb + c} \tag{6.10}$$

then the poles of this system are given by

$$P_{1,2} = -\frac{b}{2} \pm \frac{b}{2}\sqrt{1 - \frac{4c}{b^2}} \tag{6.11}$$

If the poles are well separated, then $4c/b^2 \ll 1$ Therefore,

$$P_{1,2} \approx -\frac{b}{2} \pm \frac{b}{2}\left(1 - \frac{2c}{b^2}\right) \tag{6.12}$$

and

$$P_1 \approx -\frac{c}{b} \tag{6.13}$$

$$P_2 \approx -b \tag{6.14}$$

Example 6.5 Calculation of Poles and Zeros with Simplified Expressions

With the expression as above, the poles of Example 6.4 occur at 2.60 and 120.94 GHz, which are reasonably close to the exact values.

6.1.3 The Common-Base Amplifier (Cascode)

The common-base amplifier is often combined with the common-emitter amplifier to form an LNA, but it can be used by itself as well. Since it has low input impedance when it is driven from a current source, it can pass current through it with near unity gain up to a very high frequency. Therefore, with an appropriate choice of impedance levels, it can also provide voltage gain. The small-signal model for the common-base amplifier is shown in Figure 6.5 (ignoring output impedance).

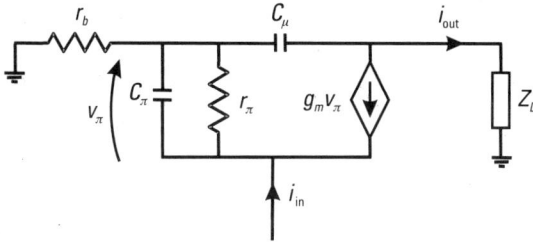

Figure 6.5 Small-signal model for the common-base amplifier.

The current gain (ignoring C_μ and r_o) for this stage can be found to be

$$\frac{i_{out}}{i_{in}} \approx \frac{1}{1 + j\omega C_\pi r_e} \approx \frac{1}{1 + j\dfrac{\omega}{\omega_T}} \quad (6.15)$$

where $\omega_T = 2\pi f_T$ [see (6.7)]. At frequencies below ω_T, the current gain for the stage is 1. Note that the pole in this equation is usually at a much higher frequency than the one in the common-emitter amplifier, since $r_e < r_b + R_S$. As mentioned above, the input impedance of this stage is low and is equal to $1/g_m$ at low frequencies. At the pole frequency, the capacitor will start to dominate and the impedance will drop.

This amplifier can be used in combination with the common-emitter amplifier (discussed in Section 6.1.1) to form a cascode LNA, as shown in Figure 6.6. In this case, the current i_{c1} through Q_1 is about the same as the current i_{c2} through Q_2, since the common-base amplifier has a current gain

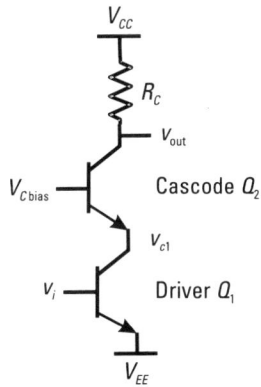

Figure 6.6 Common-base amplifier used as a cascode transistor in an LNA.

of approximately 1. Then, $i_{c1} \approx i_{c2} = g_{m1} v_i$. For the case where $R_S + r_b \ll r_\pi$ and $v_o/v_i \approx -g_m R_C$, the gain is the same as for the common-emitter amplifier. However, the cascode transistor reduces the feedback of $C_{\mu 1}$, resulting in increased high-frequency gain.

In the previous section, we showed that as the load resistance on the common-emitter amplifier is reduced, the dominant pole frequency is increased. Thus, by adding the common-base amplifier as the load of the common-emitter amplifier, the impedance seen by the collector of Q_1 is $r_e \approx 1/g_{m2}$ (a low value). Thus, the frequency response of this stage has been improved by adding the common-base amplifier.

The new estimate for the pole frequency in the common-emitter amplifier is

$$f_{P1} = \frac{1}{2\pi \left[r_{\pi 1} \| (r_{b1} + R_S) \right] \left[C_\pi + 2C_\mu \right]} \approx \left(1 + \frac{r_{\pi 1}}{R_S + r_{b1}} \right) \frac{f_T}{\beta} \quad (6.16)$$

Example 6.6 Improving the Pole Frequency of a Common-Emitter Amplifier

For the previous example with 15x npn, 5 mA, $Z_L = 100\Omega$, $C_\mu = 23.2$ fF, $C_\pi = 700$ fF, $R_S = 50\Omega$, $r_b = 5\Omega$, estimate the pole frequency for the amplifier with a cascode transistor added.

Solution

Without the cascode transistor, the estimated pole frequency was $f_{P1} \approx 2.7$ GHz. With the approximate expression (6.16), the pole frequency is 4.44 GHz.

Another advantage of the cascode amplifier is that adding another transistor improves the isolation between the two ports (very little reverse gain in the amplifier). The disadvantage is that the additional transistor adds additional poles to the system. This can become a problem for a large load resistance, leading to rapid high-frequency-gain roll-off (−12 dB/octave compared to the previous −6 dB/octave) and excess phase lag, which can cause problems if feedback is used. Also, an additional bias voltage is required, and if this cascode bias node is not properly decoupled, instability can occur. A further problem is the reduced signal swing at a given supply voltage, compared to the simple common-emitter amplifier, since the supply must now be split between two transistors instead of just one.

6.1.4 The Common-Collector Amplifier (Emitter Follower)

The common-collector amplifier is a very useful general-purpose amplifier. It has a voltage gain that is close to 1, but has a high input impedance and a low output impedance. Thus, it makes a very good buffer stage or output stage. It can also be used to do a dc level shift in a circuit.

The common-collector amplifier and its small-signal model are shown in Figure 6.7. The resistor R_E may represent a resistor or the output resistance of a current source. Note that the Miller effect is not a problem in this amplifier, since the collector is grounded. Since C_μ is typically much less than C_π, it can be left out of the analysis with little impact on the gain. The voltage gain of this amplifier is given by

$$A_v(s) = A_{vo}\left(\frac{1 - s/z_1}{1 - s/p_1}\right) \tag{6.17}$$

A_{vo} is the gain at low frequency and is given by

$$A_{vo} = \frac{g_m R_E + \dfrac{R_E}{r_\pi}}{1 + g_m R_E + \dfrac{(R_B + R_E)}{r_\pi}} \approx \frac{g_m R_E}{1 + g_m R_E} \approx 1 \tag{6.18}$$

where $R_B = R_S + r_b$.

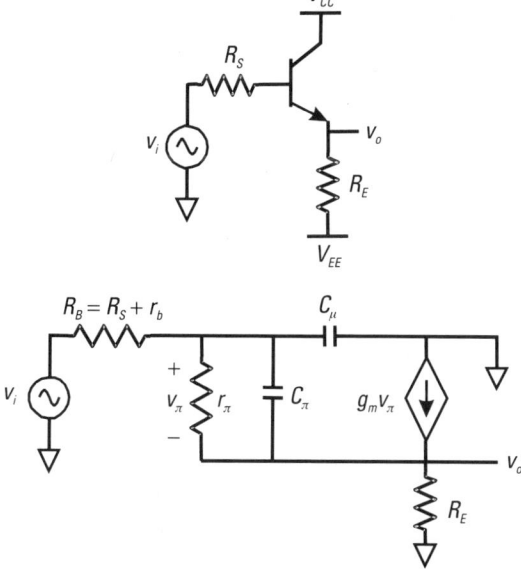

Figure 6.7 Common-collector amplifier and its small-signal model.

The pole and zero are given by

$$f_{z1} \approx \frac{g_m}{C_\pi} = -\omega_T \qquad (6.19)$$

$$f_{p1} \approx -\frac{1}{C_\pi R_A} \qquad (6.20)$$

where

$$R_A = r_\pi \left\| \frac{R_B + R_E}{1 + g_m R_E} \right. \qquad (6.21)$$

if $g_m R_E \gg 1$, then

$$R_A \approx \frac{R_B + R_E}{g_m R_E} \qquad (6.22)$$

and

$$f_{p1} \approx -\frac{R_E}{R_E + R_B} \omega_T \qquad (6.23)$$

The input impedance of this amplifier can also be determined. If $g_m R_E \gg 1$, then we can use the small-signal circuit to find Z_{in}. The input impedance, again ignoring C_μ, is given by

$$Z_{in} = Z_\pi + R_E(1 + g_m Z_\pi) \qquad (6.24)$$

Likewise, the output impedance can be found and is given by

$$Z_{out} = \frac{r_\pi + R_B + sC_\pi r_\pi R_B}{1 + g_m r_\pi + sC_\pi r_\pi} \qquad (6.25)$$

Provided that $r_\pi > R_B$ and $\omega C_\pi r_\pi > r_\pi$ at the frequency of interest, the output impedance simplifies to

$$Z_{out} \approx r_e \frac{1 + g_m R_B j\omega/\omega_T}{1 + j\omega/\omega_T} \qquad (6.26)$$

At low frequencies, this further simplifies to

$$Z_{out} \approx r_e \approx \frac{1}{g_m} \quad (6.27)$$

At higher frequencies, if $r_e > R_B$ (recalling that $R_B = R_S + r_b$), for example, at low current levels, then $|Z_{out}|$ decreases with frequency, and so the output impedance is capacitive. However, if $r_e < R_B$, then $|Z_{out}|$ increases for higher frequency and the output can appear inductive. In this case, if the circuit is driving a capacitive load, the inductive component can produce peaking (resonance) or even instability.

The output can be modeled as shown in Figure 6.8, where

$$R_1 = \frac{1}{g_m} + \frac{R_S + r_b}{\beta} \quad (6.28)$$

$$R_2 = R_S + r_b \quad (6.29)$$

$$L = \frac{1}{\omega_T}(R_S + r_b) \quad (6.30)$$

Example 6.7 Emitter Follower Example

Calculate the output impedance for the emitter follower with a transistor, as before with 5 mA, $C_\mu = 23.2$ fF, $C_\pi = 700$ fF, $r_b = 5\Omega$. Assume that both input impedance and output impedance are 50Ω.

Solution

Solving for the various components, it can be shown that low-frequency output resistance is 5.5Ω and high-frequency output impedance is 55Ω. The equivalent inductance is 0.2 nH, the zero frequency is 4.59 GHz, and the pole frequency is 45.9 GHz.

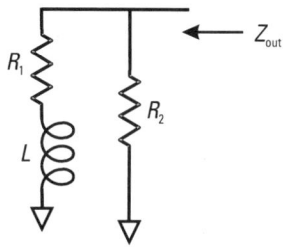

Figure 6.8 Output impedance of the common-collector amplifier.

6.2 Amplifiers with Feedback

There are numerous ways to apply feedback in an amplifier, and it would be almost a book in itself to discuss them all. Only a few of the common feedback techniques will be discussed here.

6.2.1 Common-Emitter with Series Feedback (Emitter Degeneration)

The two most common configurations for RF LNAs are the common-emitter configuration and the cascode configuration shown in Figure 6.9. In most applications, the cascode is preferred over the common-emitter topology because it can be used at higher frequencies (the extra transistor acts to reduce the Miller effect) and has superior reverse isolation (S_{12}). However, the cascode also suffers from reduced linearity due to the stacking of two transistors, which reduces the available output swing.

Most common-emitter and cascode LNAs employ the use of degeneration (usually in the form of an inductor in narrowband applications) as shown in Figure 6.9. The purpose of degeneration is to provide a means to transform the real part of the impedance seen looking into the base to a higher impedance for matching purposes. This inductor also trades gain for linearity as the inductor is increased in size.

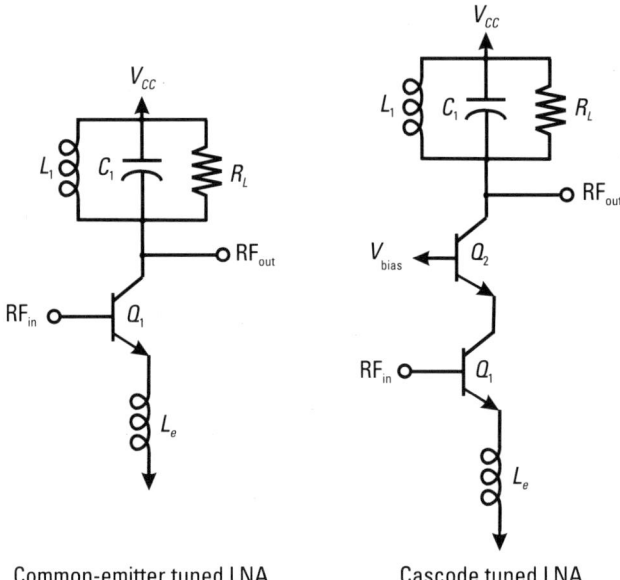

Figure 6.9 Narrowband common-emitter and cascode LNAs with inductive degeneration.

The gain of either amplifier at the resonance frequency of the tank in the collector, ignoring the effect of C_μ, is found with the aid of Figure 6.10 and is given by

$$\frac{v_{out}}{v_{in}} = \frac{-g_m R_L}{\left(1 + \frac{Z_E}{Z_\pi} + g_m Z_E\right)} \approx -\frac{R_L}{Z_E} \quad (6.31)$$

where Z_E is the impedance of the emitter degeneration. Here it is assumed that the impedance in the emitter is a complex impedance. Thus, as the degeneration becomes larger, the gain ceases to depend on the transistor parameters and becomes solely dependent on the ratio of the two impedances. This is, of course, one of the advantages of this type of feedback. This means that the circuit becomes less sensitive to temperature and process variations.

If the input impedance is matched to R_S (which would require an input series inductor), then the gain can be written out in terms of source resistance and f_T. v_{out} in terms of i_x in Figure 6.10 can be given by

$$v_{out} = -g_m v_\pi R_L = -g_m i_x Z_\pi R_L \quad (6.32)$$

Noting that i_x can also be equated to the source resistance R_S as $i_x = v_{in}/R_s$:

$$\frac{v_{out}}{v_{in}} = \frac{-g_m Z_\pi R_L}{R_S} \quad (6.33)$$

assuming that Z_π is primarily capacitive at the frequency of interest:

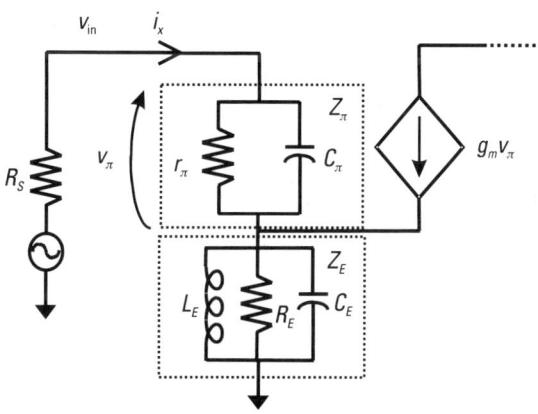

Figure 6.10 Small-signal model used to find the input impedance and gain.

$$\left|\frac{v_{out}}{v_{in}}\right| = \frac{g_m R_L}{R_S \omega_o C_\pi} = \frac{R_L \omega_T}{R_S \omega_o} \qquad (6.34)$$

where ω_o is the frequency of interest.

The input impedance has the same form as the common-collector amplifier and is also given by

$$Z_{in} = Z_\pi + Z_E(1 + g_m Z_\pi) \qquad (6.35)$$

Of particular interest is the product of Z_E and Z_π. If the emitter impedance is inductive, then when this is reflected into the base, it will become a real resistance. Thus, placing an inductor in the emitter tends to raise the input impedance of the circuit, so it is very useful for matching purposes. (Conversely, placing a capacitor in the emitter will tend to reduce the input impedance of the circuit and can even make it negative.)

6.2.2 The Common-Emitter with Shunt Feedback

Applying shunt feedback to a common-emitter amplifier is a good basic building block for broadband amplifiers. This technique allows the amplifier to be matched over a broad bandwidth while having minimal impact on the noise figure of the stage. A basic common-emitter amplifier with shunt feedback is shown in Figure 6.11. Resistor R_f forms the feedback and capacitor C_f is added to allow for independent biasing of the base and collector. C_f can normally be chosen so that it is large enough to be a short circuit over the frequency of interest. Note that this circuit can be modified to become a cascode amplifier if desired.

Ignoring the Miller effect and assuming C_f is a short circuit ($1/\omega C_f \ll R_f$), the gain is given by

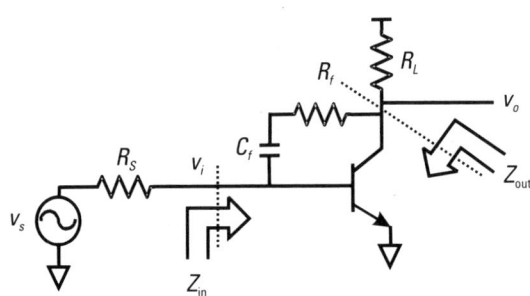

Figure 6.11 A common-emitter amplifier with shunt feedback.

$$A_v = \frac{v_o}{v_i} = \frac{\frac{R_L}{R_F} - g_m R_L}{1 + \frac{R_L}{R_f}} \approx \frac{-g_m R_L}{1 + \frac{R_L}{R_f}} \quad (6.36)$$

Thus, we see that in this case the gain without feedback $(-g_m R_L)$ is reduced by the presence of feedback.

The input impedance of this stage is also changed dramatically by the presence of feedback. Ignoring C_μ, the input admittance can be computed to be

$$Y_{in} = \frac{1}{R_f} + \frac{g_m R_L - \frac{R_L}{R_F}}{R_f + R_L} + \frac{1}{Z_\pi} \quad (6.37)$$

Alternatively, the input impedance can be given by

$$Z_{in} = \frac{Z_\pi(R_f + R_L)}{R_f + R_L + Z_\pi(1 + g_m R_L)} \approx R_f \| Z_\pi \| \frac{R_f + R_L}{g_m R_L} \approx \frac{R_f + R_L}{g_m R_L} \quad (6.38)$$

This can be seen to be the parallel combination of Z_π with R_f along with a parallel component due to feedback. The last term, which is usually dominant, shows that the input impedance is equal to $R_f + R_L$ divided by the open loop gain. As a result, compared to the open-loop amplifier, the input impedance for the shunt feedback amplifier has less variation over frequency and process.

Similarly, the output impedance can be determined as

$$Z_{out} = \frac{R_f}{1 + Z_{ip}\left(g_m - \frac{1}{R_f}\right)} \approx \frac{R_f}{1 + g_m Z_{ip}} \quad (6.39)$$

where $Z_{ip} = R_S \| R_f \| Z_\pi$.

Feedback results in the reduction of the role the transistor plays in determining the gain and therefore improves linearity, but the presence of R_f may degrade the noise depending on the choice of value for this resistor.

With this type of amplifier, it is sometimes advantageous to couple it with an output buffer, as shown in Figure 6.12. The output buffer provides some inductance to the input, which tends to make for a better match. The presence of the buffer does change the previously developed formulas somewhat. If the buffer is assumed to be lossless, the input impedance now becomes

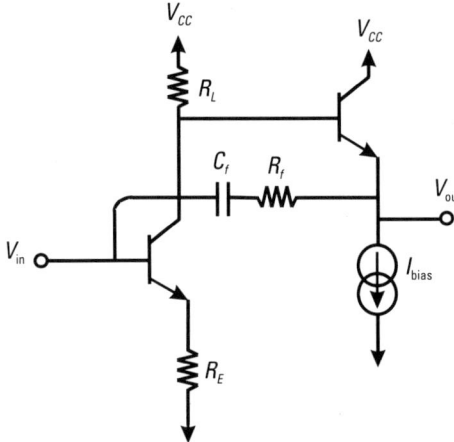

Figure 6.12 A common-emitter amplifier with shunt feedback and output common collector (CC) buffer.

$$Z_{in} = Z_\pi \left(1 + \frac{Z_\pi}{R_f} + \frac{g_m R_L Z_\pi}{R_f}\right)^{-1} = \frac{R_f}{1 + g_m R_L + \frac{R_f}{Z_\pi}} \approx \frac{R_f}{g_m R_L} \quad (6.40)$$

With the addition of a buffer, the voltage gain is no longer affected by the feedback, so it is approximately that of a common emitter amplifier given by $[R_L/(R_E + 1/g_m)]$ minus the loss in the buffer. However, if driven from a resistor, the gain will have some dependence on the feedback resistor R_f.

Example 6.8 Calculation of Gain and Input and Output Impedance with Shunt Feedback

An amplifier similar to that shown in Figure 6.11 has a load resistor of $R_L = 100\Omega$, and a source resistance of $R_S = 50\Omega$. The transistors are in a 12-GHz f_T process with 15x transistors and 5 mA for a transconductance of $g_m = 200$ mA/V. With $\beta = 100$ and $r_\pi = 500\Omega$, parasitic capacitance was estimated at $C_\pi = 2.37$ pF. Assume C_μ can be ignored. (The Miller effect can be minimized by using a cascode structure.) At 900 MHz, compare gain and input and output impedance with feedback resistor at $R_f \to \infty$ and $R_f = 500\Omega$.

Solution

Without feedback from the output to the base, the gain is $g_m R_L$ or 26 dB. From input source to output, including the effect of the source resistance and Z_π, calculated as $10.9 - j73$ (or $73.8 \angle -81.5°$), the gain A_v is equal to

$$A_v = (-g_m R_L) \frac{Z_\pi}{R_S + Z_\pi} = -20 \frac{10.9 - j73}{50 + 10.9 - j73}$$
$$= 13.26 - j8.08 = 15.52 \angle -31.34° \text{ (or 23.8 dB)}$$

Input impedance is given by Z_π, calculated as $10.9 - j73$ (or $73.8 \angle -81.5°$). The output impedance for this simple example is infinity. However, for a real circuit, the output impedance would be determined by the transistor, integrated circuit, and package parasitic impedance.

We now turn to the case when the feedback is 500Ω. From (6.36), the voltage gain is

$$A_v = \frac{v_o}{v_i} \approx \frac{-0.2 \times 100}{1 + \frac{100}{500}} = \frac{20}{1.2} = 16.67 \text{ (or 24.4 dB)}$$

(Note the exact expression from the same equation would have resulted in 16.5, so the approximate expression is sufficient.) Thus, the gain has been reduced from 26 dB to 24.4 dB.

From the source the gain is reduced much more, since the input impedance is much reduced, so there is a lot of attenuation. The gain from the source, instead of from v_i, can be shown to be

$$A_{v\text{source}} = -5.487 + j1.290 = 5.637 \angle 166.8° \text{ (or 15.02 dB)}$$

Thus, gain is reduced from 23.8 to 15.02 dB by feedback.

$$Z_{in} \approx R_f \| Z_\pi \| \frac{R_f + R_L}{g_m R_L} = 500 \| (10.9 - j73) \| \left(\frac{500 + 100}{20}\right)$$
$$= 23.89 - j8.65$$

$$Z_{out} = \frac{R_f}{1 + g_m Z_{ip}} = \frac{500}{1 + 0.2 \cdot (31.76 - j17.73)} = 55.68 + j26.82$$

where

$$Z_{ip} = R_S \| R_f \| Z_\pi = 50 \| 500 \| (10.9 - j73) = 31.76 - j17.73$$

We note that at low frequency without feedback, Z_{in} would have been 500Ω due to r_π, while with feedback it is 27Ω, so the impedance is much steadier across frequency.

Example 6.9 Simulation of Gain and Input and Output Impedance with Shunt Feedback

Simulate a cascode amplifier as shown in Figure 6.13 with a load resistor of $R_L = 50\Omega$ and source resistance of $R_S = 50\Omega$. The transistors are in a 12-GHz f_T process with 15x transistors, with $\beta = 100$ and 5 mA for a transconductance of $g_m = 200$ mA/V. Compare gain and input and output impedance with various values of feedback resistor R_f.

Solution

The results are shown in Figure 6.14. It can be seen that noise figure is not strongly affected if the feedback resistor is larger than about 500Ω. It can also be seen that for this design, linearity seems to be limited by the output, which here remains at a nearly constant OIP3 of about 14 dBm. IIP3 seems to vary significantly, but this is mainly due to the change of gain. Specifically, gain decreases by 4 dB and IIP3 improves by 5 dB, but OIP3 improves by only about 1 dB as the feedback resistance is reduced from $1,500\Omega$ to 500Ω.

In Section 6.8 we will conclude the chapter with a major example of a broadband amplifier design, including considerations of noise and linearity. Noise and linearity are topics that will be discussed in some of the following sections.

6.3 Noise in Amplifiers

When the signal is first received by the radio, it can be quite weak and can be in the presence of a great deal of interference. The LNA is the first part of the

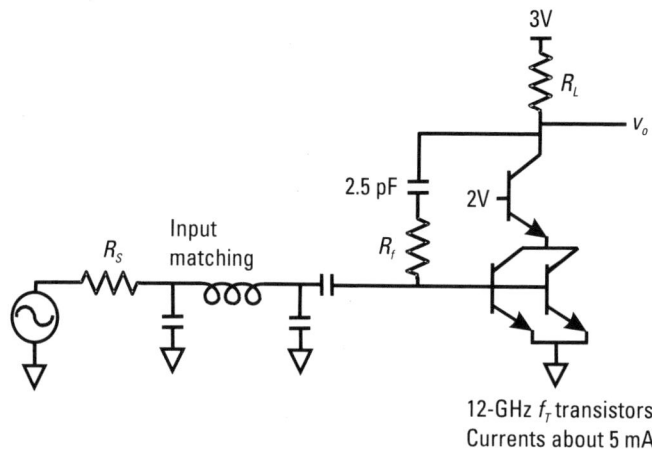

Figure 6.13 Amplifier with shunt feedback.

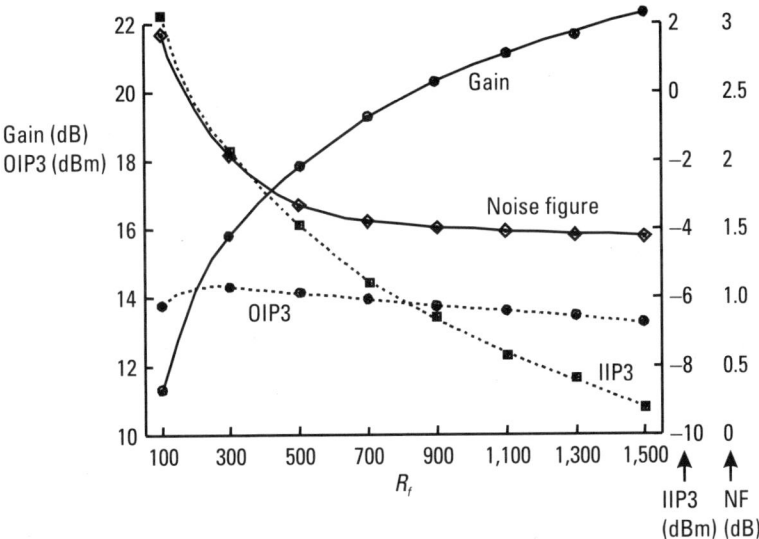

Figure 6.14 Sample plots using shunt feedback.

radio to process the signal, and it is therefore essential that it amplify the signal while adding a minimal amount of additional noise to it. Thus, one of the most important considerations when designing an LNA is the amount of noise present in the circuit. The following sections discuss this important topic.

6.3.1 Input-Referred Noise Model of the Bipolar Transistor

Note that the following two sections contain many equations, which may be tedious for some readers. Reader discretion is advised. Those who find equations disturbing can choose to go directly to Section 6.3.3.

In Chapter 2, we made use of an idealized model for an amplifier with two noise sources at the input. If the model is to be applied to an actual LNA, then all the noise sources must be written in terms of these two input-referred noise sources, as shown in Figure 6.15. Starting with the model shown in Figure 6.15(a), and assuming that the emitter is grounded with base input and collector output, the model may be determined with some analysis.

When the input is shorted in Figure 6.15(b), then v_n is the only source of noise in the model, and then the output noise current i_{on_tot} would be (assuming that r_b is small enough to have no effect on the gain)

$$\overline{i_{on\_tot}^2} = \overline{v_n^2} g_m^2 \qquad (6.41)$$

If instead the actual noise sources in the model are used as in Figure 6.15(a), then the output noise current can also be found. In this case, the effect

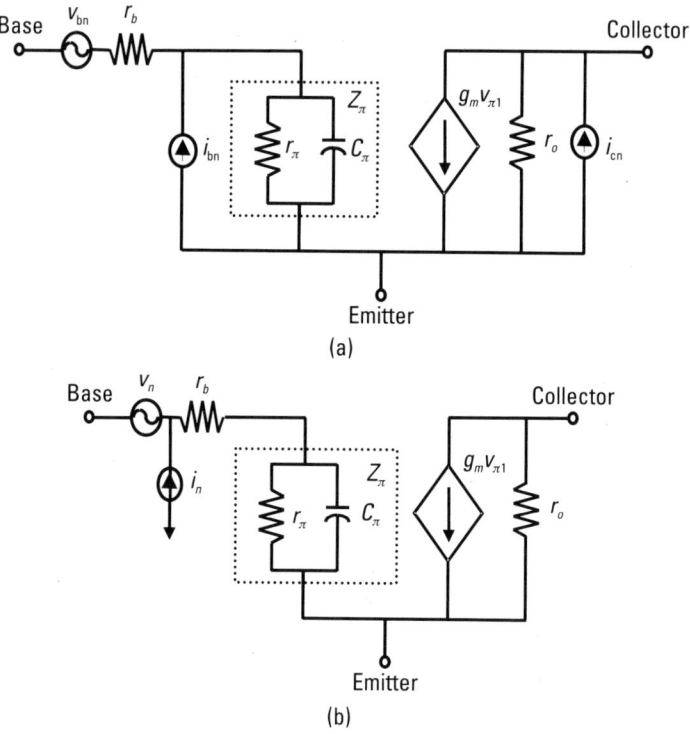

Figure 6.15 Noise model for the (a) bipolar transistor, and (b) equivalent input-referred noise model.

of the base shot noise is assumed to be shorted out (i.e., r_b is small compared to the input impedance of the transistor). That leaves the collector shot noise and base resistance noise sources in the model to be accounted for.

$$\overline{i_{on\_tot}^2} = \overline{v_{bn}^2} g_m^2 + \overline{i_{cn}^2} \tag{6.42}$$

We wish to make these two models equivalent, so we set (6.41) equal to (6.42) and solve for v_n giving

$$\overline{v_n^2} = \frac{2qI_c}{g_m^2} + 4kTr_b \tag{6.43}$$

where $\overline{v_{bn}^2} = 4kTr_b$ and $\overline{i_{cn}^2} = 2qI_c$. Now if the input is open circuited in Figure 6.15(b), then only i_n can have any effect on the circuit. In this case, the output noise is

LNA Design

$$\overline{i_{on\_tot}^2} = \overline{i_n^2} Z_\pi^2 g_m^2 \qquad (6.44)$$

Similarly, for the model in Figure 6.15(a),

$$\overline{i_{on\_tot}^2} = \overline{i_{bn}^2} Z_\pi^2 g_m^2 + \overline{i_{cn}^2} \qquad (6.45)$$

Now solving (6.44) and (6.45) for i_n gives

$$\overline{i_n^2} = 2qI_B + \frac{2qI_C}{g_m^2} Y_\pi^2 \qquad (6.46)$$

where $\overline{i_{bn}^2} = 2qI_B$ and $\overline{i_{cn}^2} = 2qI_C$.

6.3.2 Noise Figure of the Common-Emitter Amplifier

Now that the equivalent input-referred noise model has been derived, it can be applied to the results in Chapter 2 so that the optimum impedance for noise can be found in terms of transistor parameters.

The input-referred noise current has two terms. One is due to base shot noise and one is due to collector shot noise. Since collector shot noise is present for both v_n and i_n, this part of the input noise current is correlated with the input noise voltage, but the other part is not. Thus,

$$\overline{i_c^2} = \frac{2qI_C}{g_m^2} Y_\pi^2 \qquad (6.47)$$

$$\overline{i_u^2} = 2qI_B \qquad (6.48)$$

Likewise, in the case of the v_n,

$$\overline{v_c^2} = \frac{2qI_C}{g_m^2} \qquad (6.49)$$

$$\overline{v_u^2} = 4kTr_b \qquad (6.50)$$

The correlation admittance Y_c can be determined (see Section 2.2.5). It will be assumed that at the frequencies of interest the transistor looks primarily capacitive.

$$Y_c = \frac{i_c}{v_c} = \sqrt{\frac{\frac{2qI_C}{g_m^2}Y_\pi^2}{\frac{2qI_C}{g_m^2}}} = Y_\pi = j\omega C_\pi \qquad (6.51)$$

where it is assumed that r_π is not significant. Explicitly,

$$G_c \approx 0 \qquad (6.52)$$

$$B_c = \omega C_\pi \qquad (6.53)$$

Thus, the correlation admittance is just equal to the input impedance of the transistor.

R_c, R_u, and G_u can also be written down directly.

$$R_c = \frac{\overline{v_c^2}}{4kT} = \frac{2qI_C}{4kTg_m^2} = \frac{v_T}{2I_C} \qquad (6.54)$$

$$R_u = \frac{\overline{v_u^2}}{4kT} = \frac{4kTr_b}{4kT} = r_b \qquad (6.55)$$

$$G_u = \frac{\overline{i_u^2}}{4kT} = \frac{2qI_B}{4kT} = \frac{I_C}{2v_T\beta} \qquad (6.56)$$

Using these equations, an explicit expression for the noise figure can be written in terms of circuit parameters:

$$\text{NF} = 1 + \frac{\frac{I_C}{2v_T\beta} + \left[G_S^2 + (\omega C_\pi)^2\right]\frac{v_T}{2I_C} + G_S^2 r_b}{G_S} \qquad (6.57)$$

Here it is assumed that the source resistance has no reactive component.

These equations also lead to expressions for G_{opt} and B_{opt}:

$$G_{opt} = \sqrt{\frac{\frac{I_C}{2v_T\beta} + r_b\left(\frac{-\frac{v_T}{2I_C}(\omega C_\pi)}{\frac{v_T}{2I_C}+r_b}\right)^2 + \frac{v_T}{2I_C}\left(\omega C_\pi - \frac{\frac{v_T}{2I_C}(\omega C_\pi)}{\frac{v_T}{2I_C}+r_b}\right)^2}{\frac{v_T}{2I_C}+r_b}} \qquad (6.58)$$

$$B_{opt} = \frac{-\dfrac{v_T}{2I_C}(\omega C_\pi)}{\dfrac{v_T}{2I_C} + r_b} \qquad (6.59)$$

Thus, G_{opt} will vary with the size of the device used (through C_π and r_b) as well as the bias current. At some point, it would equal $1/50\Omega$. Thus, for a device of that size, matching to 50Ω would also give the best noise figure.

The expression for B_{opt} can be simplified if r_b is small compared to $1/2g_m$. This would be a reasonable approximation over most normal operating points:

$$B_{opt} = -\omega C_\pi \qquad (6.60)$$

Thus, it can be seen that the condition for maximum power transfer (resonating out the reactive part of the input impedance) is the same as the condition for providing optimal noise matching.

Hence, the method in the next section can be used for matching an LNA.

6.3.3 Input Matching of LNAs for Low Noise

Since the LNA is the first component in the receiver chain, the input must be matched to be driven by 50Ω. Many methods for matching the input using passive circuit elements are possible with varying bandwidths and degrees of complexity, many of which have already been discussed. However, one of the most elegant is described in [2]. This method requires two inductors to provide the power and noise match for the LNA, as shown in Figure 6.16.

Starting with (6.35), the input impedance for this transistor (assuming that the Miller effect is not important and that r_π is not significant at the frequency of interest) is

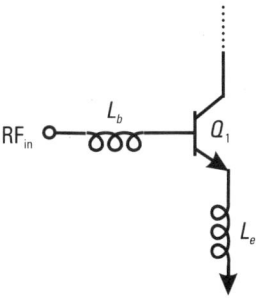

Figure 6.16 LNA driver transistor with two inductors to provide power and noise matching.

$$Z_{in} = \frac{-j}{\omega C_\pi} + j\omega L_e + \frac{g_m L_e}{C_\pi} + j\omega L_b \qquad (6.61)$$

Note that to be matched, the real part of the input impedance must be equal to the source resistance R_S so that

$$\frac{g_m L_e}{C_\pi} = R_S \qquad (6.62)$$

Therefore,

$$L_e = \frac{R_S C_\pi}{g_m} = \frac{R_S}{\omega_T} \qquad (6.63)$$

Note that if the Miller effect is considered, the value of the capacitance will be larger than C_π, and therefore a larger inductor will be required to perform the match.

Also, the imaginary part of the input impedance must equal zero. Therefore,

$$L_b = \frac{1}{C_\pi \omega^2} - \frac{R_S C_\pi}{g_m} \qquad (6.64)$$

Making use of the above analysis, as well as the discussions on noise so far, the following method for simultaneously matching an LNA for power and noise was created. It is outlined in the following steps.

1. Find the current density in the process that will provide the lowest minimum NF, and set the current density in the transistor to be this value regardless of the size of the device. The minimum NF for a process can be found from device measurements, but for the circuit designer it can be determined by the use of simulators such as HPADS or SPICE.

2. Once the current density is known, then the length of the transistor l_e (equivalently, transistor emitter area, since the width is constant) should be chosen so that the real part of the optimum source impedance for lowest noise figure is equal to 50Ω. The current must be adjusted in this step to keep the current density at its optimal level determined in step 1.

3. Size L_e, the emitter degeneration inductor, such that the real part of the input impedance is 50Ω. The use of inductive degeneration will tend to increase the real part of the input impedance. Thus, as this

inductor is increased, the impedance will (at some point) have a real part equal to 50Ω.

4. The last step in the matching is simply to place an inductor in series with the base L_b. Without this inductor, the input impedance is capacitive due to C_π. This inductor is sized so that it resonates with L_e and C_π at the center frequency of the design. This makes the resultant input impedance equal to 50Ω with no additional reactive component.

The technique above has several advantages over other matching techniques. It is simple and requires only one additional matching component in series with the input of the transistor. It produces a relatively broadband match and it achieves simultaneous noise and power matching of the transistor. This makes it a preferred method of matching.

There are many instances where this method cannot be applied without modification. For some designs, power constraints may not accommodate the necessary current to achieve the best noise performance. In this case, the design could proceed as before, except with a less than optimum current density. In other cases, the design may not provide the necessary gain required for some applications. In this case, more current may be needed or some other matching technique that does not require degeneration may have to be employed. Also, linearity constraints may demand a larger amount of degeneration than this method would produce.

Example 6.10 Simultaneous Noise and Power Matching

Design an LNA to work at 5 GHz using a 1.8-V supply with the simultaneous noise and power matching technique discussed in this chapter. For the purpose of this example, design a simple buffer so that the circuit can drive 50Ω. Assume that a 50-GHz 0.5-μm SiGe technology is available.

Solution

At 1.8V it is still possible to design an LNA using a cascode configuration. The cascode transistor can have its base tied to the supply. A simple output buffer that will drive 50Ω quite nicely is an emitter follower. Thus, the circuit would be configured as shown in Figure 6.17. In this case, V_{bias} will be set to 1.8V. The buffer should be designed to accommodate the linearity required by the circuit and drive 50Ω, but for the purposes of this example, we will set the current I_{bias} at 3 mA. This will mean that the output impedance of the circuit will be roughly 8.3Ω and there will be very little voltage loss through the follower. Note that in industry common-emitter buffers are often used to drive off-chip loads at this frequency for reasons of stability and short-circuit protection. However, if the LNA drives an on-chip mixer, this should be fine.

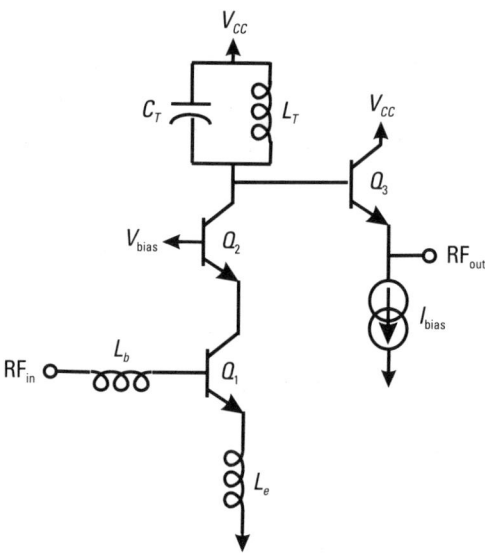

Figure 6.17 Cascode LNA with output buffer.

Next, the size of the tank capacitor and inductor must be set. We will choose an inductance of 1 nH. If the Q of this inductor is 10, then the parallel resistance of the inductor at 5 GHz is 314Ω. The inductor can be adjusted later to either move the gain up or down as required. With 1 nH of inductance, this means that a capacitance of 1 pF is required to resonate at 5 GHz.

Now the current density that results in lowest NF_{min} must be determined. We do this by sweeping the current in Q_1 and Q_2. To start, we arbitrarily chose a 20-μm emitter length. The results are shown in Figure 6.18 and show that for the technology used here, a current of 3.4 mA is optimal, or about 170 $\mu A/\mu m$.

Next, we sweep the transistor length, keeping the transistor current density constant to see where the real part of the optimal noise impedance is 50Ω. The results of this simulation are shown in Figure 6.19. From this graph, it can be seen that an emitter length of 27.4 μm will be the length required. This corresponds to a current of 4.6 mA. Now the transistor size and current are determined.

From a dc simulation, the C_π of the transistor may be found and in this case was 742.5 fF. Since the current is 4.6 mA, the g_m is 184 mA/V and thus the f_T is 39.4 GHz. This is probably a bit optimistic, but it can be used to compute initial guesses for L_e and L_b.

$$L_e = \frac{R_S}{\omega_T} = \frac{50\Omega}{248 \text{ Grad/s}} = 200 \text{ pH}$$

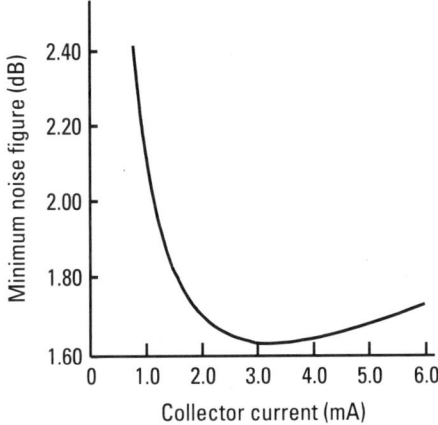

Figure 6.18 Minimum noise figure plotted versus bias current for a 20-μm transistor.

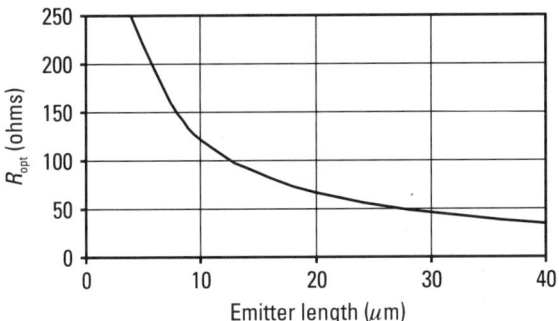

Figure 6.19 R_{opt} plotted versus emitter length with current density set for low noise.

$$L_b = \frac{1}{C_\pi \omega^2} - \frac{R_S C_\pi}{g_m} = \frac{1}{(742.5 \text{ fF})(2 \cdot \pi \cdot 5 \text{ GHz})^2} - \frac{50\Omega \ (742.5 \text{ fF})}{184 \text{ mA/V}}$$

$$= 1.16 \text{ nH}$$

These two values were refined with the help of the simulator. In the simulator, $L_e = 290$ pH and $L_b = 1$ nH were found to be the best choices.

Once these values were chosen, a final set of simulations was performed on the circuit.

First, the S_{11} of the circuit was plotted to verify that the matching was successful. A plot of S_{11} is shown in Figure 6.20. Note that the circuit is almost perfectly matched at 5 GHz as designed. Of course, in practice, loss from inductors as well as packaging and bond wires would never allow such perfect results in the lab.

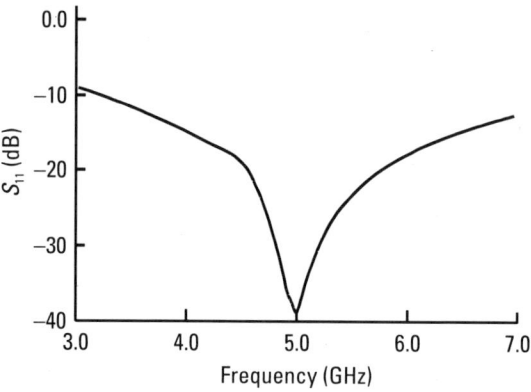

Figure 6.20 Plot showing the input matching for the LNA.

The gain is plotted in Figure 6.21. Note that it peaks about 400 MHz lower than initially calculated. This is due to the capacitance of the transistor Q_2 and the output buffer transistor Q_3. This could be adjusted by reducing the capacitor C_T until the gain is once more centered properly. The gain should have a peak value given by (6.34), which in this case would have a value of

$$\left|\frac{v_{out}}{v_{in}}\right| = \frac{R_L \omega_T}{R_S \omega_o} = \frac{314\Omega \ (2 \cdot \pi \cdot 39.4 \text{ GHz})}{50\Omega \ (2 \cdot \pi \cdot 5 \text{ GHz})} = 33.9 \text{ dB}$$

minus the loss in the buffer. The gain was simulated to have a peak value of about 29 dB, which is close to this value. If this gain is found to be too high, it can be reduced by reducing L_T or adding resistance to the resonator.

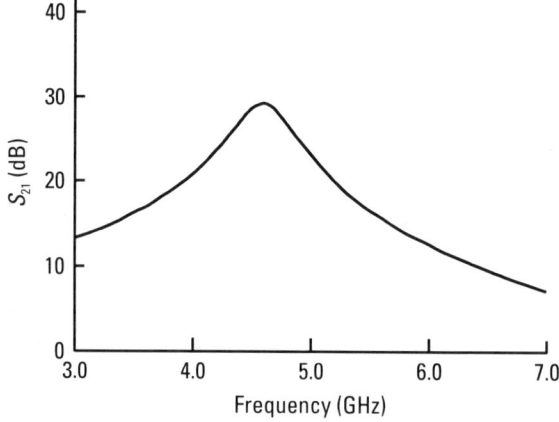

Figure 6.21 Plot showing the gain of the LNA.

LNA Design

Finally, the noise figure of the circuit was plotted as shown in Figure 6.22. The design had a noise figure of 1.76 dB at 5 GHz, which is very close to the minimum achievable noise figure of 1.74 dB, showing that we have in fact noise-matched the circuit at 5 GHz.

6.3.4 Relationship Between Noise Figure and Bias Current

Noise due to the base resistance is in series with the input voltage, so it sees the full amplifier gain. The output noise due to base resistance is given by

$$v_{no,r_b} \approx \sqrt{4kTr_b} \cdot g_{m1} R_L \quad (6.65)$$

Note that this noise voltage is proportional to the collector current, as is the signal, so the SNR is independent of bias current.

Collector shot noise is in parallel with collector signal current and is directly sent to the output load resistor.

$$v_{no,I_C} \approx \sqrt{2qI_C} R_L \quad (6.66)$$

Note that this output voltage is proportional to the square root of the collector current, and therefore, to improve the noise figure due to collector shot noise, we increase the current.

Base shot noise can be converted to input voltage by considering the impedance on the base. If Z_{eq} is the impedance on the base (formed by a combination of matching, base resistance, source resistance, and transistor input impedance), then

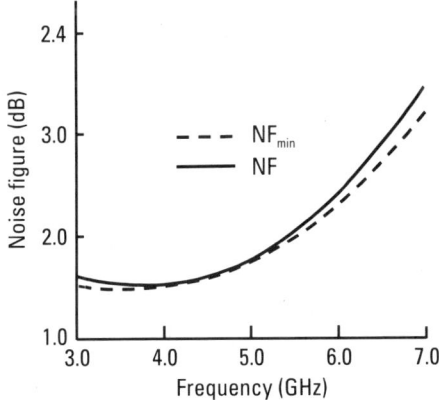

Figure 6.22 Plot showing the noise figure compared to the minimum noise figure for the design.

$$v_{no,I_B} \approx \sqrt{\frac{2qI_C}{\beta}} Z_{eq} g_m R_L \qquad (6.67)$$

Note that this output voltage is proportional to the collector current raised to the power of 3/2. Therefore, to improve the noise figure due to base shot noise, we decrease the current, because the signal-to-noise ratio (in voltage terms) is inversely proportional to the square root of the collector current.

Therefore, at low currents, collector shot noise will dominate and noise figure will improve with increasing current. However, the effect of base shot noise also increases and will eventually dominate. Thus, there will be some optimum level to which the collector current can be increased, beyond which the noise figure will start to degrade again. Note that this simple analysis ignores the fact that r_b increases at lower currents, so, in practice, thermal noise due to r_b is more important at low currents than is indicated by this analysis.

6.3.5 Effect of the Cascode on Noise Figure

As discussed in Section 6.1.3, the cascode transistor is a common-base amplifier with current gain close to 1. By *Kirchoff's current law* (KCL) of the dotted box in Figure 6.23, $i_{c2} = i_{e2} - i_{b2} \approx i_{e2}$. Thus, the cascode transistor is forced to pass the current of the driver on to the output. This includes signal and noise current. Thus, to a first order, the cascode can have no effect on the noise figure of the amplifier. However, in reality it will add some noise to the system. For this reason, the cascode LNA can never be as low noise as a common-emitter amplifier.

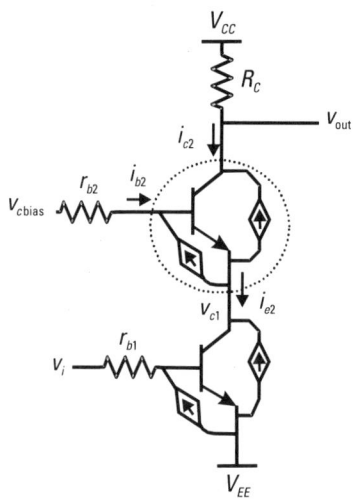

Figure 6.23 A cascode LNA showing noise sources.

6.3.6 Noise in the Common-Collector Amplifier

Since this type of amplifier is not often used as an LNA stage, but more commonly as a buffer, we will deal with its noise only briefly. The amplifier with noise sources is shown in Figure 6.24. Noise due to r_b is directly in series with noise due to R_S. If noise at the output were due only to R_S, the noise figure would be 0 dB.

The noise due to collector shot current is reduced due to negative feedback caused by R_E. For example, current added causes v_{be} to decrease (as v_e increases), and i_e decreases, counteracting the added current. Note that noise due to R_E sees the same effect (negative feedback reduction of noise).

The base shot noise current is injected into the base where an input voltage is developed across R_S: $v_{nbs} \approx i_{nbs} R_S \approx v_o$. The exact relationship between the output noise voltage v_{o_nbs} due to base shot noise current i_{nbs} is

$$v_{o\_nbs} = \frac{R_E Z_\pi (1 - g_m R_S)}{R_E (1 + g_m Z_\pi) + R_S + r_\pi} i_{nbs} \qquad (6.68)$$

Assuming that R_E is large, $g_m R_S \gg 1$, $g_m Z_\pi \gg 1$, and then

$$v_{o\_nbs} = \frac{R_E Z_\pi (1 - g_m R_S)}{R_E (1 + g_m Z_\pi)} i_{nbs} \approx \frac{R_E Z_\pi g_m R_S}{R_E g_m Z_\pi} i_{nbs} \approx R_S i_{nbs} \qquad (6.69)$$

The relationship between the collector shot noise i_{ncs} and the output noise voltage v_{o_ncs} can be shown to be

$$v_{o\_ncs} = \frac{R_E (R_S + Z_\pi)}{R_E (1 + g_m Z_\pi) + Z_\pi + R_S} i_{ncs} \qquad (6.70)$$

Assuming that R_E is large, and $R_S \gg Z_\pi$, then

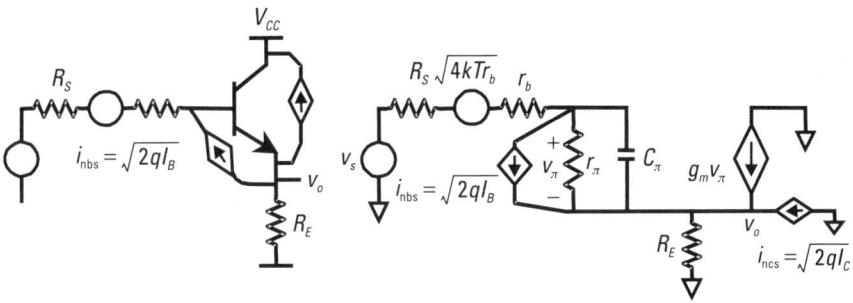

Figure 6.24 A common-collector amplifier with noise illustrated.

$$v_{o\_ncs} \approx \frac{R_S + Z_\pi}{1 + g_m Z_\pi} i_{ncs} \approx r_e i_{ncs} \quad (6.71)$$

Therefore, the collector shot noise current sees r_e, a low value, and output voltage is low. Thus, the common-collector adds little noise to the signal except through r_b.

6.4 Linearity in Amplifiers

Nonlinearity analysis will follow the same basic principles as those discussed in Chapter 2, with power series expansions and nonlinear terms present in the amplifier. These will now be discussed in detail.

6.4.1 Exponential Nonlinearity in the Bipolar Transistor

In bipolar transistors, one of the most important nonlinearities present is the basic exponential characteristic of the transistor itself, illustrated in Figure 6.25.

Source resistance improves linearity. As an extreme example, if the input is a current source, $R_S = \infty$, then $i_c = \beta i_b$. This is as linear as β is. It can be shown that a resistor in the emitter of value R_E has the same effect as a source or base resistor of value $R_E \beta$. The transistor base has a bias applied to it and an ac signal superimposed. Summing the voltages from ground to the base and assuming that $i_e = i_c$,

$$v_s + V_S = v_{be} + V_{BE} + R_E(I_C + i_c) \quad (6.72)$$

where V_{BE} and v_{be} are the dc and ac voltages across the base emitter junction of the transistor.

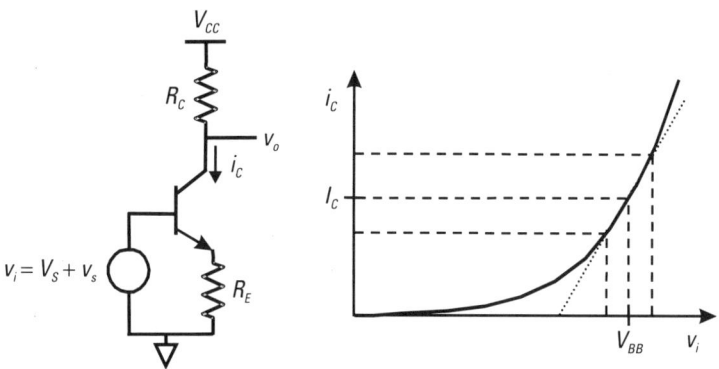

Figure 6.25 Bipolar common-emitter amplifier for linearity analysis.

Extracting only the small-signal components from this equation gives

$$v_s = v_{be} + R_E i_c \qquad (6.73)$$

Also, from the basic properties of the junction,

$$I_C + i_c = I_S e^{\frac{V_{BE}+v_{be}}{v_T}} = I_S e^{\frac{V_{BE}}{v_T}} e^{\frac{v_{be}}{v_T}} = I_C e^{\frac{v_{be}}{v_T}} \qquad (6.74)$$

where, from Chapter 3, $v_T = kT/q$. Solving for v_{be} gives

$$v_{be} = v_T \ln\left(1 + \frac{i_c}{I_C}\right) \qquad (6.75)$$

Now making use of the math identity

$$\ln(1+x) = x - \frac{1}{2}x^2 + \frac{1}{2}x^3 \ldots \qquad (6.76)$$

and expanding (6.75) using (6.76) and substituting it back into (6.73), we get

$$v_s = R_E i_c + v_T \left[\frac{i_c}{I_C} - \frac{1}{2}\left(\frac{i_c}{I_C}\right)^2 + \frac{1}{3}\left(\frac{i_c}{I_C}\right)^3 \ldots\right] \qquad (6.77)$$

Noting that $v_T/I_C = r_e$ and rearranging, we get

$$v_s = (R_E + r_e)i_c - \frac{1}{2}r_e \frac{i_c^2}{I_C} + \frac{1}{3}r_e \frac{i_c^3}{I_C^2} \ldots \qquad (6.78)$$

This can be further manipulated to give

$$\frac{v_s}{(R_E + r_e)} = i_c - \frac{1}{2}\frac{r_e}{(R_E + r_e)}\frac{i_c^2}{I_C} + \frac{1}{3}\frac{r_e}{(R_E + r_e)}\frac{i_c^3}{I_C^2} \ldots \qquad (6.79)$$

This is the equation we need, but it is in the wrong form. It needs to be solved for i_c. Thus, a few more relationships are needed. Given

$$y = a_1 x + a_2 x^2 + a_3 x^3 + \ldots \qquad (6.80)$$

The following can be found:

$$x = b_1 y + b_2 y^2 + b_3 y^3 + \ldots \qquad (6.81)$$

where

$$b_1 = \frac{1}{a_1}$$

$$b_2 = -\frac{a_2}{a_1^3} \qquad (6.82)$$

$$b_3 = \frac{1}{a_1^5}(2a_2^2 - a_1 a_3)$$

(6.79) can now be rewritten as a function of i_c:

$$i_c = \frac{v_s}{R_E + r_e} + \frac{1}{2I_C}\left(\frac{r_e}{R_E + r_e}\right)\left(\frac{v_s}{R_E + r_e}\right)^2 \qquad (6.83)$$

$$+ \left[\frac{1}{2I_C^2}\left(\frac{r_e}{R_E + r_e}\right)^2 - \frac{1}{3I_C^2}\left(\frac{r_e}{R_E + r_e}\right)\right]\left(\frac{v_s}{R_E + r_e}\right)^3$$

Now the third-order intercept voltage can be determined (note that this is the peak voltage and the rms voltage will be lower by a factor of $\sqrt{2}$):

$$v_{\text{IP3}} = 2\sqrt{\frac{k_1}{3k_3}} = 2\sqrt{\frac{1}{3}\left(\frac{1}{|R_E + r_e|}\right)\frac{6I_C^2(|R_E + r_e|)^5}{r_e|2R_E - r_e|}} \qquad (6.84)$$

$$= 2\sqrt{2}v_T \frac{|R_E + r_e|^2}{\sqrt{r_e^3|2R_E - r_e|}}$$

This very useful equation can be used to estimate the linearity of gain stages. An approximation to (6.84) that can be quite useful for hand calculations is

$$v_{\text{IP3}} = 2\sqrt{2}v_T \left(\frac{R_E + r_e}{r_e}\right)^{3/2} \qquad (6.85)$$

In the special case where there is no emitter degeneration, the above expression can be simplified to

$$v_{IP3} = 2\sqrt{2}v_T \tag{6.86}$$

Example 6.11 Linearity Calculations in Common-Emitter Amplifier

For a common-emitter amplifier with no degeneration, if the input is assumed to be composed of two sine waves of amplitude A_1 and A_2, compute the relevant frequency components to graph the fundamental and third-order products and predict what the IIP3 point will be. Assume that $I_{CA} = 1$ mA and $v_T = 25$ mV.

Solution

The first step is to calculate the coefficients k_1, k_2, and k_3 for the power series expansion from (6.83) as

$$k_1 = \frac{1}{R_E + r_e} = \frac{1}{0 + 25} = 0.04$$

$$k_2 = \frac{1}{2I_C}\left(\frac{r_e}{R_E + r_e}\right)\left(\frac{1}{R_E + r_e}\right)^2$$

$$= \frac{1}{2I_C}\frac{r_e}{(R_E + r_e)^3} = \frac{1}{2 \cdot 1m}\frac{25}{(0 + 25)^3}$$

$$= 0.8$$

$$k_3 = \left[\frac{1}{2I_C^2}\left(\frac{r_e}{R_E + r_e}\right)^2 - \frac{1}{3I_C^2}\left(\frac{r_e}{R_E + r_e}\right)\right]\left(\frac{1}{R_E + r_e}\right)^3$$

$$= \left[3\left(\frac{r_e}{R_E + r_e}\right)^2 - 2\left(\frac{r_e}{R_E + r_e}\right)\right]\left(\frac{1}{6I_C^2}\right)\left(\frac{1}{R_E + r_e}\right)^3$$

$$= \left[3\left(\frac{25}{0 + 25}\right)^2 - 2\left(\frac{25}{0 + 25}\right)\right]\left(\frac{1}{6 \cdot 1m^2}\right)\left(\frac{1}{0 + 25}\right)^3$$

$$= [3 - 2]\left(\frac{1}{6 \cdot 1m^2}\right)\left(\frac{1}{0 + 25}\right)^3$$

$$= 10.667$$

resulting in an expression for current as follows:

$$i_c = k_1 v_s + k_2 v_s^2 + k_3 v_s^3 + \ldots = 0.04 v_s + 0.8 v_s^2 + 10.667 v_s^3 + \ldots$$

The dc, fundamental, second harmonic, and intermodulation components are found in Table 6.1, and equations for them are listed for the above coefficients in Table 6.2.

The intercept point is at a voltage of 70.7 mV at the input, 2.828 mA at the output, as shown in Table 6.2 and in Figure 6.26, which agrees with (6.86). For an input of 70.7 mV, the actual output fundamental current is 11.3 mA, which illustrates the gain expansion for an exponential nonlinearity.

The voltage-versus-current transfer function is shown in Figure 6.27. Also shown are the time domain input and output waveforms demonstrating the expansion offered by the exponential nonlinearity.

This diagram illustrates a number of points about nonlinearity.

Due to the second-order term k_2, there is a dc shift. Using the dc component term shown in Table 6.2, we make the following calculations.

Table 6.1
Harmonic Components

Component	With A_1, A_2	With $A_1 = A_2 = A$	
dc	$k_o + \dfrac{k_2}{2}(A_1^2 + A_2^2)$	$k_o + k_2 A^2$	$1m + 0.8 A^2$
Fundamental	$k_1 A_1 + k_3 A_1 \left(\dfrac{3}{4} A_1^2 + \dfrac{3}{2} A_2^2\right)$	$k_1 A + \dfrac{9}{4} k_3 A^3$	$0.4 A + 24 A^3$
Second Harmonic	$\dfrac{k_2 A_1^2}{2}$	$\dfrac{k_2 A^2}{2}$	$0.4 A^2$
Intermod	$\dfrac{3}{4} k_3 A_1^2 A_2$	$\dfrac{3}{4} k_3 A^3$	$8 A^3$

Table 6.2
Values for Harmonic Components

A_1 (mV)	A_2 (mV)	IM3	2nd Harmonic	Fundamental	Ideal Fund
0.1	0.1	8 pA	4 nA	4 μA	4 μA
1	1	8 nA	400 nA	40 μA	40 μA
10	10	8 μA	40 μA	424 μA	400 μA
20	20	64 μA	160 μA	992 μA	800 μA
30	30	216 μA	360 μA	1.848 mA	1.2 mA
70.7	70.7	2.828 mA	2 mA	11.309 mA	2.828 mA

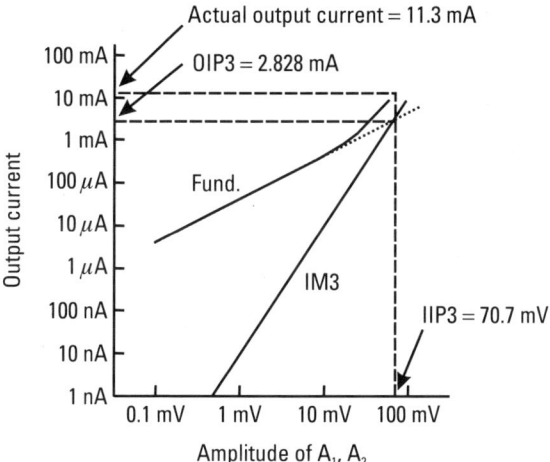

Figure 6.26 Plot of fundamental and third-order products coming out of an exponential nonlinearity.

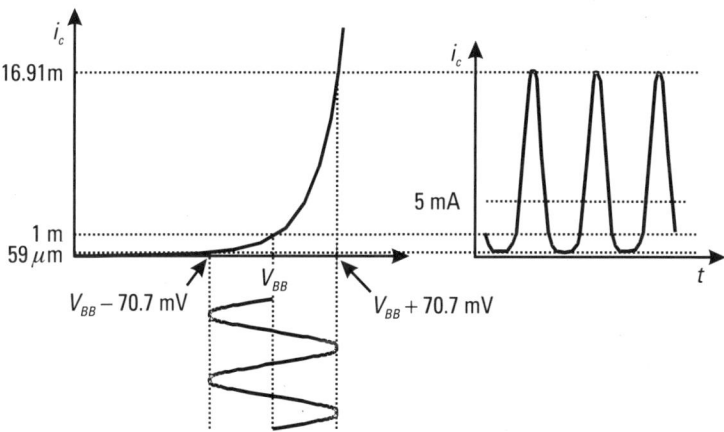

Figure 6.27 Input-output transfer function and time domain voltage and current waveforms.

With $A = 70.71$ mV, the shift is

$$\frac{1 \text{ mA}}{4 \cdot (25 \text{ mV})^2} (2 \cdot 70.71 \text{ mV}^2) = 4.0 \text{ mA}$$

Thus, the waveform is centered on 5 mA.

Because of positive coefficient k_3, the waveform is not compressed but expanded. However, either way, compression or expansion, the result is distortion.

The above calculations all assume R_C is small enough so that the transistor does not saturate. If saturation does occur, the power series is no longer valid.

Typically, inputs would not be allowed to be bigger than about 10 dB below IP3, which for this example is about 22.34 mV. Figure 6.28 shows the transfer function for an input of this amplitude. At this level, current goes from 0.409 to 2.444 mA. The dc shift is 0.4 mA, so current is about 1.4 ± 1.01 mA.

Example 6.12 Linearity Calculations in Common-Emitter Amplifier with Degeneration

Continue the previous example by determining the effect of emitter degeneration. For an input of two sine waves of amplitude A_1 and A_2 compute the IIP3 point for $R_E = 0\Omega$, 5Ω, 10Ω, 15Ω, and 20Ω. Again, assume that $I_{CA} = 1$ mA and $v_T = 25$ mV.

Solution

To determine IP3, two tones can be applied at various amplitudes and graphical extrapolations made of the fundamental and third-order tones, as previously illustrated in Figure 6.26. Instead, for this example, values for k_0, k_1, k_2, and k_3 are determined from the equations in Table 6.1. Then these are used to calculate the fundamental and *third-order intermodulation* (IM3) components from the equations in Table 6.2, and IIP3 is calculated from the given fundamental and third-order terms similar to that discussed in Section 2.3.2.

$$\text{IIP3} = 10 \log\left(\frac{A^2}{2 \times 50 \times 1 \text{ mW}}\right) + \frac{1}{2} \times 20 \log\left(\frac{\text{Fundamental}}{|\text{IM3}|}\right)$$

$$= -50 + 10 \log\left(\frac{\text{Fundamental}}{|\text{IM3}|}\right)$$

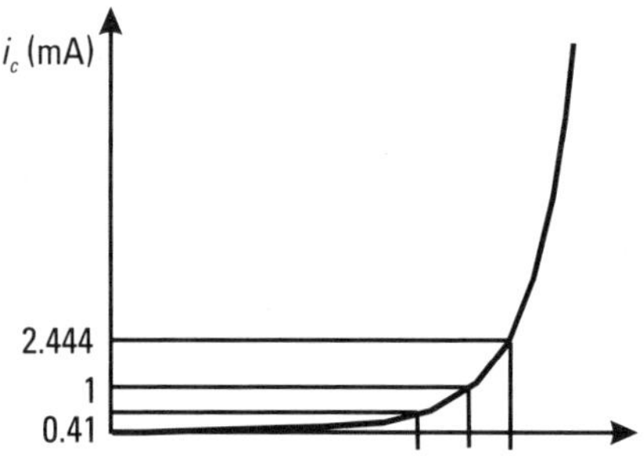

Figure 6.28 Transistor characteristic for smaller input signal.

Results are shown in Table 6.3. Figure 6.29 shows a continuous curve of IIP3 and shows the approximation of (6.85).

This example clearly shows how IP3 improves with degeneration resistance. From the fundamental equations, it can also be seen that for larger I_C and hence larger I_B, the improvement will be higher. It can also be seen from the equation that it is possible to cancel the third-order term if $R_E = r_e/2$, which in this example requires a degeneration of 12.5Ω. It can be seen that for lower degeneration, k_3 is positive, resulting in gain expansion, while for larger values

Table 6.3
Calculations of IP3 Versus Degeneration Resistance

R_E	0	5	10	15	20
A (peak) (mV)	1	1	1	1	1 mV
k_0 (mA)	1	1	1	1	1
k_1 (mA/V)	4	33.3	28.6	25	22.2
k_2 (A/V^2)	0.8	0.463	0.2915	0.1953	0.1372
k_3 (A/V^3)	10.667	2.572	0.3967	−0.2035	−0.3387
Fundamental (μA)	40	33.34	28.57	25.0	22.22
IM3 (nA)	8.0	1.929	0.297	−0.153	−0.254
IIP3 (dBm)	−13.01	−7.62	−0.175	2.144	−0.581

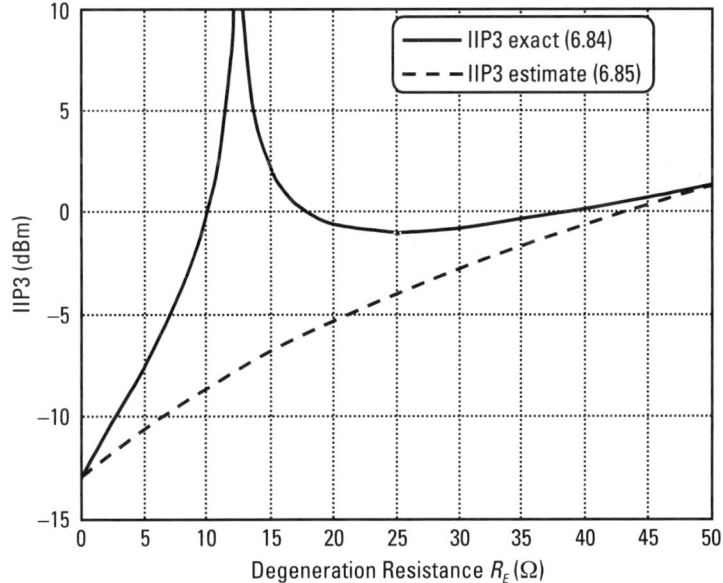

Figure 6.29 Input IP3 as a function of degeneration resistance.

of degeneration, k_3 is negative, resulting in gain compression. At exactly 12.5Ω, k_3 goes through zero and theoretical IP3 goes to infinity. In real life, if k_3 is zero, there will be a component from the k_5 term, which will limit the linearity. However, this improvement in linearity is real and can be demonstrated experimentally [3].

A related note of interest is that for a MOSFET transistor operated in subthreshold, the transistor drain characteristics are exponential and hence k_3 is positive, while for higher bias levels, k_3 is negative. Thus, by an appropriate choice of bias conditions, k_3 can be set to zero for improvements in linearity [4]. In MOSFETs, it turns out to be quite challenging to take advantage of this linearity improvement, since the peak occurs for a narrow region of bias conditions and the use of degeneration resistance or inductance reduces this linearity improvement.

6.4.2 Nonlinearity in the Output Impedance of the Bipolar Transistor

Another important nonlinearity in the bipolar or CMOS transistor is the output impedance. An example of where this may be important is in the case of a transistor being used as a current source. In this circuit, the base of the transistor is biased with a constant voltage and the current into the collector is intended to remain constant for any output voltage. Of course, the transistor has a finite output impedance, so if there is an ac voltage on the output, there is some finite ac current that flows through the transistor, as shown by r_o in Figure 6.30. Worse than this, however, is the fact that the transistor's output impedance will change with applied voltage and it can therefore introduce nonlinearity.

The dc output impedance of a transistor is given by

$$r_{o\_dc} = \frac{V_A}{I_C} \qquad (6.87)$$

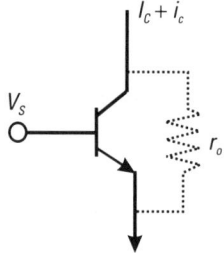

Figure 6.30 Bipolar transistor as a current source.

where V_A is the Early voltage of the transistors. An ac current into the collector can be written as a function of ac current i_c.

$$r_{o\_ac}(i_c) = \frac{V_A}{I_C + i_c} \tag{6.88}$$

Assuming for this analysis that there is no significant impedance in the circuit other than the transistor output resistance, the ac collector-emitter voltage can be written as

$$v_{ce} = i_c r_{o\_ac} = \frac{i_c V_A}{I_C + i_c} = V_A \frac{\frac{i_c}{I_C}}{1 + \frac{i_c}{I_C}} \tag{6.89}$$

Now from the relationship

$$\frac{x}{1+x} = x - x^2 + x^3 - x^4 + x^5 \ldots \tag{6.90}$$

(6.89) can be written out as a power series:

$$v_{ce} = V_A \left(\frac{i_c}{I_C}\right) - V_A \left(\frac{i_c}{I_C}\right)^2 + V_A \left(\frac{i_c}{I_C}\right)^3 = r_{o\_dc} i_c - \frac{r_{o\_dc}}{I_C} i_c^2 + \frac{r_{o\_dc}}{I_C^2} i_c^3 \tag{6.91}$$

The intermodulation current can now be easily determined.

$$i_{IP3} = 2\sqrt{\frac{k_1}{3k_3}} = 2\sqrt{\frac{1}{3} r_{o\_dc} \frac{I_C^2}{r_{o\_dc}}} = \frac{2I_C}{\sqrt{3}} \tag{6.92}$$

Thus, the output intermodulation voltage is just

$$v_{OP3} = \frac{2I_C}{\sqrt{3}} r_o \tag{6.93}$$

This is a fairly intuitive result. As the dc current is increased, the ac current is a smaller percentage of the total, and therefore the circuit behaves more linearly. Thus, designers have two choices if the current source is not linear enough. They can either increase the current or increase the output impedance.

Also, it should be noted that this relationship only holds true if the transistor does not start to saturate. If it does, the nonlinearity will get much worse.

6.4.3 High-Frequency Nonlinearity in the Bipolar Transistor

Many frequency-dependent devices can reduce the linearity of a circuit. One of the most troublesome is the base-collector junction capacitance C_μ. This capacitance is voltage dependent, which results in a nonlinearity. This nonlinearity is especially important in circuits with low supply voltages because the capacitance is largest at low reverse bias.

This capacitor's effect is particularly harmful for both frequency response and nonlinearity in the case of a standard common-emitter amplifier. In this configuration, C_μ is multiplied by the gain of the amplifier (the Miller effect) and appears across the source.

The value of C_μ as a function of bias voltage is given by

$$C_\mu(V) = \frac{C_{\mu o}}{\left(1 - \frac{V}{\psi_o}\right)^{(1/n)}} \quad (6.94)$$

where $C_{\mu o}$ is the capacitance of the junction under zero bias, ψ_o is the built-in potential of the junction, and n is usually between 2 and 5. Since this capacitor's behavior is highly process dependent and hard to model, there is little benefit in deriving detailed equations for it. Rather, the designer must rely on simulation and detailed models to predict its behavior accurately.

6.4.4 Linearity in Common-Collector Configuration

The common-collector amplifier is often called the *emitter-follower* because the emitter voltage "follows" the base voltage. However, the amplifier cannot do this over all conditions. If the current is constant, v_{BE} is constant and the transfer function will be perfectly linear. However, as v_o changes, and $i_{out} = v_o/R_{out}$ will change as shown in Figure 6.31. Thus, i_E will change and so will v_{BE}, and there will be some nonlinearity.

If R_{out} is large so that i_{out} is always much less than I_B, the linearity will be good, as the operating point will not change significantly over a cycle of the signal. It is important to keep the peak output current less than the bias current. This means that

$$\frac{|v_{o,\text{peak}}|}{R_{out}} < I_B \quad (6.95)$$

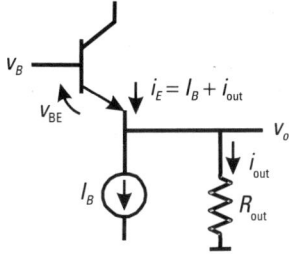

Figure 6.31 Illustration of nonlinearity in the common-collector amplifier.

If this is the case, then there will be no clipping of the waveform.

The linearity can be improved by increasing I_B or R_{out}. This will continue to improve performance as long as power supply voltage is large enough to allow this swing. Thus, for large R_{out}, the power supply limits the voltage swing and therefore the linearity. In this case, the current is not a limiting factor.

6.5 Differential Pair (Emitter-Coupled Pair) and Other Differential Amplifiers

Any of the amplifiers that have already been discussed can be made differential by adding a mirrored copy of the original and connecting them together at the points of symmetry so that voltages are no longer referenced to ground, but rather swing plus or minus relative to each other. While this is hard to describe, it is easy to show an example of a differential common-emitter amplifier (more commonly called a *differential pair* or *emitter-coupled pair*) in Figure 6.32. Here the bias for the stage is supplied with a current source in the emitter. Note that

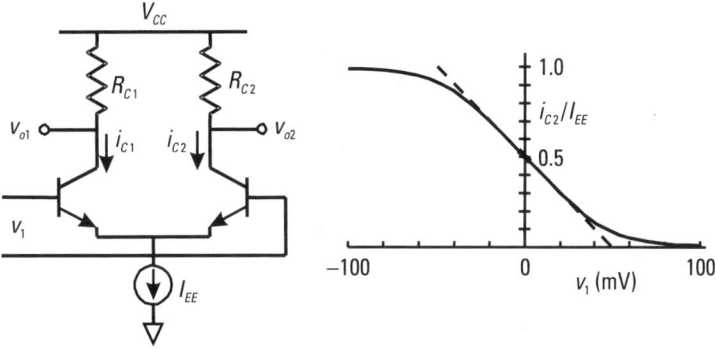

Figure 6.32 Differential common-emitter amplifier or emitter-coupled pair.

when the bias is applied this way, the emitter is at a virtual ground. This means that for small-signal differential inputs, this voltage never moves from its nominal voltage.

This stage can be used in many circuits such as mixers, oscillators, or dividers. If an input voltage is applied larger than about $5v_T$, then the transistors will be fully switched and they can act as a limiting stage or "square wave generator" as well.

All the equations already developed are still valid for the differential amplifier. The large signal current and voltage relationships are often written as hyperbolic tangents. The currents are given by

$$i_{C1} = \frac{I_{EE}}{1 + e^{-(v_1/v_T)}} = \frac{I_{EE}}{2}\left[1 + \tanh\left(\frac{v_1}{2v_T}\right)\right] \qquad (6.96)$$

$$i_{C2} = \frac{I_{EE}}{1 + e^{(v_1/v_T)}} = \frac{I_{EE}}{2}\left[1 - \tanh\left(\frac{v_1}{2v_T}\right)\right] \qquad (6.97)$$

and the differential output voltage is given by

$$v_{o2} - v_{o1} = I_{EE}R_C \tanh\left(\frac{v_1}{2v_T}\right) \qquad (6.98)$$

Note that there will only be even-order terms in a power series expansion of this nonlinearity and hence no dc shifts or even harmonics in $v_{o2} - v_{o1}$ as v_1 grows.

The slope at $v_1 = 0$ will be

$$\frac{\partial i_{C2}}{\partial v_1} = \frac{\partial i_{C2}}{2\partial v_{BE2}} = -\frac{1}{2}g_{m2} = -\frac{1}{2}\frac{I_{C2}}{v_T} = -\frac{I_{EE}}{4v_T} \qquad (6.99)$$

This can be found directly by taking the derivative of the above equation for i_{C2} and setting v_1 to 0.

6.6 Low-Voltage Topologies for LNAs and the Use of On-Chip Transformers

Of the configurations described so far, the common-emitter amplifier would seem ideally suited to low-voltage operation. However, if the improved properties of the cascode are required at lower voltage, then the topology must be modified slightly. This has led some designers to "fold" the cascode as shown in Figure

6.33(a) [5]. With the use of two additional LC tanks and one very large coupling capacitor, the cascode can now be operated down to a very low voltage. This approach does have drawbacks, however, as it uses two additional inductors, which will use a lot of die area. The other drawback present with any folding scheme is that both transistors can no longer reuse the current. Thus, this technique will use twice the current compared to an unfolded cascode, although it could possibly be used at half the voltage to result in comparable power consumption.

An alternative to this topology involves using a transformer to produce magnetic rather than electric coupling between the two stages, as shown in Figure 6.33(b). In this circuit, L_p and L_s form the primary and secondary windings of an on-chip transformer, respectively. Note that there is no longer any need for the coupling capacitor. The transformer, although slightly larger than a regular inductor, will nevertheless use much less die area than two individual inductors.

Typically, LNAs as already discussed make use of inductors for many reasons, including low-loss biasing, maximized signal swing for high dynamic range, and simultaneous noise and power matching. It is also possible to replace the collector and emitter inductors with a transformer as shown in Figure 6.34

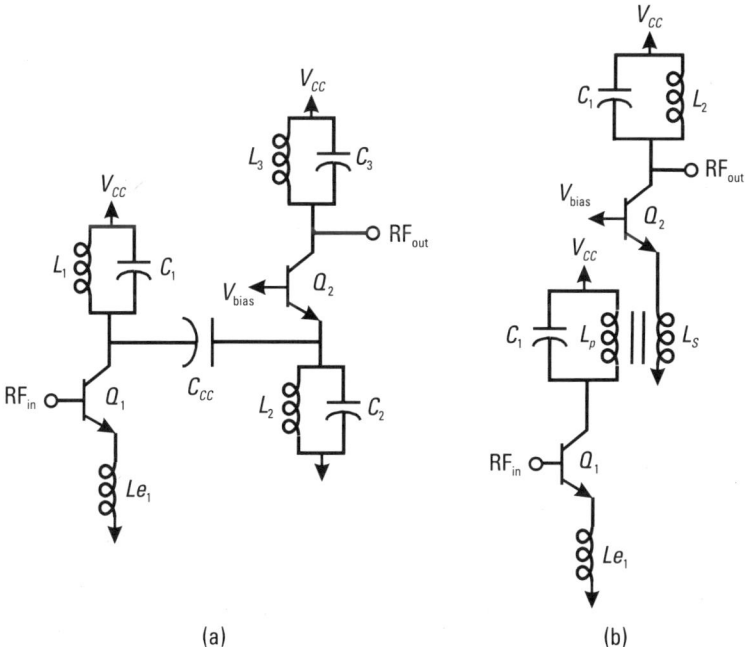

Figure 6.33 A folded cascode LNA with (a) capacitive coupling, and (b) inductive coupling.

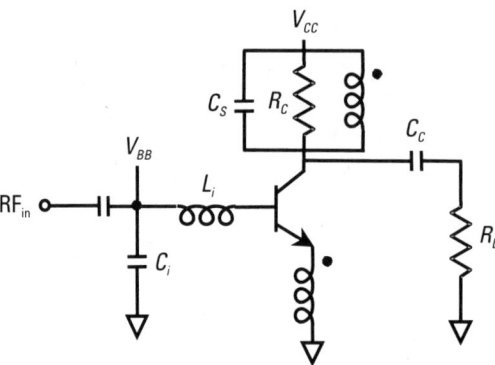

Figure 6.34 LNA with transformer coupling collector to emitter.

[6]. This circuit has all the same useful properties as the previously discussed LNA but adds some additional benefits. From [6], the gain of this circuit is given by

$$S_{21} = \frac{-g_m Z_L}{A_{BJT} + g_m Z_L \left[\frac{1}{n} + j\omega r_b C_\mu \left(\frac{1}{n} + 1\right) - \omega^2 L_i C_\mu \left(\frac{1}{n} + 1\right)\right]} \quad (6.100)$$

where

$$A_{BJT} = 1 + j\omega r_b (C_\pi + C_\mu) - \omega^2 L_i (C_\pi + C_\mu) \quad (6.101)$$

At low frequencies the gain is given by

$$S_{21} \approx \frac{-g_m Z_L}{1 + g_m Z_L \left[\frac{1}{n}\right]} \quad (6.102)$$

Under many circumstances, $g_m Z_L$ is large and the gain is approximately equal to n, the turns ratio. This means that there is very little dependence on transistor parameters.

Considering the redrawn circuit in Figure 6.35, a simplistic description of this circuit can be provided. The transistor acts as a source follower to the input of the transformer. A transformer by itself cannot provide power gain, since, if the voltage is increased by a factor of n, the current is decreased by a factor of n. However, in this circuit, the transistor feeds the primary current

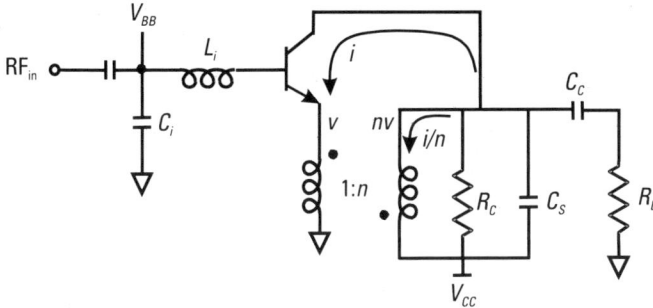

Figure 6.35 Redrawn transformer-coupled LNA.

into the secondary, adding it to the secondary current, but also allowing a lower impedance to be driven. The net result is that the gain S_{21} is approximately equal to n. Thus, with a turns ratio of 4:1, the amplifier can achieve a gain of 12 dB.

The advantage of this circuit is that the gain is determined largely by the transformer turns ratio, thus minimizing the dependence on transistor parameters. The transformer has high linearity and low noise; thus, the amplifier also has high linearity and low noise. This circuit has not been widely used due to the complexity of designing with monolithic transformers. However, with good transformer design techniques and good models now more widely available, this circuit is expected to become more popular.

6.7 DC Bias Networks

A number of circuits have already been discussed in this text, and it is probably appropriate to say at least a few words about biasing at this point. Bias networks are used in all types of circuits and are not unique to LNAs.

The most common form of biasing in RF circuits is the current mirror. This basic stage is used everywhere and it acts like a current source. Normally, it takes a current as an input and this current is usually generated, along with all other references on the chip, by a circuit called a *bandgap reference generator*. A bandgap reference generator is a temperature-independent bias generating circuit. The bandgap reference generator balances the V_{BE} dependence on temperature, with the temperature dependence of v_T to result in a voltage or current nearly independent of temperature. Design details for the bandgap reference generator can be found in [7].

Perhaps the most basic current mirror is shown in Figure 6.36(a). In this mirror, the bandgap reference generator produces current I_{bias} and forces this

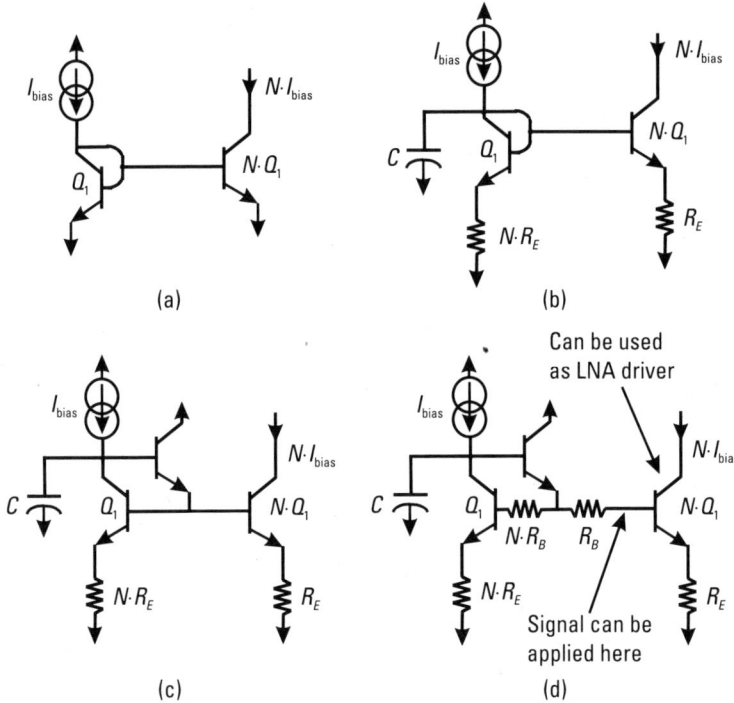

Figure 6.36 Various current mirrors: (a) simple mirror; (b) mirror with improved noise performance; (c) mirror with improved current matching; and (d) mirror with transistor doing double duty as current source and driver.

current through Q_1. Scaling the second transistor allows the current to be multiplied up and used to bias working transistors. One major drawback to this circuit is that it can inject a lot of noise at the output due primarily to the high g_m of the transistor $N \cdot Q_1$ (larger than Q_1 by a factor of N), which acts like an amplifier for noise. A capacitor can be used to clean up the noise, and degeneration can be put into the circuit to reduce the gain of the transistor, as shown in Figure 6.36(b). If Q_1 is going to drive many current stages, then base current can affect the matching, so an additional transistor can be added to provide the base current without affecting I_{bias}, as shown in Figure 6.36(c).

Another useful technique for an LNA design is to make the $N \cdot Q_1$ transistor function both as a mirror transistor and as the LNA driver transistor, as shown in Figure 6.36(d). In this case, resistors have to be added in the base to isolate the input from the low impedance of Q_1. Provided that R_B is big compared to the input impedance of the transistor $N \cdot Q_1$, little noise is injected here.

With any of these mirrors, a voltage at the collector of $N \cdot Q_1$ must be maintained above a minimum level or else the transistor will go into saturation. Saturation will lead to bad matching and nonlinearity.

6.7.1 Temperature Effects

For transistor current given by

$$i_C \approx I_S e^{\frac{V_{BE}}{v_T}} \tag{6.103}$$

the temperature affects parameters such as I_S and V_{BE}. Also, current gain β is affected by temperature. I_S doubles for every 10°C rise in temperature, while the relationship for V_{BE} and β with temperature is shown in (6.104) and (6.105).

$$\left. \frac{\Delta V_{BE}}{\Delta T} \right|_{i_C = \text{constant}} \approx -2 \text{ mV/°C} \tag{6.104}$$

$$\frac{\Delta \beta}{\Delta T} \approx +0.5\%/\text{°C} \tag{6.105}$$

A typical temperature range for an integrated circuit might be 0 to 85°C. Thus, for a constant voltage bias, if the current is 1 mA at 20°C, then it will change to 0.2 mA at 0°C and to 1.65 mA at 85°C. Thus, the current changes by more than eight times over this temperature range. This illustrates why constant current biasing (for example, with the current mirrors discussed in Section 6.7) is used. If both transistors in the current mirror are at the same temperature, then output current is roughly independent of temperature.

6.8 Broadband LNA Design Example

As a final major design example, we will design a broadband LNA to work from 50 to 900 MHz and with input matched to 75Ω with an S_{11} better than −10 dB over this range. The gain must be more than 12 dB, noise figure less than 5 dB, and the IIP3 must be greater than 6 dBm. The circuit must operate with a 3.3V supply and consume no more than 8 mA of current. Assume that there is a suitable 50-GHz process available for this design. Assume that the LNA will drive an on-chip mixer with an input impedance of 5 kΩ.

This is going to be a high-linearity part and it needs to be broadband. Therefore, the matching and the load cannot make use of reactive components.

This means that, of the designs presented so far, an LNA with shunt feedback will have to be used. This can be combined with a common-base amplifier to provide better frequency response and an output buffer to avoid the problem of loading the circuit. A first cut at a topology that could satisfy the requirements is shown in Figure 6.37. Note that emitter degeneration has been added to this circuit. Degeneration will almost certainly be required due to the linearity requirement. We have left the current source as ideal for this example. Also, we are not including the bias circuitry that will be needed at the base of Q_1.

The first specification we will satisfy is the requirement of 8-mA total supply current. It may even be possible to do this design with less current. The trade-off is that as the current is decreased, R_E must be increased to deliver the same linearity, and this will affect noise. We will begin this design using all the allowed current, and at the end, we will consider the possibility of reducing the current in a second iteration.

The total current must be divided between the two stages of this amplifier. The buffer must have enough current so that it continues to operate properly even when it has to deliver a lot of current in the presence of large signals. Since the load resistance is large, the buffer will have to drive an effective resistance of approximately $R_f + 75\Omega$. This is expected to be a few hundred ohms and will require several hundreds of microamps of ac current. We will start with 3 mA in the buffer and 5 mA in the driver stage.

We can now start to size the resistors and capacitor in the circuit by considering linearity. An IIP3 of 6 dBm in a 75Ω system, assuming that the input is matched, means that the IIP3 in terms of voltage will be 546 mV$_{rms}$. Since we have assumed a current of 5 mA in the driver, we can now use (6.85) to determine the size of R_E:

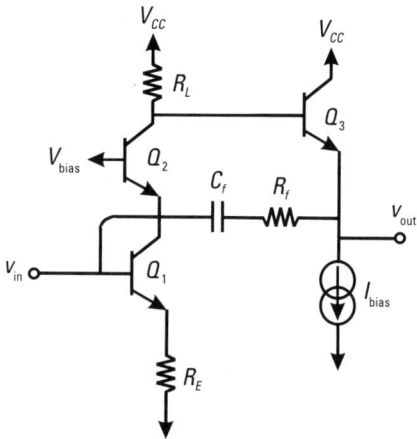

Figure 6.37 Broadband LNA sample circuit.

$$R_E = r_e \left(\frac{v_{IP3}}{2v_T}\right)^{2/3} - r_e = 5\Omega \left(\frac{546 \text{ mV}}{2 \cdot 25 \text{ mV}}\right)^{2/3} - 5\Omega = 19.6\Omega$$

This is a rough estimate for what the linearity should be. Also, there are many other factors that can limit the linearity of the circuit. We will start with $R_E = 20\Omega$.

The gain can also be found now. Knowing that we want 12 dB of voltage gain means a gain of 4 V/V. We will assume that the buffer has a voltage gain A_{BO} of about 0.9 (they will always have a bit of loss). Thus, the load resistance can be obtained:

$$G = \frac{R_L}{R_E + r_e} A_{BO} \Rightarrow R_L = \frac{G}{A_{BO}}(R_E + r_e) = \frac{4}{0.9}(20\Omega + 5\Omega) \approx 115\Omega$$

Now we need to set the feedback resistor. Knowing that the input impedance needs to be 75Ω (we approximate that the input impedance is R_f divided by the gain),

$$Z_{in} \approx \frac{R_f}{\frac{R_L}{R_E + r_e}} \Rightarrow R_f = \frac{Z_{in} R_L}{R_E + r_e} = \frac{75\Omega \cdot 115\Omega}{20\Omega + 5\Omega} = 345\Omega$$

The other thing that must be set is the value of C_f. Since the LNA must operate down to 50 MHz, this capacitor will have to be fairly large. At 50 MHz, if it has an impedance that is 1/20th of R_f, then this would make it approximately 50 pF. We will start with this value. It can be refined in simulation later.

The only thing left to do in this example is to size the transistors. With all the feedback around this design, the transistors will have a much smaller bearing on the noise figure than in a tuned LNA. Thus, we will make the input transistor fairly large (60 μm) and the other two transistors will be sized to be 30 μm fairly arbitrarily. Having high f_T is important, but in a 50-GHz process, this will probably not be an issue. The other last detail that needs to be addressed is the bias level at the base of Q_2. Given that the emitter of Q_1 is at 100 mV, the base will have to sit at about 1V. The collector of Q_1 should be higher than this, for example, about 1.2V. This means that the base of Q_2 will need to be at about 2.2V, and since its collector will sit at about 2.7V, this transistor will have plenty of headroom.

The noise figure of this design can now be estimated. First, the noise voltage produced by the source resistance is

$$v_{ns} = \sqrt{4kTR_S} = \sqrt{4 \cdot (4 \cdot 10^{-21}) \cdot 75\Omega} = 1.1 \text{ nV}/\sqrt{\text{Hz}}$$

Since the input is matched, this voltage is divided by half to the input of the driver transistor and then sees the full voltage gain of the amplifier. Thus, the noise at the output due to the source resistance is

$$v_{o(source)} = \frac{1}{2} v_{ns} G = \frac{1}{2} \cdot 1.1 \text{ nV}/\sqrt{\text{Hz}} \cdot 4 = 2.2 \text{ nV}/\sqrt{\text{Hz}}$$

The current produced by the degeneration resistor is

$$i_{n_E} = \sqrt{\frac{4kT}{R_E}} = \sqrt{\frac{4 \cdot (4 \cdot 10^{-21})}{20\Omega}} = 28.3 \text{ pA}/\sqrt{\text{Hz}}$$

This current is split between the degeneration resistor and the emitter of the driver transistor. The fraction that enters the driver transistor develops into a voltage at the collector of the cascode transistor and is then passed to the output through the follower:

$$v_{on_E} = i_{n_E} \cdot \left(\frac{R_E}{r_e + R_E}\right) R_L \cdot A_{BO}$$

$$= 28.3 \text{ pA}/\sqrt{\text{Hz}} \cdot \left(\frac{20\Omega}{5\Omega + 20\Omega}\right) \cdot 115\Omega \cdot 0.9$$

$$= 2.3 \text{ nV}/\sqrt{\text{Hz}}$$

If we assume that the source resistance and the emitter degeneration resistor are the two dominant noise sources, then the noise figure is

$$\text{NF} = 10 \log \left(\frac{v_{on_E}^2 + v_{ons}^2}{v_{ons}^2}\right)$$

$$= 10 \log \left(\frac{(2.3 \text{nV}/\sqrt{\text{Hz}})^2 + (2.2 \text{nV}/\sqrt{\text{Hz}})^2}{(2.2 \text{nV}/\sqrt{\text{Hz}})^2}\right)$$

$$= 3.2 \text{ dB}$$

The performance of this design is now verified by simulation. The voltage gain is shown in Figure 6.38 and is between 12.3 and 12.4 dB over the frequency

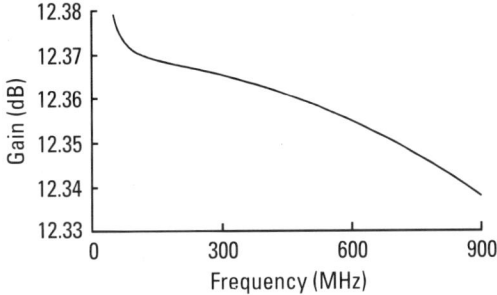

Figure 6.38 Simulated voltage gain of the broadband LNA sample circuit.

range of interest. This is very close to the value predicted by our calculations and is very constant. The magnitude of S_{11} is shown in Figure 6.39 and is less than −19 dB over the whole range. Thus, the circuit is almost perfectly matched to 75Ω over all frequencies. The noise figure was also simulated and is shown in Figure 6.40. The noise figure was less than 3.5 dB and only slightly higher than our calculated value. We could have gotten closer to the right value by

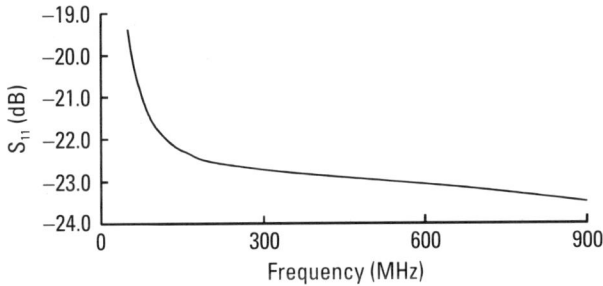

Figure 6.39 Simulated S_{11} of the broadband LNA sample circuit.

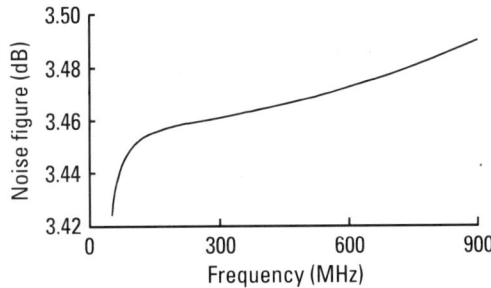

Figure 6.40 Simulated noise figure of the broadband LNA sample circuit.

considering more noise sources. Since this is lower than required, a second iteration of this example could reduce the current in the driver stage, and a larger value for R_E could be used to maintain the linearity. In order to test the linearity of the circuit, two tones were fed into the circuit. One was at a frequency of 400 MHz and one at a frequency of 420 MHz, each having an input power of −20 dBm. The *fast Fourier transform* (FFT) of the output is shown in Figure 6.41. The two tones at the fundamental have an rms amplitude of −19.5 dBV, and the amplitude of the intermodulation tones have an amplitude of −73.4 dBV. Using (2.55) in Chapter 2, this means that at the input this corresponds to an IIP3 of 572 mV or 6.4 dBm. Thus, the specification for linearity is met for this part.

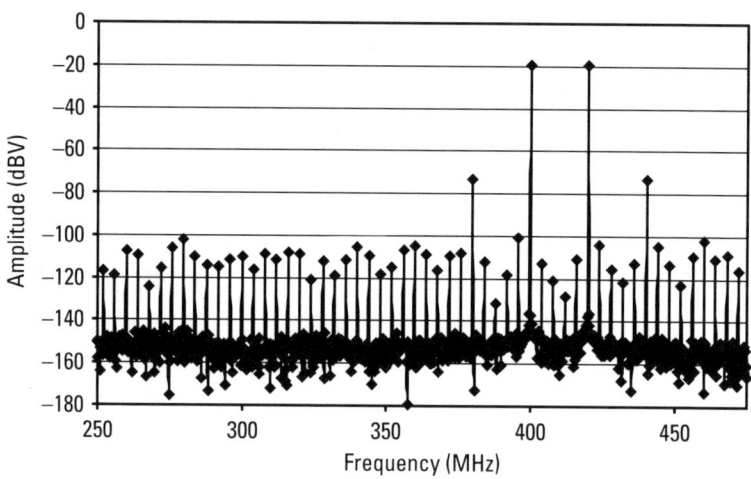

Figure 6.41 FFT of the broadband LNA with two tones applied at the input.

References

[1] Johns, D. A., and K. Martin, *Analog Integrated Circuit Design,* New York: John Wiley & Sons, 1997.

[2] Voinigescu, S. P., et al., "A Scalable High-Frequency Noise Model for Bipolar Transistors with Applications to Optimal Transistor Sizing for Low-Noise Amplifier Design," *IEEE J. Solid-State Circuits,* Vol. 32, Sept. 1997, pp. 1430–1439.

[3] van der Heijden, M. P., H. C. de Graaf, and L. C. N. de Vreede, "A Novel Frequency Independent Third-Order Intermodulation Distortion Cancellation Technique for BJT Amplifiers," *Proc. BCTM,* Sept. 2001, pp. 163–166.

[4] Toole, B., C. Plett, and M. Cloutier, "RF Circuit Implications of a Low-Current Linearity 'Sweet Spot' in MOSFETs," *Proc. ESSCIRC,* Sept. 2002, pp. 619–622.

[5] Ray, B., et al., "A Highly Linear Bipolar 1V Folded Cascode 1.9GHz Low Noise Amplifier," *Proc. BCTM,* Sept. 1999, pp. 157–160.

[6] Long, J. R., and M. A. Copeland, "A 1.9GHz Low-Voltage Silicon Bipolar Receiver Front-End for Wireless Personal Communications Systems," *IEEE J. Solid-State Circuits,* Vol. 30, Dec. 1995, pp. 1438–1448.

[7] Gray, P. R., et al., *Analysis and Design of Analog Integrated Circuits,* 4th ed., New York: John Wiley & Sons, 2001.

Selected Bibliography

Abou-Allam, E., J. J. Nisbet, and M. C. Maliepaard, "A 1.9GHz Front-End Receiver in 0.5μm CMOS Technology," *IEEE J. Solid-State Circuits,* Vol. 36, Oct. 2001, pp. 1434–1443.

Baumberger, W., "A Single-Chip Rejecting Receiver for the 2.44 GHz Band Using Commercial GaAs-MESFET-Technology," *IEEE J. Solid-State Circuits,* Vol. 29, Oct. 1994, pp. 1244–1249.

Copeland, M. A., et al., "5-GHz SiGe HBT Monolithic Radio Transceiver with Tunable Filtering," *IEEE Trans. on Microwave Theory and Techniques,* Vol. 48, Feb. 2000, pp. 170–181.

Harada, M., et al., "2-GHz RF Front-End Circuits at an Extremely Low Voltage of 0.5V," *IEEE J. Solid-State Circuits,* Vol. 35, Dec. 2000, pp. 2000–2004.

Krauss, H. L., C. W. Bostian, and F. H. Raab, *Solid State Radio Engineering,* New York: John Wiley & Sons, 1980.

Long, J. R., "A Low-Voltage 5.1–5.8GHz Image-Reject Downconverter RFIC," *IEEE J. Solid-State Circuits,* Vol. 35, Sept. 2000, pp. 1320–1328.

Macedo, J. A., and M. A. Copeland, "A 1.9 GHz Silicon Receiver with Monolithic Image Filtering," *IEEE J. Solid-State Circuits,* Vol. 33, March 1998, pp. 378–386.

Razavi, B., "A 5.2-GHz CMOS Receiver with 62-dB Image Rejection," *IEEE J. Solid-State Circuits,* Vol. 36, May 2001, pp. 810–815.

Rogers, J. W. M., J. A. Macedo, and C. Plett, "A Completely Integrated Receiver Front-End with Monolithic Image Reject Filter and VCO," *Proc. IEEE RFIC Symposium,* June 2000, pp. 143–146.

Rudell, J. C., et al., "A 1.9-GHz Wide-Band IF Double Conversion CMOS Receiver for Cordless Telephone Applications," *IEEE J. Solid-State Circuits,* Vol. 32, Dec. 1997, pp. 2071–2088.

Samavati, H., H. R. Rategh, and T. H. Lee "A 5-GHz CMOS Wireless LAN Receiver Front End," *IEEE J. Solid-State Circuits,* Vol. 35, May 2000, pp. 765–772.

Schmidt, A., and S. Catala, "A Universal Dual Band LNA Implementation in SiGe Technology for Wireless Applications," *IEEE J. Solid-State Circuits,* Vol. 36, July 2001, pp. 1127–1131.

Schultes, G., P. Kreuzgruber, and A. L. Scholtz, "DECT Transceiver Architectures: Superheterodyne or Direct Conversion?" *Proc. 43rd Vehicular Technology Conference,* Secaucus, NJ, May 18–20, 1993, pp. 953–956.

Steyaert, M., et al., "A 2-V CMOS Transceiver Front-End," *IEEE J. Solid-State Circuits,* Vol. 35, Dec. 2000, pp. 1895–1907.

Yoshimasu, T., et al., "A Low-Current Ku-Band Monolithic Image Rejection Down Converter," *IEEE J. Solid-State Circuits,* Vol. 27, Oct. 1992, pp. 1448–1451.

7

Mixers

7.1 Introduction

The purpose of the mixer is to convert a signal from one frequency to another. In a receiver, this conversion is from radio frequency to intermediate frequency. Mixing requires a circuit with a nonlinear transfer function, since nonlinearity is fundamentally necessary to generate new frequencies. As described in Chapter 2, if an input RF signal and a local oscillator signal are passed through a system with a second-order nonlinearity, the output signals will have components at the sum and difference frequencies. A circuit realizing such nonlinearity could be as simple as a diode followed by some filtering to remove unwanted components. On the other hand, it could be more complex, such as the double-balanced cross-coupled circuit, commonly called the *Gilbert cell*. In an integrated circuit, the more complex structures are often preferred, since extra transistors can be used with little extra cost but with improved performance. In this chapter, the focus will be on the cross-coupled double-balanced mixer. Consideration will also be given as to how to design a mixer for low-voltage operation.

7.2 Mixing with Nonlinearity

A diode or a transistor can be used as a nonlinearity. The two signals to be mixed are combined and applied to the nonlinear circuit. In the transistor, they can be applied separately to two control inputs, for example, to the base and emitter in a bipolar transistor or to the gate and source in a field-effect transistor. If a diode is the nonlinear device, then signals might be combined with additional circuitry. As described in Chapter 2, two inputs at ω_1 and ω_2, which are passed through a nonlinearity that multiplies the two signals together will produce mixing terms at $\omega_1 \pm \omega_2$. In addition, other terms (harmonics, feed-through, intermodulation) will be present and will need to be filtered out.

7.3 Basic Mixer Operation

Mixers can be made from the LNAs that have already been discussed and some form of controlled inverter. One of the simplest forms of this type of mixer is shown in Figure 7.1. The input of the mixer is simply a gain stage like one that has already been considered. The amplified current from the gain stage is then passed into the switching stage. This stage steers the current to one side of the output or the other depending on the value of v_2 (this provides the nonlinearity just discussed). If the control signal is assumed to be a periodic one, then this will have the effect of multiplying the current coming out of the gain stage (Q_1, Q_2) by ±1 (a square wave). Multiplying a signal by another signal will cause the output to have components at various frequencies. Thus, this can be used to move the signal v_1 from one frequency to another.

7.4 Controlled Transconductance Mixer

Figure 7.2 shows a transconductance-controlled mixer made from a bipolar differential pair. In this case, the current is related to the input voltage v_2 by the transconductance of the input transistors Q_1 and Q_2. However, the

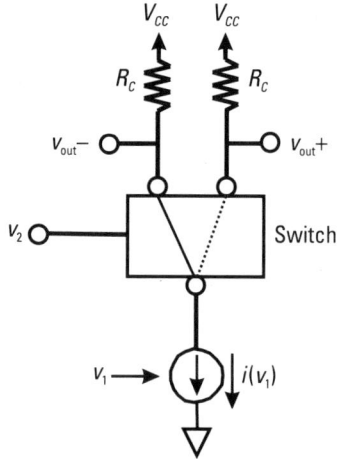

Figure 7.1 Simple conceptual schematic of a mixer.

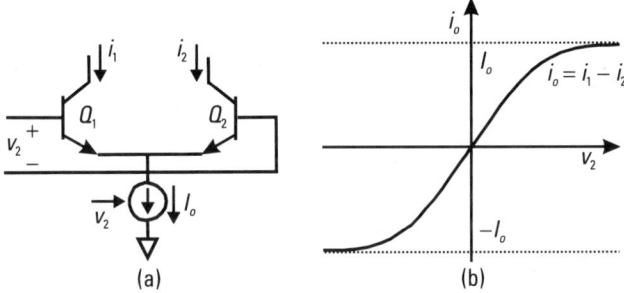

Figure 7.2 Transconductance controlled mixer: (a) basic circuit, and (b) output current waveform.

transconductance is controlled by the current I_o, which in turn is controlled by the input voltage v_1. Thus, the output current will be dependent on both input voltages v_1 and v_2.

Let us now look in detail at the operation of this mixer. As shown in Chapter 6, the current in a differential pair is related to the voltage by the following equation:

$$i_1 = \frac{I_o}{1 + e^{-v_2/v_T}} \tag{7.1}$$

$$i_2 = \frac{I_o}{1 + e^{v_2/v_T}}$$

Thus, the difference in the output currents from the mixer is given by

$$i_o = i_1 - i_2 = I_o \left(\frac{1}{1 + e^{-v_2/v_T}} - \frac{1}{1 + e^{v_2/v_T}} \right) = I_o \tanh \frac{v_2}{2v_T} \tag{7.2}$$

This can be converted to a differential voltage with equal load resistors in the collectors.

For small input signals, if $v_2 \ll v_T$, then

$$i_o \approx I_o \frac{v_2}{2v_T} \tag{7.3}$$

Now, if we assume the current source is modulated by v_1 so that the current I_o is replaced with $I_o + g_{mc} v_1$, where g_{mc} is the transconductance of the current source, then

$$i_o = (I_o + g_{mc}v_1)\tanh\frac{v_2}{2v_T} = \underbrace{I_o \tanh\frac{v_2}{2v_T}}_{v_2 \text{ feedthrough}} + \underbrace{g_{mc}v_1 \tanh\frac{v_2}{2v_T}}_{\text{multiplication (mixing)}} \quad (7.4)$$

We note that feedthrough appear equally in the output voltages above (common mode), and so does not appear in the differential output voltage.

7.5 Double-Balanced Mixer

In order to eliminate the v_2 feed-through, it is possible to combine the output of this circuit with another circuit driven by $-v_2$, as shown in Figure 7.3. This circuit has four switching transistors known as the *switching quad*. The output current from the second differential pair is given by

$$i_o' = i_6 - i_5 = I_o \tanh\frac{v_2}{2v_T} - g_{mc}v_1 \tanh\frac{v_2}{2v_T} \quad (7.5)$$

Therefore the total differential current is

$$i_{ob} = i_o - i_o' = 2g_{mc}v_1 \tanh\frac{v_2}{2v_T} \quad (7.6)$$

This removes the v_2 feed-through term that was present in (7.4).

The last step in making this circuit practical is to replace the ideal current sources with an actual amplifier stage, as shown in Figure 7.4. Now v_1 is applied

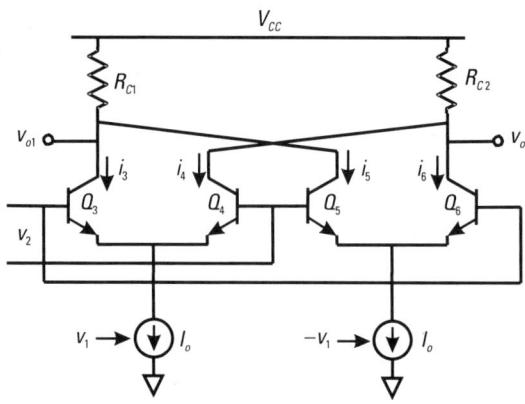

Figure 7.3 Double-balanced transconductance-controlled mixer.

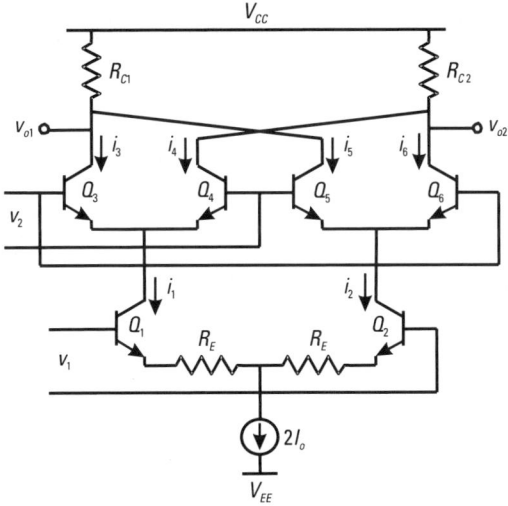

Figure 7.4 Double-balanced mixer with amplifier implemented as a differential pair with degeneration.

to a differential pair so that the small-signal component of i_1 and i_2 are the inverse of each other, that is,

$$i_1 \approx I_o + \frac{v_1}{2} \frac{1}{r_e + R_E} \quad (7.7)$$

and

$$i_2 = I_o - \frac{v_1}{2} \frac{1}{r_e + R_E} \quad (7.8)$$

Currents from the switching quad are related to v_2, i_1, and i_2 by (7.2) through (7.6).

$$i_3 = \frac{i_1}{1 + e^{-v_2/v_T}}$$

$$i_4 = \frac{i_1}{1 + e^{v_2/v_T}} \quad (7.9)$$

$$i_5 = \frac{i_2}{1 + e^{v_2/v_T}}$$

$$i_6 = \frac{i_2}{1 + e^{-v_2/v_T}}$$

Then, assuming that the amplifier formed by Q_1 and Q_2 is linear,

$$(i_3 + i_5) - (i_4 + i_6) = \tanh\left(\frac{v_2}{2v_T}\right)(i_1 - i_2) = \tanh\left(\frac{v_2}{2v_T}\right)\frac{v_1}{r_e + R_E} \quad (7.10)$$

The output differential voltage is

$$v_o = -\tanh\left(\frac{v_2}{2v_T}\right)\frac{v_1}{r_e + R_E}R_C \quad (7.11)$$

Thus, the gain of the circuit relative to v_1 can be determined:

$$\frac{v_o}{v_1} = -\tanh\left(\frac{v_2}{2v_T}\right)\frac{R_C}{r_e + R_E} \quad (7.12)$$

With $R_E = 0$, a general large-signal expression for the output can also be written:

$$v_o = -2R_C I_o \tanh\left(\frac{v_1}{2v_T}\right)\tanh\left(\frac{v_2}{2v_T}\right) \quad (7.13)$$

This takes into account nonlinearity from the bottom pair as well as from the top quad. Previously, with R_E present, the bottom was treated in a linear fashion.

7.6 Mixer with Switching of Upper Quad

Usually, a downconverting mixer is operated with v_1 as the RF signal and v_2 as the local oscillator. The RF input must be linear, and linearity is improved by the degeneration resistors. Why must the RF input be linear? In a communications receiver application, the RF input can have several channels at different frequencies and different amplitudes. If the RF input circuitry were nonlinear, adjacent channels could intermodulate and interfere with the desired channel. For example, with inputs at 900, 900.2, and 900.4 MHz, if 900 MHz is desired, the third-order intermodulation term from the two other signals occurs at $2 \cdot 900.2 - 900.4 = 900$ MHz, which is directly on top of the desired signal.

The LO input need not be linear, since the LO is clean and of known amplitude. In fact, the LO input is usually designed to switch the upper quad so that for half the cycle Q_3 and Q_6 are on and taking all of the current i_1 and i_2. For the other half of the LO cycle, Q_3 and Q_6 are off and Q_4 and Q_5 are on. More formally, if $v_2 \gg 2v_T$, then (7.12) can be approximated as

$$\frac{v_o}{v_1} = u(v_2)\frac{R_C}{r_e + R_E} \tag{7.14}$$

where

$$u(v_2) = \begin{cases} 1 & \text{if } v_2 \text{ is positive} \\ -1 & \text{if } v_2 \text{ is negative} \end{cases} \tag{7.15}$$

This is equivalent to alternately multiplying the signal by 1 and −1. This can also be expressed as a Fourier series. If v_2 is a sine wave with frequency ω_{LO}, then

$$u(v_2) = \frac{4}{\pi}\sin(\omega_{LO}t) + \frac{4}{3\pi}\sin(3\omega_{LO}t) + \frac{4}{5\pi}\sin(5\omega_{LO}t) \tag{7.16}$$
$$+ \frac{4}{7\pi}\sin(7\omega_{LO}t) + \ldots$$

7.6.1 Why LO Switching?

For small LO amplitude, the amplitude of the output depends on the amplitude of the LO signal. Thus, gain is larger for larger LO amplitude. For large LO signals, the upper quad switches and no further increases occur. Thus, at this point, there is no longer any sensitivity to LO amplitude.

As the LO is tuned over a band of frequencies, for example, to pick out a channel in the 902- to 928-MHz range, the LO amplitude may vary. If the amplitude is large enough, the variation does not matter.

For image reject mixers (to be discussed in Section 7.10), matching two LO signals in amplitude and phase is important. By using a switching modulator and feeding the LO signal into the switching input, amplitude matching is less important.

Noise is minimized with large LO amplitude. With large LO, the upper quad transistors are alternately switched between completely off and fully on. When off, the transistor contributes no noise, and when fully on, the switching transistor behaves as a cascode transistor, which, as described in Chapter 6, does not contribute significantly to noise.

7.6.2 Picking the LO Level

The differential pair will require an input voltage swing of about 4 to 5 v_T for the transistors to be hard-switched one way or the other. Therefore, the LO input to the mixer should be at least 100 mV peak for complete switching. At 50Ω, 100-mV peak is −10 dBm. Small improvements in noise figure and conversion gain can be seen for larger signals; however, for LO levels larger than about 0 dBm, there is minimal further improvement. Thus, −10 to 0 dBm (100 to 300 mV) is a reasonable compromise between noise figure, gain, and required LO power. If the LO voltage is made too large, then a lot of current has to be moved into and out of the bases of the transistors during transitions. This can lead to spikes in the signals and can actually reduce the switching speed and cause an increase in LO feed-through. Thus, too large a signal can be just as bad as too small a signal.

Another concern is the parasitic capacitance on node V_d, as shown in Figure 7.5. The transistors have to be turned on and off, which means that any capacitance in the emitter has to be charged and discharged. Essentially, the input transistors behave like a simple rectifier circuit, as shown in Figure

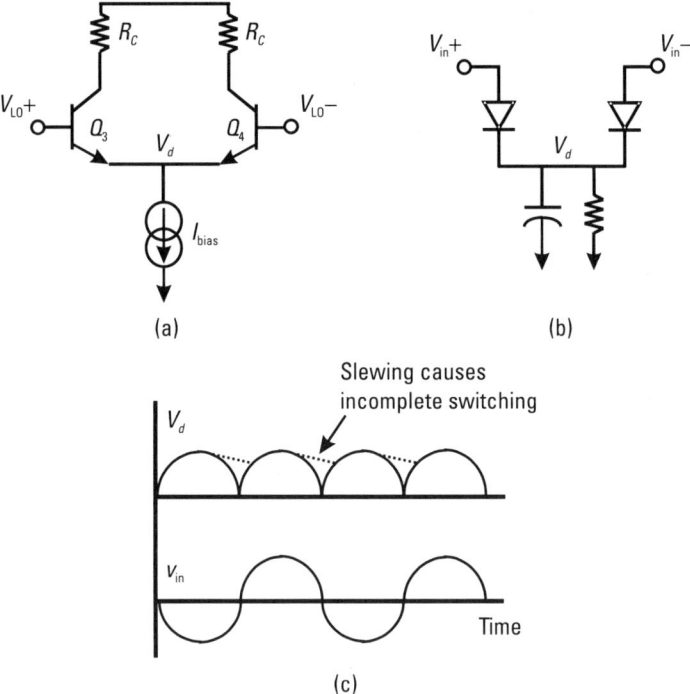

Figure 7.5 Large-signal behavior of the differential pair: (a) schematic representation; (b) diode rectifier model; and (c) waveforms illustrating the problem of slewing.

7.5(b). If the capacitance on the emitters is too large, then V_d will stop following the input voltage and the transistors will start to be active for a smaller portion of the cycle, as shown in Figure 7.5(c). Since V_d is higher than it should be, it takes longer for the transistor to switch and it switches for a smaller portion of the cycle. This will lead to waveform distortion.

7.6.3 Analysis of Switching Modulator

The top switching quad alternately switches the polarity of the output signal as shown in Figure 7.6.

The LO signal has the effect of multiplying the RF input by a square wave going from −1 to +1. In the frequency domain, this is equivalent to a convolution of RF and LO signals, which turns out to be a modulation of the RF signal with each of the Fourier components of the square wave.

As can be seen from Figure 7.7, the output amplitude of the product of the fundamental component of square wave is

$$v_o = \frac{4}{\pi} v_{RF} \sin(\omega_{RF} t) \sin(\omega_{LO} t) \tag{7.17}$$

$$= \frac{1}{2} \cdot \frac{4}{\pi} v_{RF} \sin[(\omega_{RF} + \omega_{LO})t] + \frac{1}{2} \cdot \frac{4}{\pi} v_{RF} \sin[(\omega_{RF} - \omega_{LO})t]$$

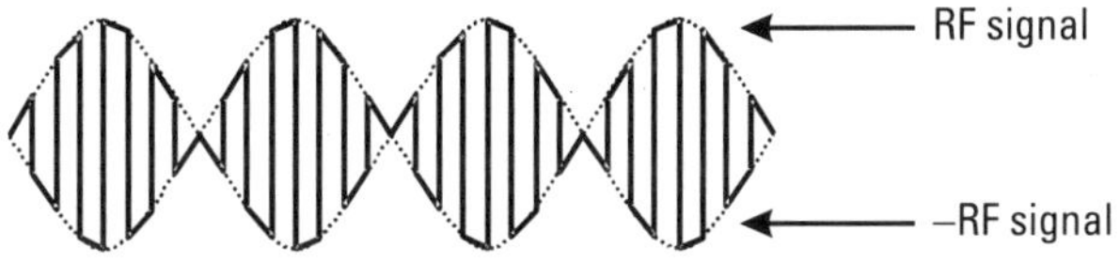

Figure 7.6 Switching waveform.

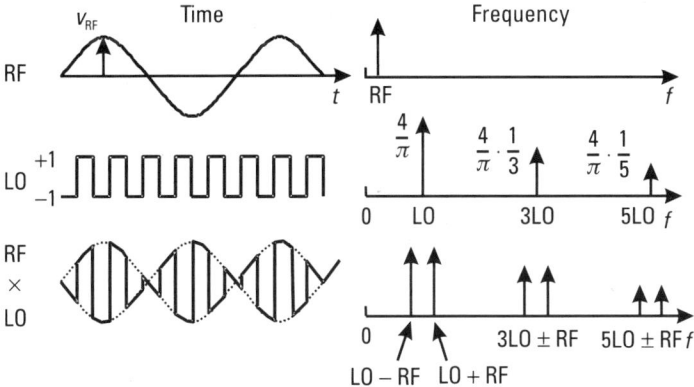

Figure 7.7 Analysis of the switching mixer in the frequency and time domain.

where v_{RF} is the output voltage obtained without the switching (i.e., for a differential amplifier). This means that because of the frequency translation, the amplitude of each mixed frequency component is

$$v_o = \frac{2}{\pi} v_{RF} = v_{RF_{dB}} - 3.9 \text{ dB} \tag{7.18}$$

As a result of mixing, gain is modified by a factor of 1/2 or −6 dB, but a square wave has a fundamental larger by 2.1 dB, for a net change of −3.9 dB. Third harmonic terms are down by 1/3 or −9.5 dB, while fifth harmonics are 1/5 or −14 dB. Intermodulation (other than mixing between RF and LO) is often due to the RF input and its nonlinearity. Thus, the analysis of the differential pair may be used here. From Chapter 6, the gain for the differential amplifier with load resistors of R_C and emitter degeneration resistors of R_E per side was given by:

$$v_{RF} = \frac{R_C}{r_e + R_E} v_{in} \tag{7.19}$$

Thus, a final useful estimate of gain in a mixer such as the one shown in Figure 7.4 (at one output frequency component) is the following:

$$v_o = \frac{2}{\pi} \frac{R_C}{r_e + R_E} v_{in} \tag{7.20}$$

We note that this is voltage gain from the base of the input transistors to the collector of the switching quads. In an actual implementation with matching circuits, these also have to be taken into account.

In Figures 7.6 and 7.7, the LO frequency is much greater than the RF frequency (it is easier to draw the time domain waveform). This is upconversion, since the output signal is at a higher frequency than the input signal. Downconversion is shown in Figure 7.8. This is downconversion because the output signal of interest is at a lower frequency than the input signal. The other output component which appears at higher frequency will be removed by the IF filter.

Note also that any signals close to the LO or its multiples can mix into the IF. These signals can be other signals at the input, intermodulation between input signals (this tells us we need linear RF inputs), noise in the inputs, or noise in the mixer itself.

7.7 Mixer Noise

Mixer noise figure is somewhat more complicated to define compared to that of an LNA, because of the frequency translation involved. Therefore, for mixers,

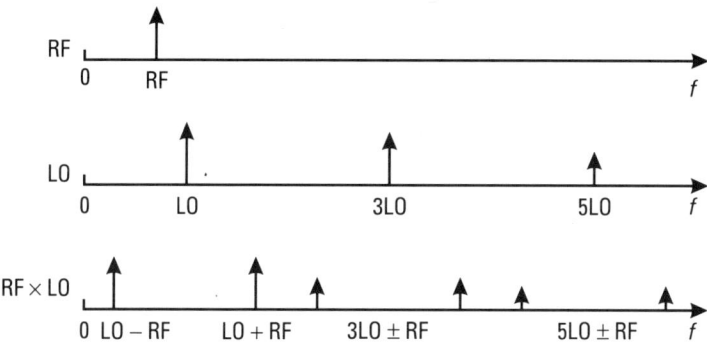

Figure 7.8 Downconversion frequency domain plot.

a slightly modified definition of noise figure is used. Noise factor for a mixer is defined as

$$F = \frac{N_{o\,tot}(\omega_{IF})}{N_{o(source)}(\omega_{IF})} \qquad (7.21)$$

where $N_{o\,tot}(\omega_{IF})$ is the total output noise at the IF and $N_{o(source)}(\omega_{IF})$ is the output noise at the IF due to the source. The source and all circuit elements generate noise at all frequencies, and many of these frequencies will produce noise at the output IF due to the mixing action of the circuit. Usually the two dominant frequencies are the input frequency and the image frequency.

To make things even more complicated, *single-sideband* (SSB) noise figure or *double-sideband* (DSB) noise figure is defined. The difference between the two definitions is the value of the denominator in (7.21). In the case of double-sideband noise figure, all the noise due to the source at the output frequency is considered (noise of the source at the input and image frequencies). In the case of single-sideband noise figure, only the noise at the output frequency due to the source that originated at the RF frequency is considered. Thus, using the single-sideband noise figure definition, even an ideal noiseless mixer would have a noise figure of 3 dB. This is because the noise of the source would be doubled in the output due to the mixing of both the RF and image frequency noise to the intermediate frequency. Thus, it can be seen that

$$N_{o(source)\,DSB} = N_{o(source)\,SSB} + 3 \text{ dB} \qquad (7.22)$$

and

$$NF_{DSB} = NF_{SSB} - 3 \text{ dB} \qquad (7.23)$$

This is not quite correct, since an input filter will also affect the output noise, but this rule is usually used. Usually, single-sideband noise figure is used for a mixer in a superheterodyne radio receiver, since an image reject filter preceding the mixer removes noise from the image.

Largely because of the added complexity and the presence of noise that is frequency translated, mixers tend to be much noisier than LNAs. The differential pair that forms the bottom of the mixer represents an unattainable lower bound on the noise figure of the mixer itself. Mixer noise will always be higher because noise sources in the circuit get translated to different frequencies and this often "folds" noise into the output frequency. Generally, mixers have three frequency bands where noise is important:

1. *Noise already present at the IF:* The transistors and resistors in the circuit will generate noise at the IF. Some of this noise will make it to the output and corrupt the signal. For example, the collector resistors will add noise directly at the output IF.

2. *Noise at the RF and image frequency:* Any noise present at the RF and image frequency will also be mixed down to the IF. For instance, the collector shot noise of Q_1 at the RF and at the image frequency will both appear at the IF at the output.

3. *Noise at multiples of the LO frequencies:* Any noise that is near a multiple of the LO frequency can also be mixed down to the IF, just like the noise at the RF.

Also note that noise over a cycle of the LO is not constant, as illustrated in Figure 7.9. At large negative or positive voltage on the LO, dominant noise comes from the bottom transistors. This is the expected behavior, as the LO is causing the upper quad transistors to be switched between cutoff and saturation. In both of these two states, the transistors will add very little noise because they have no gain. We also note that gain from the RF input is maximum when the upper quad transistors are fully switched one way or the other. Thus, a large LO signal that switches rapidly between the two states is ideal to maximize the signal-to-noise ratio.

However, for the finite rise and fall time in the case of a square wave LO signal, or for a sinusoidal LO voltage, the LO voltage will go through zero. During this time, these transistors will be on and in the active region. Thus, in this region they behave like an amplifier and will cause noise in the LO and in the upper quad transistors to be amplified and passed on to the output. As shown in Figure 7.9, for LO voltages near zero, noise due to the upper quad transistors is dominant. At the same time, in this region, gain from the RF is very low; thus for small LO voltages, the signal-to-noise ratio is very poor, so time spent in this region should be minimized.

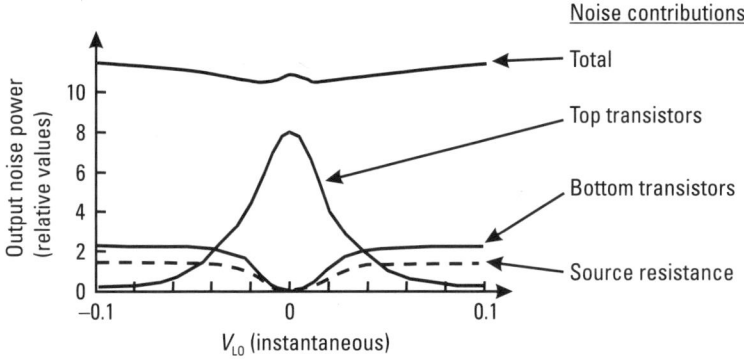

Figure 7.9 Mixer noise shown at various LO levels.

In order to determine noise figure from Figure 7.9, the relative value of the total noise compared to the noise from the source must be determined. Very conveniently, the calculated noise figure is approximately the same when calculated using a slowly swept dc voltage at the LO input or with an actual LO signal. With a slowly swept dc voltage, the mixer becomes equivalent to a cascode amplifier and the LO input serves as a gain-controlling signal. When used as a mixer, any noise (or signal) is mixed to two output frequencies, thus reducing the output level. This results in the mixer having less gain than the equivalent differential pair. However, noise from both the RF and the image frequency is mixed to the IF, resulting in a doubling of noise power at the output. Thus, the noise prediction based on a swept LO analysis is very close to that predicted using an actual LO signal.

There are a few other points to consider. One is that with a mixer treated as an amplifier, output noise is calculated at f_{RF}, so if some output filtering is included, for example, with capacitors across R_C, the noise will be reduced. However, the ratio of total noise to noise due to the source can still be correct. Another point is that in this analysis, noise that has been mixed from higher LO harmonics has not been included. However, as will be shown in Example 7.3, these end up not being very important, so the error is not severe. Some of these issues and points will be illustrated in the next three examples.

Example 7.1 Mixer Noise Figure Determination

For the mixer simulation results shown in Figure 7.9, estimate the mixer noise figure.

Solution

The noise figure is given by the ratio of total noise to noise from the source. In this example, while individual components of noise from most sources are strongly LO voltage dependent, total noise happens to be roughly independent of instantaneous LO voltage, and the relative value is approximately 11 (arbitrary units). Noise from the source is dependent on instantaneous LO voltage varying from 0 to about 1.5 (arbitrary units). This plot illustrates why maximum signal gain and minimum noise figure is realized for a sufficiently large LO signal such that minimal time is spent around 0V. Thus, minimum double-sideband noise figure is

$$\text{NF} = 10 \log \frac{N_{o\,tot}}{N_{o(\text{source})}} = 10 \log \frac{11}{1.5} = 8.65 \text{ dB}$$

For a real, sinusoidal signal, some time is inevitably spent at 0V with a resulting time domain waveform as shown in Figure 7.10.

In the diagram, the effective input noise is reduced down to about 1.1, and as a result the noise figure is increased to about 10 dB.

Example 7.2 Mixer Noise and Gain with Degeneration

In this example, some equations and simulation results will be shown for noise and gain versus degeneration.

Without consideration of input and output matching, the noise sources associated with R_S and R_E of Figure 7.11 both have gain approximately equal to the signal gain given by

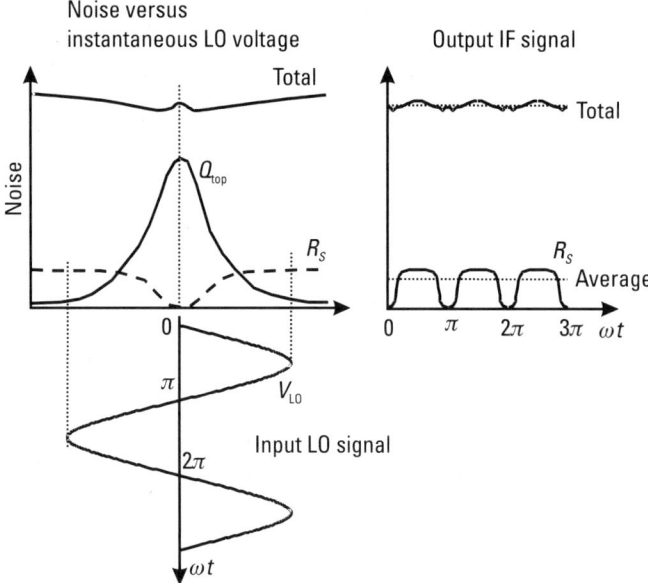

Figure 7.10 Noise calculations.

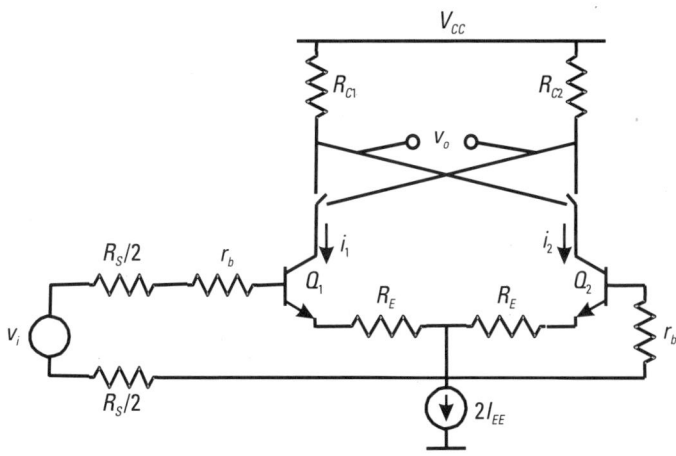

Figure 7.11 Mixer with switching.

$$\frac{v_{no,R_E}}{v_{n,R_E}} = \frac{v_{no,R_S}}{v_{n,R_S}} = \frac{v_o}{v_{RF}} \approx \frac{2}{\pi} \frac{R_C}{r_e + R_E}$$

where v_{no,R_E} and v_{no,R_S} are the output noise due to R_S and R_E and v_{n,R_E} and v_{n,R_S} are the noise voltage associated with each resistor. Thus, with degeneration, gain will decrease unless R_C is increased to match the increase of R_E. Similarly, the noise figure will degrade with increasing R_E, since the noise due to R_E is given by

$$v_{n,R_E} = \sqrt{4kTR_EB}$$

Simulation results are shown in Figure 7.12. It can be seen that gain decreases rapidly for increased values of R_E, from 12 dB to about 0 dB for R_E from 0Ω to 100Ω. Noise figure increases by about 4 dB over the same change of R_E. We note that matching will make a difference and adding degeneration resistance changes the input impedance, which will indirectly change the noise due to the effect of input impedance on base shot noise.

We also note that, as predicted by theory, gain is about 4 dB lower for a mixer compared to a differential pair. Noise is higher for the mixer by about 3 dB to 5 dB due to noise mixing from other frequencies and noise from the switching quad.

Example 7.3 Mixer Noise Components—Sources and Frequencies

Noise in a mixer, such as that shown in Figure 7.13, comes from a variety of sources and is mixed to the output from a variety of frequencies. In this

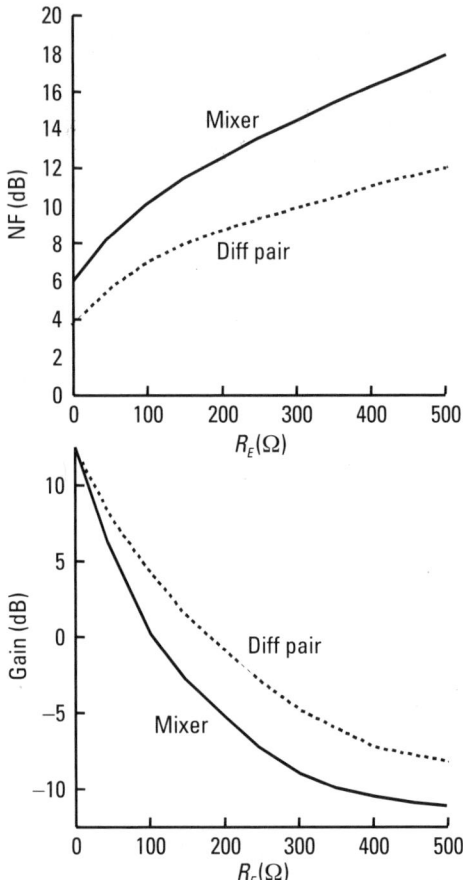

Figure 7.12 Noise simulation results.

example, we will show the typical relative levels of these noise components. Many simulators could provide the information for this analysis, but instead of discussing the simulation, the results will be discussed here.

Table 7.1 shows the noise from various sources and from various frequencies. Noise has been expressed as a voltage or as a squared voltage instead of a density by assuming a bandwidth of 1 MHz. Noise from the bottom components (RF input transistors, current sources, input resistors, and bias resistors) has been further broken down in Table 7.2. Noise from the source resistor (R_{SR}, top line of Table 7.1) has been shown in brackets, as its effect is also included with total bottom noise.

The final row in Table 7.1 shows for each specified input frequency, the total number of equivalent frequency bands that have approximately the same noise. For example, for every noise input at the RF, there is an approximately

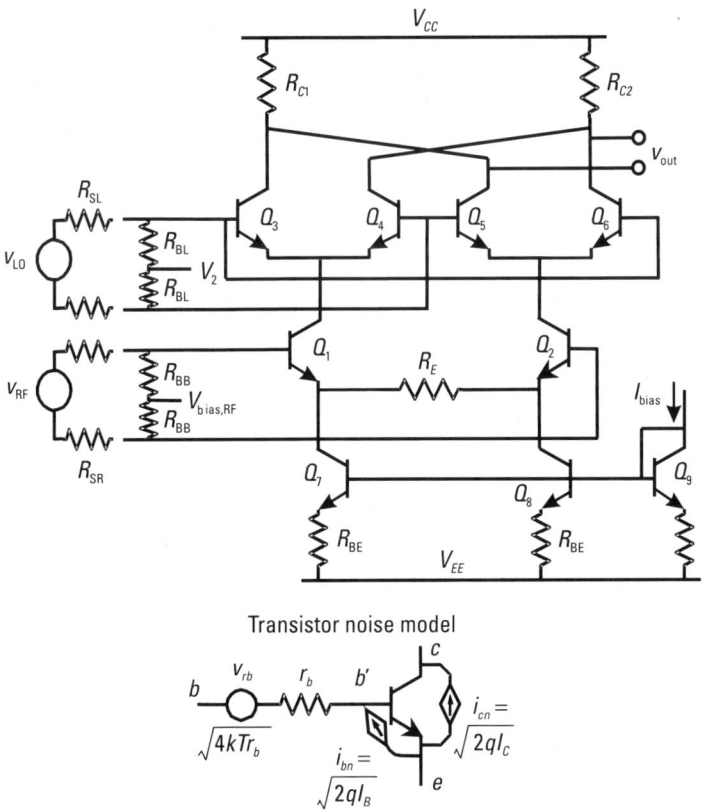

Figure 7.13 Mixer for noise analysis and transistor noise model.

Table 7.1
Mixer Noise and Dominant Sources of Noise

Source	Differential Output Voltage at ω_{IF}				Total $V^2 \times 10^{-6}$
	Input at ω_{IF}	Input at ω_{RF}	Input at $2\omega_{LO} - \omega_{IF}$	Input at $3\omega_{LO} - \omega_{IF}$	
(R_{SR})	(0.7 nV)	(0.72 mV)	(0.2 nV)	(0.11 mV)	(1.08)
Bottom	0.9 nV	1.77 mV	4.0 nV	0.29 mV	6.44
v_{rb}quad	0.25 mV	0.03 mV	0.20 mV	0.02 mV	0.15
$i_c$quad	0.26 mV	0.15 mV	1.5 μV	0.03 mV	1.17
$i_b$quad	36 nV	0.13 mV	25 nV	0.03 mV	0.04
v_{RL}	1.77 mV	16 nV	4.4 μV	1.3 nV	3.13
Freq. bands	1	2	2	2	10.93

Table 7.2
Breakdown of Noise in the RF Stage

Source	Noise Out V^2	Source	Noise Out V^2
$r_{b,Q_{1,2}}$	0.51×10^{-6}	R_{SR}	1.08×10^{-6}
$r_{e,Q_{1,2}}$	0.08×10^{-6}	R_{BB}	1.08×10^{-6}
i_b	0.06×10^{-6}	R_{BE}	0.20×10^{-6}
i_c	1.21×10^{-6}	R_E	2.22×10^{-6}
Total $Q_{1,2}$	1.86×10^{-6}	Total	6.44×10^{-6}

equal input at the image frequency. Thus, for an input at the RF or around other harmonics of the LO frequency, the multiplier is 2, while for noise inputs at the IF, the multiplier is 1. Different frequencies are dominant for different parts of the mixer. For the bottom circuitry, the most important input frequency is the RF (and image frequency). For the load resistor, the only important component occurs at the IF, while for the quad switching transistors, both the RF and IF are important. Generally, noise around the LO second harmonic or higher harmonics is not important; however, the third harmonic will add a bit of noise from the bottom circuitry.

As for sources of noise, the bottom circuitry is seen to be the dominant factor (true only if the LO amplitude is large enough to switch the quad transistors fully). Thus, to minimize noise, the RF input stage must be optimized, similar to that of an LNA design. Of the bottom noise sources, degeneration can quickly become the dominant noise source for high-linearity design. If linearity allows, it is possible to use inductor degeneration for optimal noise and power matching. Also, in this example, bias resistors contributed significantly to the noise. In an inductively degenerated mixer, bias resistors can be made significantly larger to minimize the noise contribution from them.

Double-sideband noise figure can be calculated as follows:

$$\text{DSB noise figure} = 10 \log_{10} \left(\frac{10.93}{1.08} \right) = 10.1 \text{ dB}$$

Minimum noise figure from the RF stage can be calculated as

$$\text{DSB NF}_{\min} = 10 \log_{10} \left(\frac{6.44}{1.08} \right) = 7.75$$

Note that single-sideband noise figure would consider source noise in the RF band only; thus, noise figure would have been higher by about 3 dB.

7.8 Linearity

Mixers have both desired and undesired nonlinearity. The mixing action of the switching quad is what is necessary for the operation of the circuit. However, mixers also contain amplifiers that can be nonlinear. Just as explained in Chapter 6, these amplifiers have linearity requirements.

7.8.1 Desired Nonlinearity

A mixer is inherently a nonlinear device. Linear components have output frequencies equal to the input frequencies, so no mixing action can take place for purely linear circuits. This *desired nonlinearity* comes from the switching action of the quad transistors as determined by the nonlinear exponential characteristics of the transistor. Thus, if we have two tones at the input, the desired outputs for a switching mixer are as shown in Figure 7.14.

The only desired output may be at the IF, but the other components are far enough away that they can easily be filtered out.

7.8.2 Undesired Nonlinearity

Undesired nonlinearity can occur in several places. One is at the RF input, which converts the input signal into currents i_1 and i_2 (see Figure 7.11). The reason for adding degeneration resistors R_E is to keep this conversion linear, just as in the case of an LNA. However, some nonlinearity will still be present. The resulting output is shown in Figure 7.15. Thus, the only difference in this circuit compared to those considered in Chapter 6 is that now there is a frequency translation. Thus, just as before, the input IP3 can be approximated as

$$v_{\text{IP3}} = 4\sqrt{2} v_T \left(\frac{R_E + r_e}{r_e} \right)^{3/2} \tag{7.24}$$

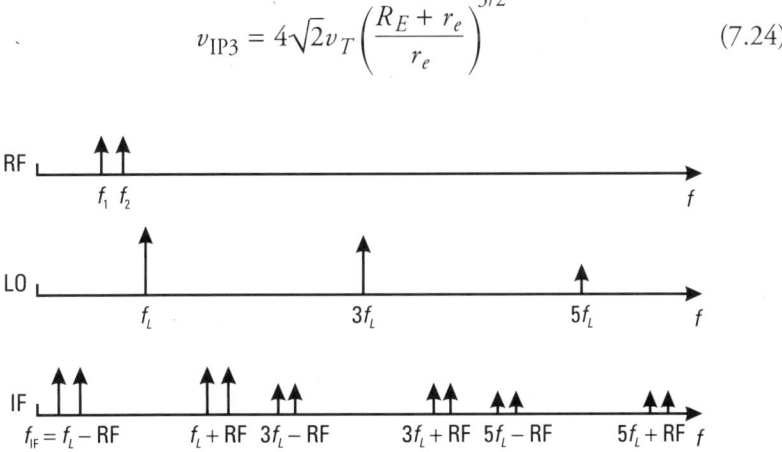

Figure 7.14 Mixer expected outputs in the frequency domain.

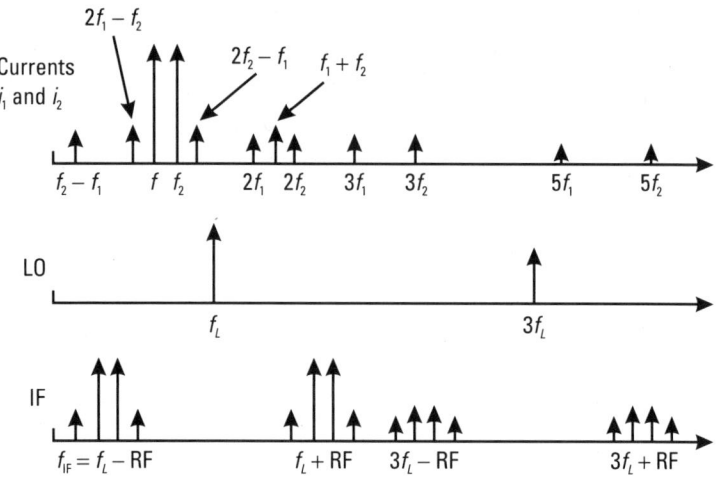

Figure 7.15 Mixer with nonlinearity in the RF input stage.

where, in this case, there is an extra factor of 2 because the circuit is differential. From here, the 1-dB compression point can also be computed using the relationships between IP3 and 1-dB compression developed in Chapter 2.

With nonlinearity in the RF input stage, the currents i_1 and i_2 are composed of a large number of frequency components. Each of these frequencies is then mixed with each of the LO harmonics, producing the IF output. Many of the intermediate frequencies (such as mixing of harmonics of the radio frequencies with harmonics of the LO) have not been shown.

Finally, if the RF input were perfectly linear, mixing action would proceed cleanly, with the result as previously shown in Figure 7.14. However, undesirable frequency components can be generated because of nonlinearity in the output stage, for example, due to limiting action. In such a case, the two IF tones at $f_L - f_1$ and $f_L - f_2$ would intermodulate, producing components at $f_L - (2f_1 - f_2)$ and $f_L - (2f_2 - f_1)$, in addition to harmonics of $f_L - f_1$ and $f_L - f_2$. This is shown in Figure 7.16.

In some cases, the saturation of the switching quad may be the limiting factor in the linearity of the mixer. For example, if the bias voltage on the base of the quadrature switching transistors v_{quad} is 2V and the power supply is 3V, then the output can swing from about 3V down to about 1.5V, for a total 1.5V peak swing. If driving 50Ω, this is 13.5 dBm. For larger swings, a tuned circuit load can be used as shown in Figure 7.17. Then the output is nominally at V_{CC} with equal swing above and below V_{CC}, for a new swing of 3V$_p$ or 6V$_{\text{p-p}}$, which translates to 18.5 dBm into 50Ω. We note that an on-chip tuned circuit may be difficult to realize for a down-converting mixer, since the output frequency is low, and therefore the inductance needs to be large.

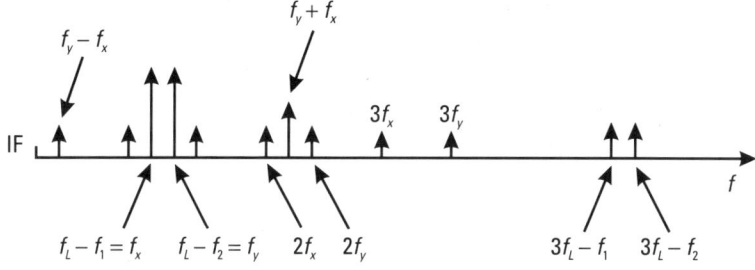

Figure 7.16 Mixer with nonlinearity in switching quad.

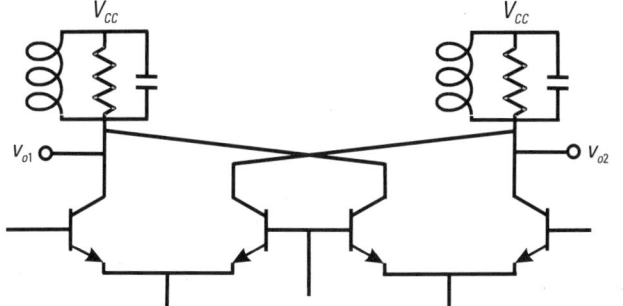

Figure 7.17 Tuned load on a mixer.

If the output needs to drive a low impedance such as 50Ω, often an emitter follower is used at the output. This can be a fairly broadband circuit, since no tuned components need be used.

7.9 Improving Isolation

It is also possible to place an inductor-capacitor (LC) series circuit across the outputs, as shown in Figure 7.18, to reduce LO or RF feed-through or to get rid of some upconverted component. This can also be accomplished with the use of a lowpass filter by placing a capacitor in parallel with the load, as shown in Figure 7.19, or using a tuned load, as shown in Figure 7.17.

7.10 Image Reject and Single-Sideband Mixer

Mixing action as shown in Figure 7.20 always produces two sidebands: one at $\omega_1 + \omega_2$ and one at $\omega_1 - \omega_2$ by multiplying $\cos \omega_1 t \times \cos \omega_2 t$. It is possible

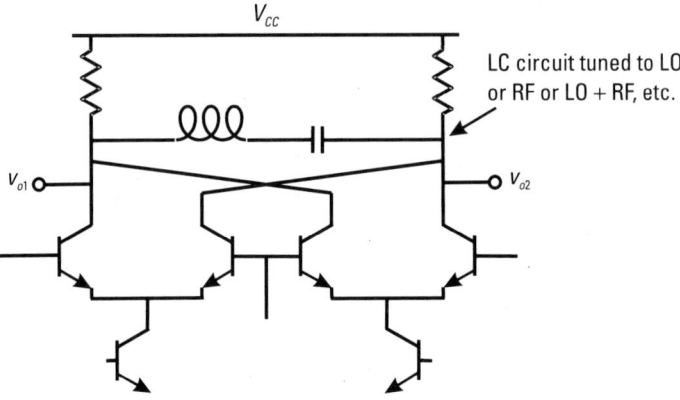

Figure 7.18 Series LC between mixer outputs.

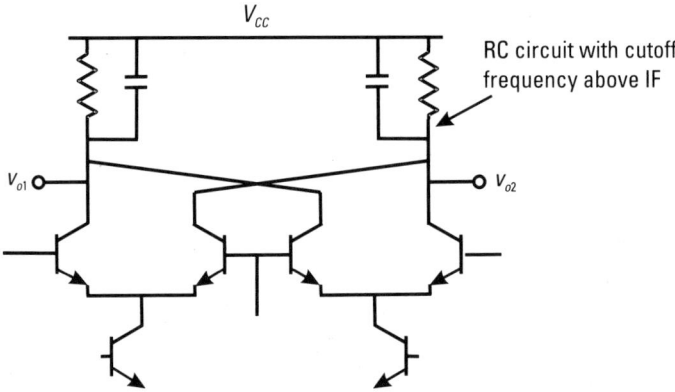

Figure 7.19 Parallel RC circuit across the output.

to use a filter *after* the mixer in the transmitter to get rid of the unwanted sideband for the up-conversion case. Similarly, it is possible to use a filter *before* the mixer in a receiver to eliminate unwanted signals at the image frequency for the down-conversion case. Alternatively, a single-sideband mixer for the transmit path, or an image reject mixer for the receive path can be used.

An example of a single-sideband up-conversion mixer is shown in Figure 7.21. It consists of two basic mixer circuits, two 90° phase shifters, and a summing stage. As can be shown, the use of the phase shifters and mixers will cause one sideband to add in phase and the other to add in antiphase, leaving only the desired sideband at the output. Which sideband is rejected depends on the placement of the phase shifts or the polarity of the summing block. By moving the phase shift from the input to the output, as shown in Figure 7.22,

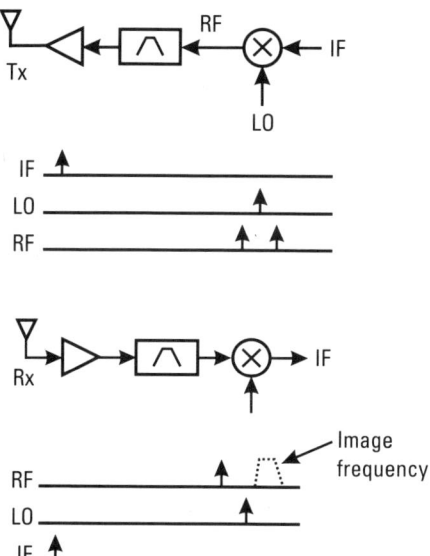

Figure 7.20 Sidebands in upconversion, image in downconversion.

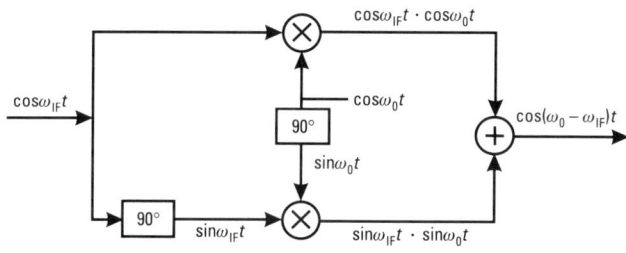

Figure 7.21 A single-sideband mixer.

an image reject mixer is formed. In this circuit, at the output, the RF signal adds in phase while the image adds in antiphase.

7.10.1 Alternative Single-Sideband Mixers

The image reject configuration in Figure 7.22 is also known as the *Hartley architecture*.

Another possible implementation of an image reject receiver is known as the *Weaver architecture,* shown in Figure 7.23. In this case, the phase shifter after the mixer in Figure 7.22 is replaced by another set of mixers to perform an equivalent operation. The advantage is that all phase shifting takes place only in the LO path and there are no phase shifters in the signal path. As a

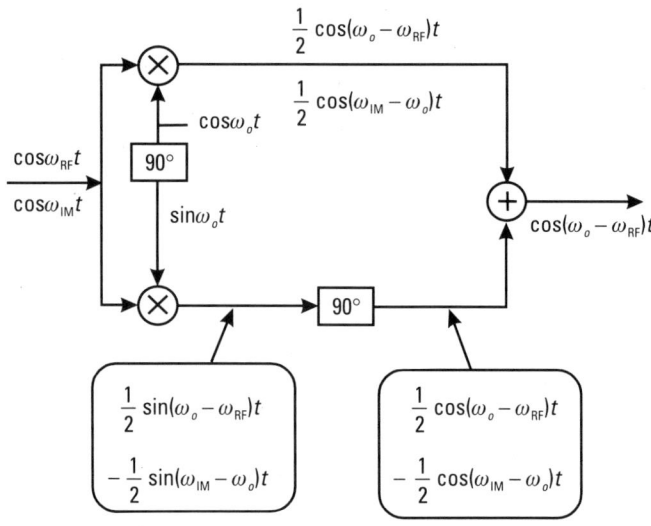

Figure 7.22 An image reject mixer.

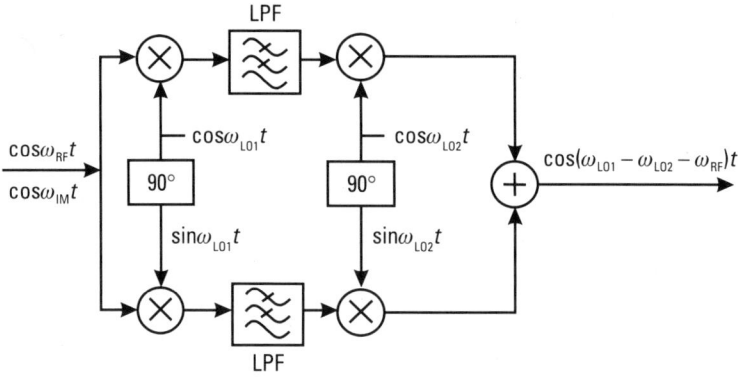

Figure 7.23 Weaver image reject mixer.

result, this architecture is less sensitive to amplitude mismatch in the phase-shifting networks and so image rejection is improved. The disadvantage is the additional mixers required, but if the receiver has a two-stage downconversion architecture, then these mixers are already present and so there is no penalty.

7.10.2 Generating 90° Phase Shift

Several circuits can be used to generate the phase shifts as required for single-sideband or image reject mixers. Some of the simplest are the RC circuits shown in Figure 7.24. The transfer functions for the two networks are simply

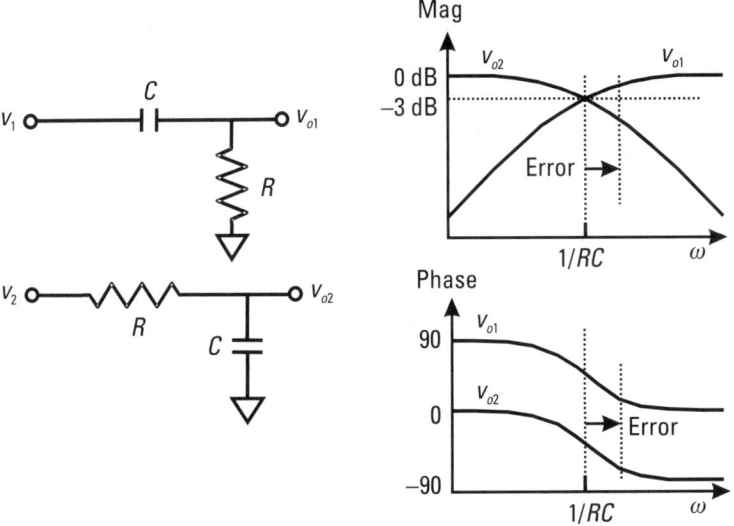

Figure 7.24 RC networks to produce phase shift.

$$\frac{v_{o1}}{v_1} = \frac{sCR}{1+sCR} = \frac{j\omega/\omega_o}{1+j\omega/\omega_o} \qquad (7.25)$$

$$\frac{v_{o2}}{v_2} = \frac{1}{1+sCR} = \frac{1}{1+j\omega/\omega_o}$$

where $\omega_o = 1/CR$.

It can be seen that at the center frequency, where $\omega = \omega_o$, the output of the lowpass filter is at $v_{o1}/v_1 = 1/\sqrt{2} \angle 45°$ and the output of the highpass filter is at $v_{o2}/v_2 = 1/\sqrt{2} \angle -45°$. Thus, if $v_1 = v_2$, then v_{o1} and v_{o2} are 90° out of phase. In a real circuit, the amplitude or phase may be shifted from their ideal value. Such mismatch between the amplitude or phase can come from a variety of sources. For example, R and C can be poorly matched, and the time constant could be off by a large percentage. As shown in Figure 7.24, such an error will cause an amplitude error, but the phase difference between the two signals will remain at approximately 90°. If the phase-shifted signals are large and fed into the switching quad of a mixer, amplitude mismatch is less important. However, in any configuration requiring a phase shifter in the signal path, such as those shown in Figures 7.21 and 7.22, the sideband cancellation or image rejection will be sensitive to amplitude and phase mismatch. Even if the phase shifter is perfect at the center frequency, there will be errors at other frequencies and this will be important in broadband designs.

Example 7.4 Calculation of Amplitude and Phase Error of Phase-Shifting Network
Calculate the amplitude and phase error for a 1% component error.

Solution
Gains are calculated as

$$\frac{v_{o1}}{v_i} = \frac{j1.01}{1 + j1.01} = 0.7106 \angle 44.71°$$

$$\frac{v_{o2}}{v_i} = \frac{1}{1 + j1.01} = 0.7036 \angle -45.29°$$

In this case, the phase difference is still 90°, but the amplitude now differs by about 1%. It will be shown later that such an error will limit the image rejection to about 40 dB.

A differential implementation of a simple phase-shifting circuit is shown in Figure 7.25. In order to function properly, the RC network must not load the output of the differential amplifier. It may also be necessary to buffer the phase shift output.

This circuit is sometimes known as a first-order polyphase filter. The polyphase filter will be discussed in the next section.

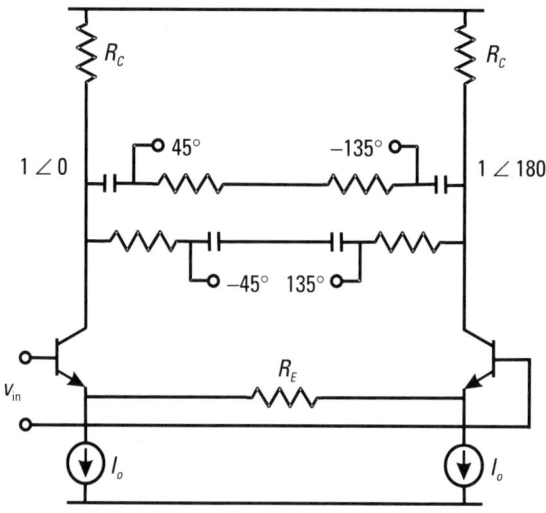

Figure 7.25 Differential circuit to produce phase shift.

Polyphase Filters

A multistage polyphase filter [1] is a circuit that improves performance in the presence of component variations and mismatches over a broader band of frequencies. All polyphase filters are simple variations or extensions of the polyphase filters shown in Figure 7.26. One of the variations is in how the input is driven. The inputs can be driven with four phases, or simple differential

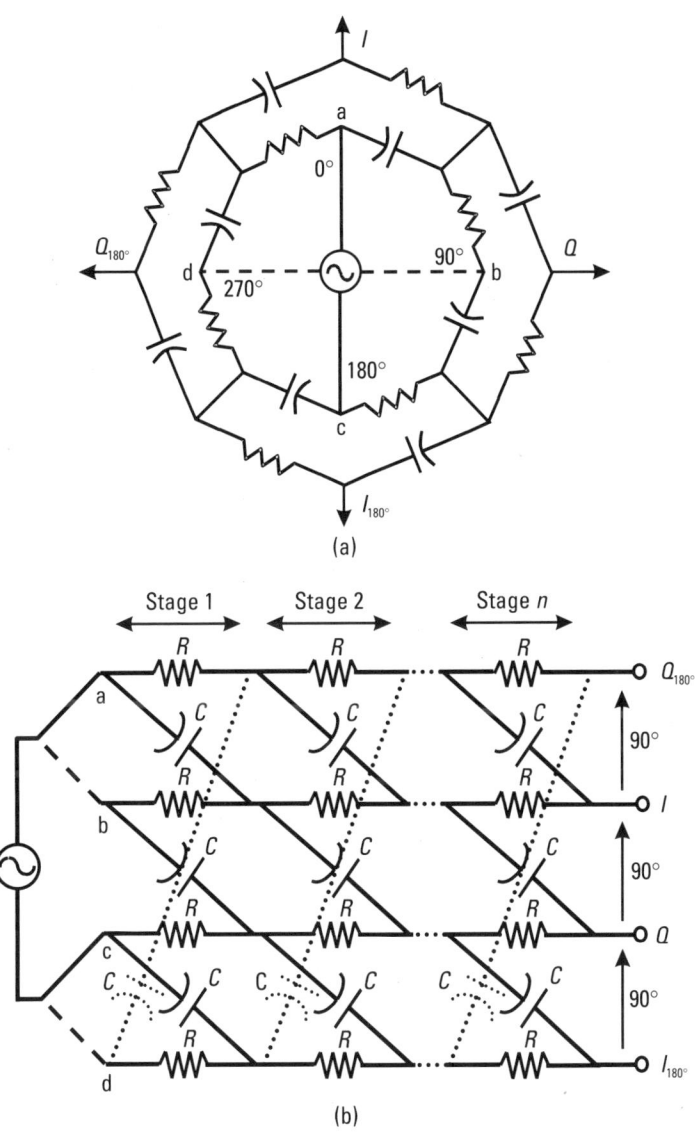

Figure 7.26 Polyphase filters: (a) two stage, and (b) n stage.

inputs can be applied at nodes "a" and "c." With the simple differential inputs, the other nodes, "b" and "d," can be connected to "a" and "c," left open, or grounded.

The polyphase filter is designed such that at a particular frequency (nominally at $\omega = 1/RC$), all outputs are 90° out of phase with each other. The filter also has the property that with each additional stage, phase shifts become more precisely 90°, even with a certain amount of tolerance on the parts. Thus, when they are used in an image reject mixer, if more image rejection is required, then polyphase filters with more stages can be employed. The drawback is that with each additional stage there is an additional loss of about 3 dB through the filter. This puts a practical upper limit on the number of stages that can be used.

7.10.3 Image Rejection with Amplitude and Phase Mismatch

The ideal requirements are that a phase shift of exactly 90° is generated in the signal path and that the LO has perfect quadrature output signals. In a perfect system, there is also no gain mismatch in the signal paths. In a real circuit implementation, there will be imperfections as shown in Figure 7.27. Therefore, an analysis of how much image rejection can be achieved for a given phase and amplitude mismatch is now performed.

The analysis proceeds as follows:

1. The input signal is mixed with the quadrature LO signal through the I and Q mixers to produce signals V_1 and V_2 after filtering. V_1 and V_2 are given by

$$V_1 = \frac{1}{2} \sin(\omega_{LO} - \omega_{RF})t - \frac{1}{2} \sin(\omega_{IM} - \omega_{LO})t \quad (7.26)$$

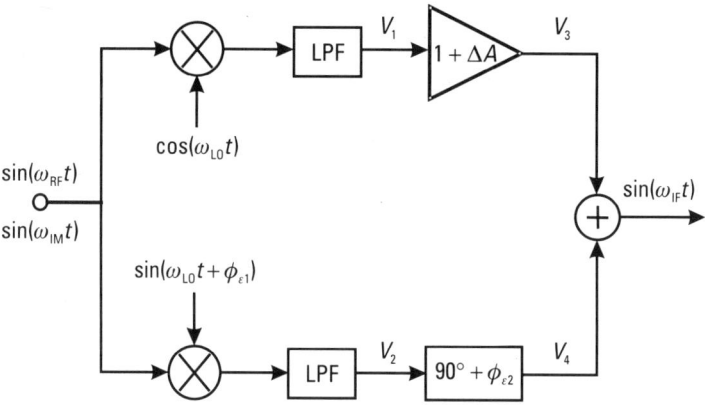

Figure 7.27 Block diagram of an image reject mixer, including phase and gain errors.

$$V_2 = \frac{1}{2}\cos[(\omega_{LO} - \omega_{RF})t + \phi_{\epsilon 1}] + \frac{1}{2}\cos[(\omega_{IM} - \omega_{LO})t - \phi_{\epsilon 1}]$$
(7.27)

2. Now V_1 experiences an amplitude error relative to V_2, and V_2 experiences a phase shift that is not exactly 90° to give V_3 and V_4, respectively.

$$V_3 = \frac{1}{2}(1 + \Delta A)\sin(\omega_{LO} - \omega_{RF})t + \frac{1}{2}\sin(\omega_{IM} - \omega_{LO})t \quad (7.28)$$

$$V_4 = \frac{1}{2}\sin[(\omega_{LO} - \omega_{RF})t + \phi_{\epsilon 1} + \phi_{\epsilon 2}] + \frac{1}{2}\sin[(\omega_{IM} - \omega_{LO})t - \phi_{\epsilon 1} + \phi_{\epsilon 2}]$$
(7.29)

3. Now V_3 and V_4 are added together. The component of the output due to the RF signal is denoted V_{RF} and is given by

$$V_{RF} = \frac{1}{2}(1 + \Delta A)\sin(\omega_{IF}t) + \frac{1}{2}\sin(\omega_{IF}t + \phi_{\epsilon 1} + \phi_{\epsilon 2}) \quad (7.30)$$

$$V_{RF} = \frac{1}{2}(1 + \Delta A)\sin(\omega_{IF}t) + \frac{1}{2}\sin(\omega_{IF}t)\cos(\phi_{\epsilon 1} + \phi_{\epsilon 2}) \quad (7.31)$$
$$+ \frac{1}{2}\cos(\omega_{IF}t)\sin(\phi_{\epsilon 1} + \phi_{\epsilon 2})$$

4. The component due to the image is denoted V_{IM} and is given by

$$V_{IM} = -\frac{1}{2}(1 + \Delta A)\sin(\omega_{IF}t) + \frac{1}{2}\sin(\omega_{IF}t)\cos(\phi_{\epsilon 2} - \phi_{\epsilon 1})$$
$$+ \frac{1}{2}\cos(\omega_{IF}t)\sin(\phi_{\epsilon 1} - \phi_{\epsilon 2}) \quad (7.32)$$

5. Only the ratio of the magnitudes is important. The magnitudes are given by

$$|V_{RF}|^2 = \frac{1}{4}\{[\sin(\phi_{\epsilon 1} + \phi_{\epsilon 2})]^2 + [(1 + \Delta A) + \cos(\phi_{\epsilon 1} + \phi_{\epsilon 2})]^2\}$$
(7.33)

$$|V_{RF}|^2 = \frac{1}{4}\{1 - [\cos(\phi_{\epsilon 1} + \phi_{\epsilon 2})]^2 + (1 + \Delta A)^2$$
$$+ 2(1 + \Delta A)\cos(\phi_{\epsilon 1} + \phi_{\epsilon 2}) + [\cos(\phi_{\epsilon 1} + \phi_{\epsilon 2})]^2\}$$
(7.34)

$$|V_{RF}|^2 = \frac{1}{4}[1 + (1 + \Delta A)^2 + 2(1 + \Delta A)\cos(\phi_{\epsilon 1} + \phi_{\epsilon 2})] \quad (7.35)$$

$$|V_{IM}|^2 = \frac{1}{4}\{[\sin(\phi_{\epsilon 1} - \phi_{\epsilon 2})]^2 + [-(1 + \Delta A) + \cos(\phi_{\epsilon 2} - \phi_{\epsilon 1})]^2\}$$
(7.36)

$$|V_{IM}|^2 = \frac{1}{4}\{1 - [\cos(\phi_{\epsilon 2} - \phi_{\epsilon 1})]^2 + (1 + \Delta A)^2$$
$$- 2(1 + \Delta A)\cos(\phi_{\epsilon 2} - \phi_{\epsilon 1}) + [\cos(\phi_{\epsilon 1} + \phi_{\epsilon 2})]^2\}$$
(7.37)

$$|V_{IM}|^2 = \frac{1}{4}[1 + (1 + \Delta A)^2 - 2(1 + \Delta A)\cos(\phi_{\epsilon 2} - \phi_{\epsilon 1})] \quad (7.38)$$

6. Therefore, the image rejection ratio is given by

$$\text{IRR} = 10 \log \frac{|V_{RF}|^2}{|V_{IM}|^2} \quad (7.39)$$

$$= 10 \log \left[\frac{1 + (1 + \Delta A)^2 + 2(1 + \Delta A)\cos(\phi_{\epsilon 1} + \phi_{\epsilon 2})}{1 + (1 + \Delta A)^2 - 2(1 + \Delta A)\cos(\phi_{\epsilon 2} - \phi_{\epsilon 1})}\right]$$

If there is no phase imbalance or amplitude mismatch, then this equation approaches infinity, and so ideally this system will reject the image perfectly, and it is only the nonideality of the components that causes finite image rejection. Figures 7.28 and 7.29 show a contour plot and a three-dimensional plot of how much image rejection can be expected for various levels of phase and amplitude mismatch. An amplitude error of about 20% is acceptable for 20 dB of image rejection, but more like 2% is required for 40 dB of image rejection. Likewise, phase mismatch must be held to less than 1.2° for 40 dB of image rejection, and phase mismatch of less than 11.4° can be tolerated for 20 dB of image rejection.

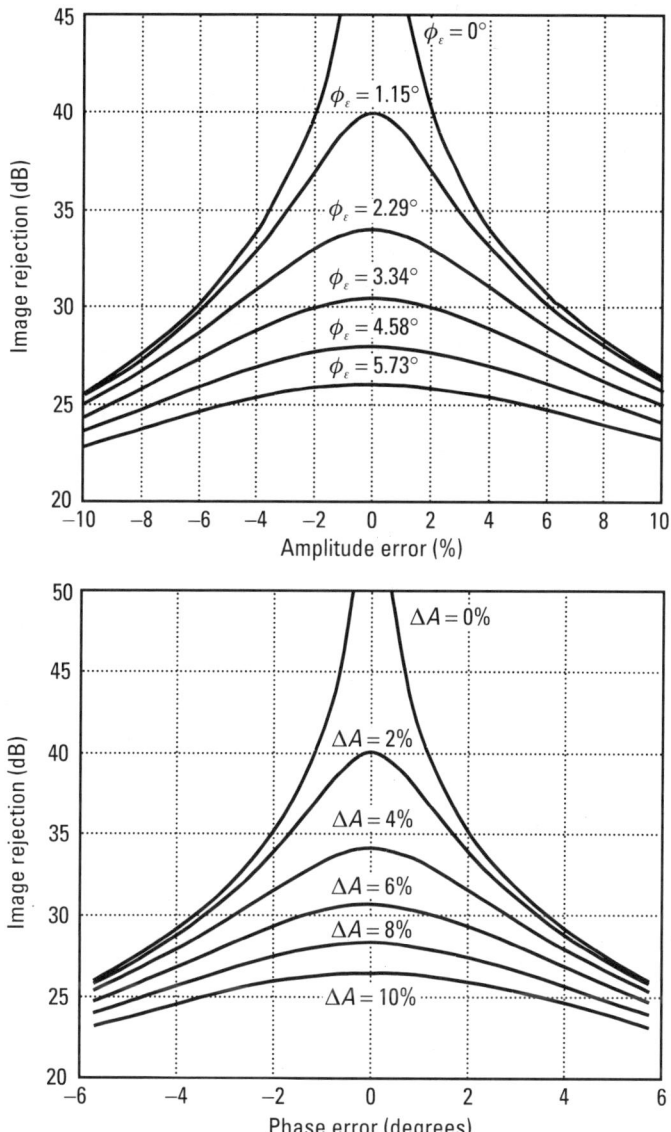

Figure 7.28 Plot of image rejection versus phase and amplitude mismatch.

7.11 Alternative Mixer Designs

In the following section, some variations of mixers will be mentioned briefly, including the Moore mixer, which rejects image noise from a degeneration resistor, mixers that make use of inductors and transformers, and some other low-voltage mixers.

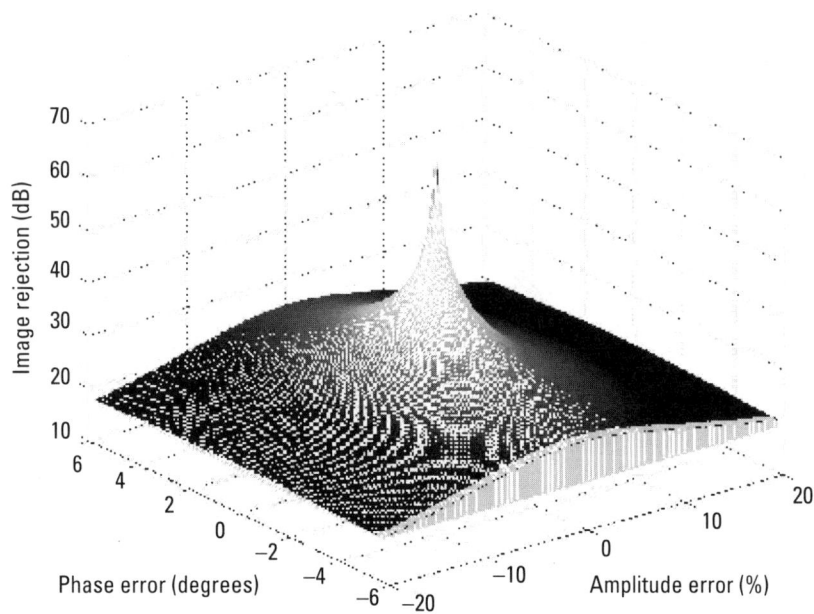

Figure 7.29 Three-dimensional plot of image rejection versus phase and amplitude mismatch.

7.11.1 The Moore Mixer

In a receiver, the noise produced by the mixer is sometimes very important. If the mixer is to use resistive degeneration and it is to have its phase shifts in the IF and LO paths, then there is a way to interleave the mixers, as shown in Figure 7.30, such that the noise produced by the degeneration resistors R_E is also image rejected. Here, the noise due to these resistors is fed into both paths of the mixer rather than just one; thus, it gets image rejected, and its effect is reduced by 3 dB. Since noise due to degeneration resistors is often very important, this can have a beneficial effect on the noise figure of the mixer.

7.11.2 Mixers with Transformer Input

Figure 7.31 shows a mixer with a transformer-coupled input and output [2]. Such a mixer has the potential to be highly linear, since a transformer is used in place of the input transistors. In addition, this mixer can operate from a low power supply voltage, since the number of stacked transistors is reduced compared to that of a conventional mixer. We note that for a downconversion mixer, the input transformer could be on-chip, but for low IF, the output transformer would have to be off-chip.

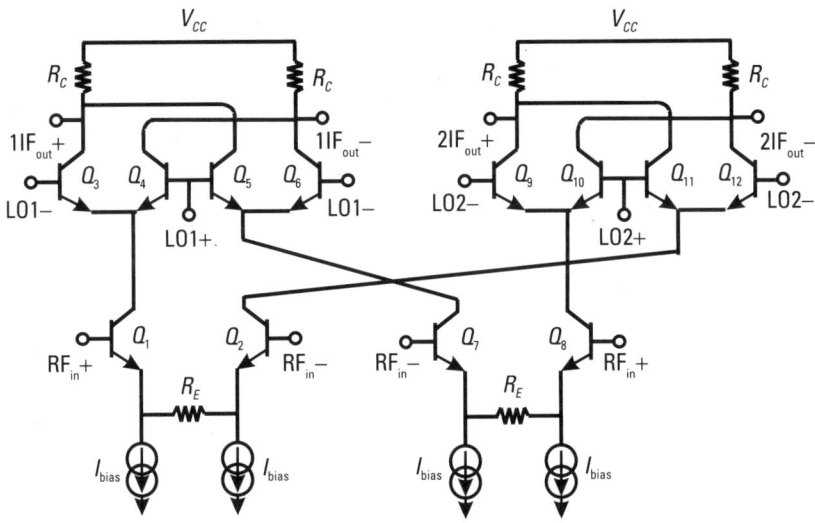

Figure 7.30 The Moore mixer.

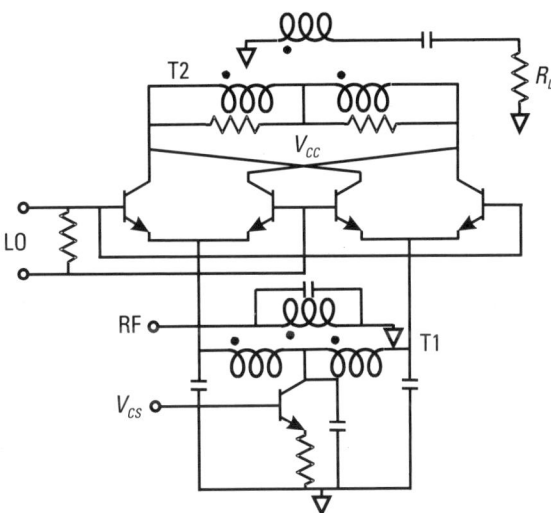

Figure 7.31 Mixer with transformer input.

7.11.3 Mixer with Simultaneous Noise and Power Match

Figure 7.32 shows a mixer with inductor degeneration and inductor input achieving simultaneous noise and power matching similar to that of a typical LNA [3].

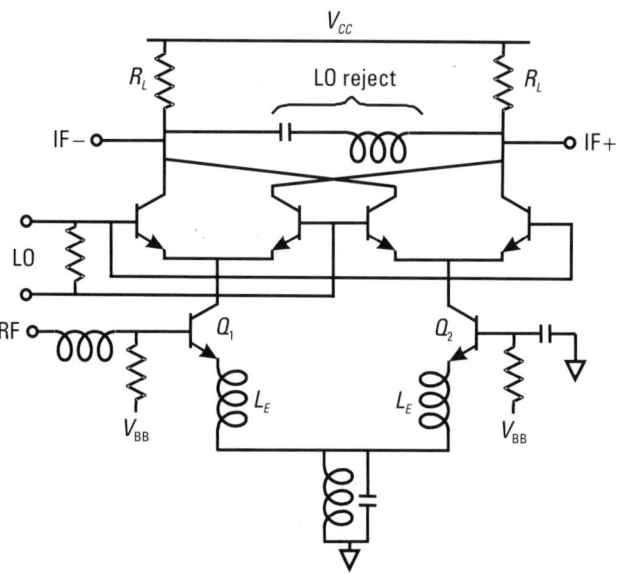

Figure 7.32 Mixer with simultaneous noise and power match.

To achieve matching, the same conditions as for an LNA are required, starting with

$$L_E = \frac{Z_0}{2\pi f_T} \tag{7.40}$$

The resulting linearity is approximately given by [3]

$$\text{IIP3} \approx \frac{\omega g_m Z_0}{2\pi f_T} \tag{7.41}$$

Noise matching is achieved by sizing L_E, selecting transistor size, and operating the RF transistors at the current required for minimum noise figure. The quad switching transistors are sized for maximum f_T, which typically means they will end up being about five to ten times smaller than the RF transistors.

7.11.4 Mixers with Coupling Capacitors

If headroom is a problem, but due to space or bandwidth constraints it is not possible to use transformers or inductors, then the circuit shown in Figure 7.33 may be one alternative. In this figure, the differential amplifier is coupled into the switching quad through the capacitors C_{cc}. Resistors R_{cc} provide a high

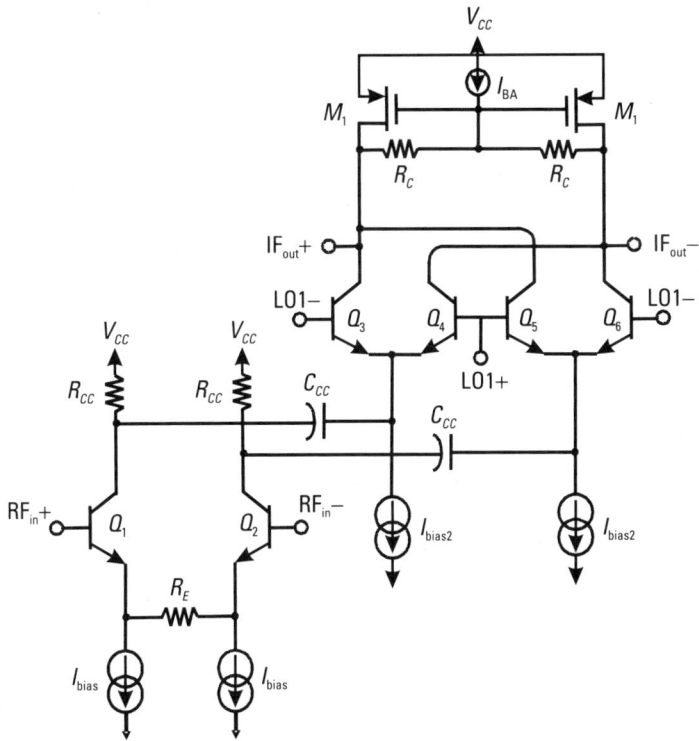

Figure 7.33 Mixer with folded switching stage and current steering PMOS transistors for high gain at low supply voltage.

impedance, so that most of the small-signal current will flow through the capacitors and up into the quad. Usually it is sufficient for R_{cc} to be about ten times the impedance of the series combination C_{cc} and input impedance of the switching quad transistors. Also, the current is steered away from the load resistors using two PMOS transistors, which act like diodes. Thus, R_C can be large to give good gain without using up as much headroom as would otherwise be required. A current source I_{BA} can also be included for bias adjustment if needed.

7.12 General Design Comments

So far in this chapter, we have discussed the basic theory of the operation of mixers. Here we will provide a summary of some general design guidelines to help with the trade-offs of optimizing a mixer for a particular application.

7.12.1 Sizing Transistors

The differential pair that usually forms the bottom of a double-balanced mixer is basically an LNA stage, and the transistors and associated passives can be optimized using the techniques of the previous chapter. The switching quad transistors are the parts of the circuit unique to the mixer. Usually, these transistors are sized so that they operate close to their peak f_T at the bias current that is optimal for the differential pair amplifier. In a bipolar design, if the differential pair transistors are biased at their minimum noise current, then the switching transistors end up being about one-eighth the size.

7.12.2 Increasing Gain

As shown previously in (7.20), without matching considerations and assuming full switching of the upper quad, voltage gain is estimated by

$$v_o = \frac{2}{\pi} \frac{R_C}{r_e + R_E} v_{\text{in}} \qquad (7.42)$$

To increase the gain, the choices are to increase the load resistance R_C, to reduce degeneration resistance R_E, or to increase the bias current I_B. Since the output bias voltage is approximately equal to $V_C \approx V_{CC} - I_B R_C$, increased gain will be possible only if adjustments to R_C, R_E, or I_B do not cause the switching transistors to become saturated.

7.12.3 Increasing IP3

How to increase IP3 depends on which part of the circuit is compressing. Compression can be due to overdriving of the lower differential pair, clipping at the output, or the LO bias voltage being too low, causing clipping at the collectors of the bottom differential pair. After a problem has been identified, IP3 can be improved by one or more of the following:

1. If the compression is due to the bottom differential pair (RF input), then linearity can be improved by increasing R_E or by increasing bias current. We note from (7.42) that increased R_E will result in decreased gain. Increased bias current will increase the gain slightly through a reduction of r_e, although this effect will be small if degeneration resistance is significantly larger than r_e.
2. Compression caused by clipping at the output is typically due to the quad transistors going into saturation. Saturation can be avoided by reducing the bias current or reducing the load resistors. Either technique

will move dc output voltage to a higher level; however, reduced load resistance will also reduce the gain. Another possibility, although not usually a practical one, is to increase V_{CC}. The use of a tuned output circuit is equivalent to raising the supply voltage.

3. If compression is caused by clipping at the collector of the RF input differential pair, then increasing the LO bias voltage will improve linearity; however, this may result in clipping at the output.

It is possible to conduct a series of tests on the actual circuit or in a circuit simulator to determine where the compression is coming from. First, it is possible to increase the power supply voltage to some higher value; for example, in a simulation this might be 5V or even 10V. If the compression point is increased, then output clipping is a problem. If the compression is unchanged, then the problem is not at the output. Next, one can determine if the LO bias voltage is sufficiently high by increasing it further. If linearity is not improved, then this was not the cause of the original linearity problem. Then, having eliminated output clipping or LO biasing problems, one can concentrate on the lower differential pair. As discussed in the previous paragraph, its linearity can be improved by increasing current or by increasing R_E.

7.12.4 Improving Noise Figure

Noise figure will be largely determined by the choice of topology, with the opportunity for the lowest noise provided by the simultaneous matched design of Section 7.11.3. The next most important factor is the value of the emitter degeneration resistor. To minimize noise, the emitter degeneration resistor should be kept as small as possible. However, with less degeneration, getting the required linearity will require more current.

7.12.5 Effect of Bond Pads and the Package

In a single-ended circuit, such as an LNA, the effect of the bond pads and the package is particularly important for the emitter, since this is a low-impedance node and has a strong influence on the gain and noise. For a differential circuit, such as the mixer, the ground is a virtual ground and the connection to the external ground is typically through a current source. Thus, the bond pad on the ground node here has minimal impact on gain and noise. At the other nodes, such as inputs and outputs, the bond pads will have some effect, since they add a series inductor. However, this can be incorporated as part of the matching network.

7.12.6 Matching, Bias Resistors, and Gain

If the base of the RF transistors were biased using a voltage divider with an equivalent resistance of 50Ω, the input would be matched over a broad band. However, the gain would have dropped by about 6 dB compared to the gain achievable when matching the input reactively with an LC network or with a transformer. For a resistively degenerated mixer, the RF input impedance (at the base of the input transistors) will be fairly high; for example, with R_E = 100Ω, Z_{in} can be of the order of a kilohm. In such a case, a few hundred ohms can make it easier to match the input; however, there will be some signal attenuation and noise implications.

At the output, if matched, the load resistor R_o is equal to the collector resistor R_C, and the voltage gain is modified by a factor of 0.5. Furthermore, to convert from voltage gain A_v to power gain P_o/P_i, one must consider the input resistance R_i and load resistance $R_o = R_C$ as follows:

$$\frac{P_o}{P_i} = \frac{\frac{v_o^2}{R_o}}{\frac{v_i^2}{R_i}} = \frac{v_o^2}{v_i^2}\frac{R_i}{R_o} = A_v^2 \frac{R_i}{R_o} \approx \left(\frac{2}{\pi}\frac{R_C/2}{R_E}\right)^2 \frac{R_i}{R_C} \quad (7.43)$$

Example 7.5 High-Linearity Mixer

Design a mixer to downconvert a 2-GHz RF signal to a 50-MHz IF. Use a low-side-injected LO at 1.95 GHz. Design the mixer to have an IIP3 of 8 dBm at 15 dB of voltage gain. The mixer must operate from a 3.3-V supply and draw no more than 12 mA of current. Determine the noise figure of the design as well. Determine what aspects of the design dominate the noise figure. Do not use any inductors in the design and match the input to 100Ω differentially.

Solution

Since inductors are not allowed in the design, the linearity must be achieved with resistor degeneration. Since current sources require at least 0.7V and the differential pair and quad will both require 1V, this would leave only 0.6V for the load resistors. A design that stacks the entire circuit is unlikely, therefore, to fit into the 3.3-V supply requirement; thus, it will have to be folded. Also, since we are using resistive degeneration, we can probably match the input with a simple resistor. Thus, the mixer topology shown in Figure 7.34 will probably be adequate for this design.

We can now begin sizing components and determining bias currents. First, we are told that we can use 12 mA in this design. There are two sources of nonlinearity of concern, one is the exponential nonlinearity of the differential pair and the other is the exponential nonlinearity of the quad. We note that

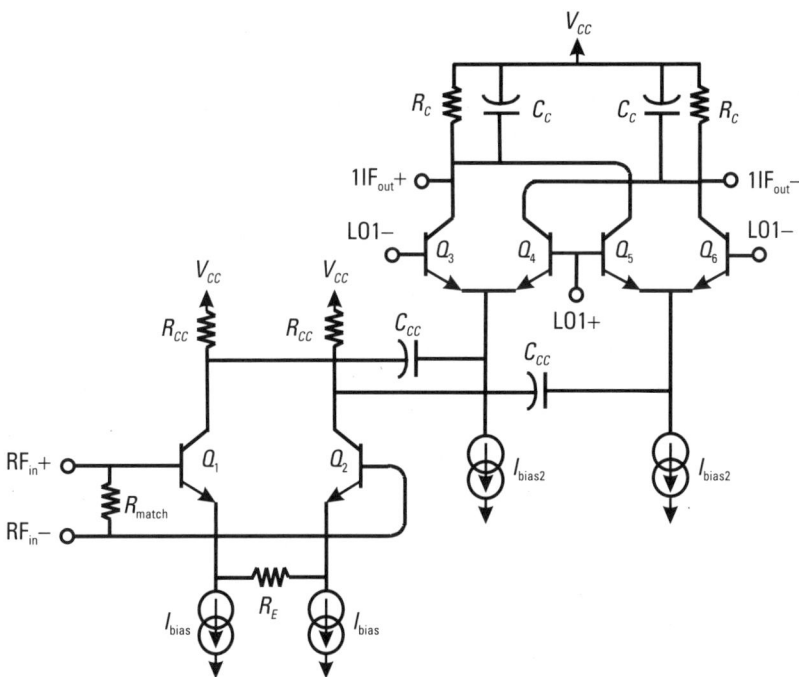

Figure 7.34 Mixer with folded switching stage and resistive input matching.

the quad nonlinearity for the folded cascode configuration is slightly more complex than the standard configuration because the current applied to it is no longer exactly equal to the current from the RF stage. We can start by assuming that each nonlinearity contributes equally to the linearity of the circuit (assuming that the circuit now has enough headroom that the output does not clip or saturate the quad). Then the input should be designed for 11-dBm IIP3 rather than 8-dBm.

The quad will be more linear as more current is passed through it because of the reduction in the emitter resistance r_e of the four transistors and the resulting reduction of the voltage swing at the emitters. Thus, as a first cut we will split the available current equally between both stages, allotting 6 mA to the driver and 6 mA to the quad. We can now start to size the degeneration resistor R_E. An IIP3 of 11 dBm at 100Ω corresponds to a signal swing of 1.12 V_{rms} at the input of the mixer or, equivalently, 0.561 V_{rms} per side for the differential circuit. Using (6.85) from the previous chapter, R_E can be determined to be

$$R_E = 2\left[r_e\left(\frac{v_{IP3}}{2v_T}\right)^{2/3} - r_e\right] = 2\left[8.3\Omega\left(\frac{561 \text{ mV}}{2 \cdot 25 \text{ mV}}\right)^{2/3} - 8.3\Omega\right] = 66.6\Omega$$

Since this formula is an approximation and we have one other nonlinearity to worry about, we will choose $R_E = 70\Omega$ for this design.

Next, we can find the load resistor, noting that we want 15 dB of voltage gain or 5.6 volts per volt (V/V). Using (7.20) (omitting output matching),

$$R_C = \frac{\pi}{2} A_v \left(r_e + \frac{R_E}{2} \right) = \frac{\pi}{2} 5.6(8.3\Omega + 35\Omega) = 380\Omega$$

Note that there will be losses due to the r_o of the quad transistors and some loss of current between the stages, but we will start with an R_C value of 400Ω. We also include capacitors C_C in parallel with the resistors R_C to filter out high-frequency signals coming out of the mixer. We will choose the filter to have a corner frequency of 100 MHz; therefore, the capacitors should be sized to be

$$C_C = \frac{1}{2\pi f_c R_C} = \frac{1}{2\pi(100 \text{ MHz})(400\Omega)} = 4 \text{ pF}$$

Now the coupling network needs to be designed. The current sources I_{bias} will need about 0.7V across them to work properly, and the differential pair should have roughly 1.5V to avoid nonlinearity. This leaves the resistors R_{cc} with about 0.8V; thus, a value between 200Ω and 400Ω would be appropriate for these resistors. We choose 300Ω.

The quad transistors will each have an r_e of 16.7Ω. This is less than 1/10th of the value of R_{cc}, and if they are placed in series with a 3-pF capacitor, they still have an impedance with a magnitude of 31.5Ω or about 1/10th that of R_{cc}. Thus, little current will be lost through the resistors R_{cc}.

The quad transistors themselves were sized so that when operated at 1.5 mA each, they were at the current for peak f_T. For minimum noise, the differential pair transistors were sized somewhat larger than the quad transistors. However, since the noise will be dominated by R_E, exact sizing for minimum noise was not critical.

The circuit also needs to be matched. Since inductors have not been allowed, we do this in a crude manner by placing a 100Ω resistor across the input.

Next, we can estimate the noise figure of this design. The biggest noise contributors will be R_E, R_{Match}, and the source resistance. The noise spectral density produced by both the matching resistor and the source $v_{n(source)}$ will be

$$v_{n(source)} = \sqrt{4kTR_{Match}} = 1.29 \frac{\text{nV}}{\sqrt{\text{Hz}}}$$

These two noise sources are voltage-divided at the input by the source and matching resistors. They will also see the same gain to the output. Thus, the output noise generated by each of these two noise sources $v_{\text{on(source)}}$ is

$$v_{\text{on(source)}} = \frac{v_{n\text{(source)}}}{2} A_v = \frac{1.29 \frac{\text{nV}}{\sqrt{\text{Hz}}}}{2} \cdot 5.6 \frac{\text{V}}{\text{V}} = 3.6 \frac{\text{nV}}{\sqrt{\text{Hz}}}$$

The other noise source of importance is R_E. It produces a current $i_{n(R_E)}$ of

$$i_{n(R_E)} = \sqrt{\frac{4kT}{R_E}} = 15.3 \frac{\text{pA}}{\sqrt{\text{Hz}}}$$

This current produces an output voltage $v_{\text{no}(R_E)}$ of

$$v_{\text{no}(R_E)} = \frac{2}{\pi} i_{n(R_E)} \cdot 2R_L = 7.81 \frac{\text{nV}}{\sqrt{\text{Hz}}}$$

Now the total output noise voltage $v_{\text{no(total)}}$ (assuming these are the only noise sources in the circuit) is

$$v_{\text{no(total)}} = \sqrt{(v_{\text{no}(R_E)})^2 + (v_{\text{no(source)}})^2 + (v_{\text{no\_source}})^2} = 9.32 \frac{\text{nV}}{\sqrt{\text{Hz}}}$$

Thus, the single-sideband noise figure can be calculated by

$$\text{NF} = 20 \log \left(\frac{v_{\text{no(total)}}}{\frac{v_{\text{on(source)}}}{\sqrt{2}}} \right) = 20 \log \left(\frac{9.32}{2.54} \right) = 11.3 \text{ dB}$$

Note that in the single-sideband noise figure, only the source noise from one sideband is considered; thus, we divided by $\sqrt{2}$.

Now the circuit is simulated. The results are summarized in Table 7.3. The voltage gain was simulated to be 13.6 dB, which is 1.4 dB lower than what was calculated. The main source of error in this calculation is ignoring current lost into R_{cc}. Since the impedance of R_{cc} is about ten times that of the path leading into the quad, it draws 1/10th of the total current causing a 1-dB loss in gain. Thus, in a second iteration, R_C could be raised to a higher

Table 7.3
Results of the Simulation of the Mixer Circuit

Parameter	Value
Gain	13.6 dB
(SSB) NF	12.9 dB
IIP3	8.1 dBm
Voltage	3.3V
Current	12 mA
LO frequency	1.95 GHz
RF	2 GHz
IF	50 MHz

value. The noise figure was also simulated and found to be 12.9 dB. This is close to what was calculated. Most noise came either from R_E or from both the source and the input-matching resistor. A more refined calculation taking more noise sources into account would have made the calculation agree much closer with simulation. To determine the IIP3, the LO was set to be 400 mV$_{pp}$ at 1.95 GHz, and two RF signals were injected at 2.0 and 2.001 GHz. The *fast Fourier transform* (FFT) of the output voltage is plotted in Figure 7.35. From this figure, using a method identical to that used in the broadband LNA example in Section 6.8, it can be found that the IIP3 is 8.1 dBm. Thus, simulations are in good agreement with the calculations.

Example 7.6 Image Reject Mixer

Take the balanced mixer cell designed in the last example and use it to construct an image reject mixer as shown in Figure 7.22. Place a simple lowpass-highpass phase shifter in the LO path. Place the second phase shifter in the IF path and

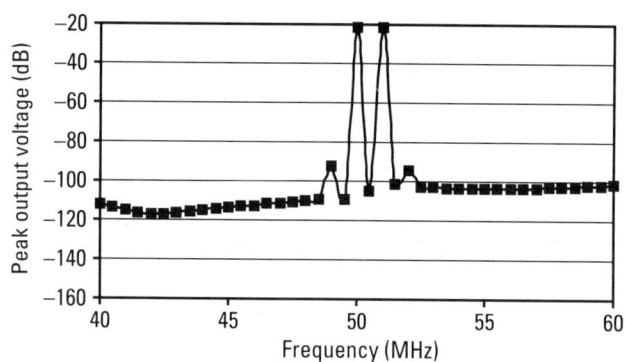

Figure 7.35 FFT of a transient simulation with two input tones used to find the IIP3.

make this one a second-order polyphase filter. Compare the design to one using only a first-order polyphase filter. Design the mixer so that it is able to drive 100-Ω output impedance. Explore the achievable image rejection over process tolerances of 20%.

Solution

For the LO path there is little additional design work to be done. For this example, we will ignore the square wave buffers that would normally be used to guarantee that the mixer is driven properly. We add a simple phase-shifting filter to the circuit like the one shown in Figure 7.25 to provide quadrature LO signals. Since this filter must be centered at 1.95 GHz, we choose $R = 300\Omega$ fairly arbitrarily, and this makes the capacitors 272 fF. Both of these are easily implemented in most technologies.

Next we must design the IF stage that will follow the mixers using the polyphase filter to achieve the second phase shifter. In order to prevent loading of the mixers by the polyphase filter, we need buffers at the input and we will need buffers at the output to drive the 100Ω load impedance. A polyphase filter with buffers is shown in Figure 7.36. Note that this circuit implements both IF paths as well as the summing-component shown in Figure 7.22. The polyphase filter components must be sized so that the impedance is large to minimize buffer current. However, if the impedance is made too large, then it

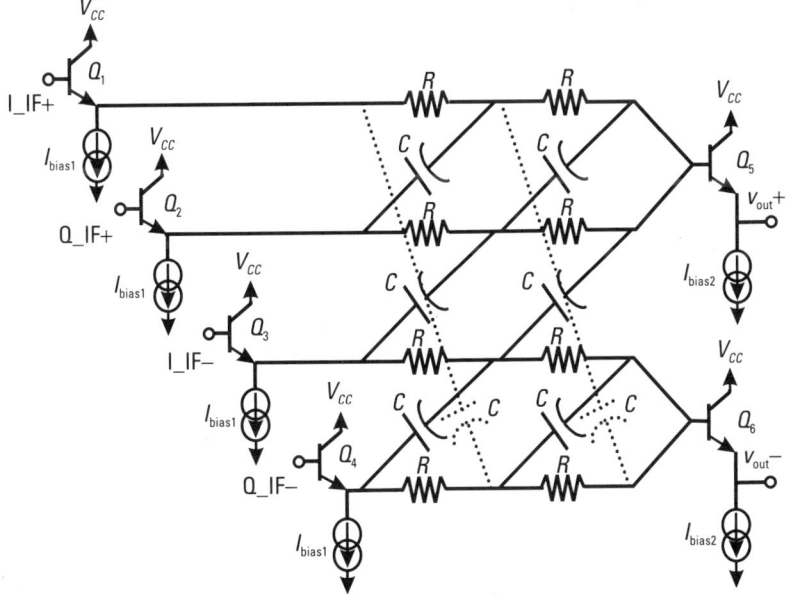

Figure 7.36 IF stage of an image reject mixer.

will form a voltage divider with the output stage, resulting in a loss of gain. Thus, we choose through trial and error a resistance of $R = 2\text{k}\Omega$, and this will make the capacitors $C = 1.6$ pF (centered at 50 MHz in this case).

The mixer of the previous example had an IIP3 of 8.1 dBm. As there are now two mixers, we can expect this system to have an IIP3 of more like 5 dBm. This means that it will have a 1-dB compression point of −5 dBm. At this power level, the input will have a peak voltage swing of 250 mV. With a gain of 13.6 dB or 4.8 V/V, this means that the buffers will have to swing 1.2V peak. If we assume that they drive a series combination of 2 kΩ and 1.6 pF, then the total impedance will be about 2.8 kΩ. This means that the transistor needs to accommodate an ac current of 429 μA. Thus, a bias current of 750 μA for this stage should be safe.

If we assume now that the polyphase filter has a loss of 3 dB per stage, then the voltage gain from input to output will drop to 7.6 dB or 2.4 V/V. Thus, the output voltage will be 600-mV peak. Into 100Ω, this will be a current of 6 mA. This large value demonstrates how hard it is to drive low impedances with high-linearity systems. We will start with a current of 5 mA in each transistor and refine this number as needed.

The circuit was then simulated. The basic circuit parameters are shown in Table 7.4. The gain and IIP3 have dropped as expected. The noise figure has also risen due to reduced gain, but not too much, as now the noise due to the input has been image rejected as well.

The components in the filters were then adjusted to show the effect of circuit tolerance on the image rejection. Table 7.5 shows how the LO phase shifter affects image rejection. Note that this port is very insensitive to amplitude changes, which is why the highpass-lowpass filter was chosen for the 90° phase

Table 7.4
Results of the Simulation of the Image Reject Mixer Circuit

Parameter	Value
Gain	7.4 dB
NF	16.3 dB
IIP3	6.9 dBm
Voltage	3.3V
Current	37 mA
Image rejection	69 dB
RF	2 GHz
LO frequency	1.95 GHz
IF	50 MHz
Image frequency	1.9 GHz

Table 7.5
Image Rejection for LO Phase Shifter

Tolerance Level for Resistance and Capacitance	Image Rejection
±20%	>20 dB
±10%	>27.4 dB
Nominal	69 dB

shift. It still provides image rejection of 20 dB even at 20% tolerance in the values.

Table 7.6 shows that the polyphase filter with two stages also does an excellent job at keeping the image suppressed, so this was a good choice for the IF filter. If this filter is reduced to a first order as shown in Table 7.7, then the image rejection suffers greatly. Thus, a second-order filter is required in this case.

Example 7.7 Image Reject Mixer with Improved Gain

The gain of the image reject mixer has been reduced by 6 dB due to the presence of the IF polyphase filter. Modify it to get the 6 dB of gain back.

Table 7.6
Image Rejection for IF Phase Shifter

Tolerance Level for Resistance and Capacitance	Image Rejection
±20%	>25 dB
±10%	>36.7 dB
Nominal	69 dB

Table 7.7
Image Rejection for IF Phase Shifter (First Order)

Tolerance Level for Resistance and Capacitance	Image Rejection
±20%	>12.5 dB
±10%	>18.2 dB
±5%	>23.2 dB
Nominal	35 dB

Solution

With the current flowing in the quad stage of the mixer, it would be impossible, due to headroom constraints, to raise the resistance, so we must now employ the PMOS current steering technique shown in Figure 7.33. The PMOS will now make up the capacitor that was placed in the tank to remove high-frequency feed-through of RF and LO signals. The PMOS must be made large to ensure that they are not noisy and that they have a low saturation voltage. A device with a length of 2 μm and a width of 800 μm was chosen through simulation. No current source was needed in this case, as the voltage levels seemed to be fine without it.

The resistors were then doubled to 800Ω to restore the gain of the circuit. Also, the buffers are all doubled in current because they will now have to handle signals that are twice as large. The results of this new SSB mixer are shown in Table 7.8.

Note that the NF has dropped due to the increased gain. The linearity has been degraded slightly due to the additional nonlinearity of the output resistance of the PMOS transistors.

One more improvement can be made to this circuit. The mixer can be put into a Moore configuration to reduce the effect of R_E on the noise figure. When this was done, the noise due to R_E reduced to about half its previous value, but because it was responsible for only a small percentage of the total noise, the new noise figure was lowered by only 0.5 to 13.0 dB. This is not a dramatic improvement, but as it comes at no additional cost, it is worthwhile. If the gain of this mixer were increased further, then the importance of R_E on the noise figure would increase and a greater improvement would be seen.

7.13 CMOS Mixers

Most of the circuits, techniques, and analyses used for bipolar mixers can also be used for CMOS mixers. For example, one can realize single-balanced and double-balanced CMOS mixers as shown in Figure 7.37.

Table 7.8
Results of the Image Reject Mixer with PMOS Current Steering Transistors

Parameter	Value
Gain	13.6 dB
NF	13.5 dB
IIP3	5.7 dBm
Voltage	3.3V
Current	50 mA
Image rejection	69 dB

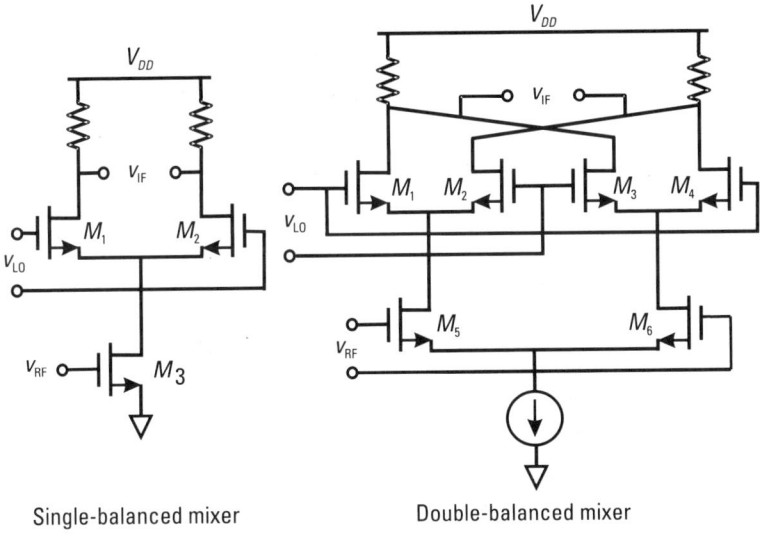

Figure 7.37 Single-balanced and double-balanced CMOS mixers.

Compared to bipolar, for the MOS mixer, the LO voltage is typically required to be larger to ensure there is complete switching of the quad network. In order to minimize the amount of extra LO voltage, the switching transistors usually have a large W/L in order to switch with minimal overdrive. (Here overdrive refers to $V_{GS} - V_T$). For the RF port, one can design with a larger overdrive in order to linearize the input. However, this will reduce the transconductance and hence will reduce the gain and increase the noise figure.

Another opportunity with CMOS is to replace an NMOS differential pair with a PMOS differential pair in the RF input and the quad network, which allows them to be stacked and the current to be reused as shown in Figure 7.38 [4]. In such a case, the output is potentially a high-gain node, so some form of common mode feedback is required for this circuit.

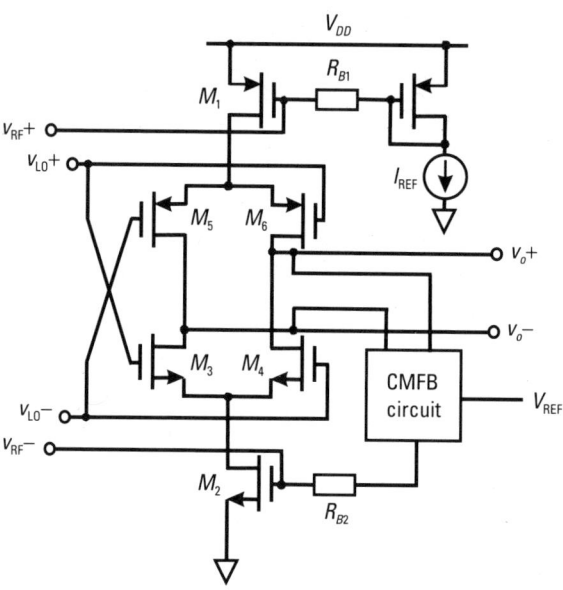

Figure 7.38 CMOS mixer with NMOS and PMOS differential pairs.

References

[1] Gingell, M. J., "Single Sideband Modulation Using Sequence Asymmetric Polyphase Networks," *Electrical Communications*, Vol. 48, 1973, pp. 21–25.

[2] Long, J. R., "A Low-Voltage 5.1–5.8GHz Image-Reject Downconverter RFIC," *IEEE J. Solid-State Circuits*, Vol. 35, Sept. 2000, pp. 1320–1328.

[3] Voinigescu, S. P., and M. C. Maliepaard, "5.8GHz and 12.6GHz Si Bipolar MMICs," *Proc. ISSCC*, 1997, pp. 372, 373.

[4] Karanicolas, A. N., "A 2.7-V 900-MHz CMOS LNA and Mixer," *IEEE J. Solid-State Circuits*, Vol. 31, Dec. 1996, pp. 1939–1944.

Selected Bibliography

Larson, L. E., (ed.), *RF and Microwave Circuit Design for Wireless Communications*, 2nd ed., Norwood, MA: Artech House, 1997.

Maas, S. A., *Microwave Mixers*, 2nd ed., Norwood, MA: Artech House, 1993.

Rudell, J. C., et al., "A 1.9-GHz Wide-Band IF Double Conversion CMOS Receiver for Cordless Telephone Applications," *IEEE J. Solid-State Circuits*, Vol. 32, Dec. 1997, pp. 2071–2088.

8
Voltage-Controlled Oscillators

8.1 Introduction

An oscillator is a circuit that generates a periodic waveform whether it be sinusoidal, square, triangular as shown in Figure 8.1, or, more likely, some distorted combination of all three. Oscillators are used in a number of applications in which a reference tone is required. For instance, they can be used as the clock for digital circuits or as the source of the LO signal in transmitters. In receivers, oscillator waveforms are used as the reference frequency to mix down the received RF to an IF or to baseband. In most RF applications, sinusoidal references with a high degree of spectral purity (low phase noise) are required. Thus, this chapter will focus on LC-based oscillators, as they are the most prominent form of oscillator used in RF applications.

In this chapter, we will first look at some general oscillator properties and then examine the resonator as a fundamental building block of the oscillator. Different types of oscillators will then be examined, but most emphasis will be on the Colpitts oscillator and the negative transconductance oscillator. Both single-ended and double-ended designs will be considered. This chapter will also include discussions of the theoretical calculations of the amplitude of oscillation and the phase noise. Finally, there will be a section on automatic amplitude control circuitry for oscillators.

8.2 Specification of Oscillator Properties

Perhaps the most important characteristic of an oscillator is its phase noise. In other words, we desire accurate periodicity with all signal power concentrated

Figure 8.1 Example of periodic waveforms.

in one discrete oscillator frequency and possibly at multiples of the oscillator frequency. A signal with power at only one discrete frequency would correspond to an impulse function if plotted in the frequency domain. However, all real oscillators have less than perfect spectral purity and thus they develop "skirts" as shown in Figure 8.2. These skirts are undesirable, and we would like to minimize them as much as possible. Power in the skirts is evidence of phase noise, which has resulted in oscillator power bands around the intended discrete spectral lines. Phase noise is any noise that changes the frequency or phase of the oscillator waveform. Phase noise is given by

$$\mathrm{PN} = \frac{P_o}{N_o} \quad (8.1)$$

where P_o is the power in the tone at the frequency of oscillation and N_o is the noise power spectral density at some specified offset from the carrier. Phase noise is usually specified in dBc/Hz, meaning noise in a 1-Hz bandwidth measured in decibels with respect to the carrier.

Since oscillators are designed to run at particular frequencies of interest, long-term stability is of concern, especially in products that are expected to function for many years. Thus, we would like to have minimum drift of oscillation frequency due to such things as aging or power supply variations. In addition, oscillators must produce sufficient output voltage amplitude for the intended application. For instance, if the oscillator is used to drive the LO switching transistors in a double-balanced mixer cell, then the voltage swing must be large enough to switch the mixer.

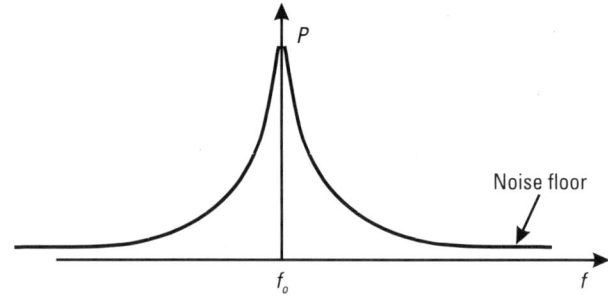

Figure 8.2 Spectrum of a typical oscillator.

8.3 The LC Resonator

At the core of almost all integrated RF oscillators is an LC resonator that determines the frequency of oscillation and often forms part of the feedback mechanism used to obtain sustained oscillations. Thus, the analysis of an oscillator begins with the analysis of a damped LC resonator such as the parallel resonator shown in Figure 8.3.

Since there are two reactive components, this is a second-order system, which can exhibit oscillatory behavior if the losses are low or if positive feedback is added. It is useful to find the system's response to an impulse of current, which in a real system could represent noise. If $i(t) = I_{\text{pulse}} \delta(t)$ is applied to the parallel resonator, the time domain response of the system can be found as

$$v_{\text{out}}(t) = \frac{\sqrt{2} I_{\text{pulse}} e^{\frac{-t}{2RC}}}{C} \cos\left(\sqrt{\left(\frac{1}{LC} - \frac{1}{4R^2 C^2}\right)} \cdot t\right) \quad (8.2)$$

From this equation, it is easy to see that this system's response is a sinusoid with exponential decay whose amplitude is inversely proportional to the value of the capacitance of the resonator and whose frequency is given by

$$\omega_{\text{osc}} = \sqrt{\frac{1}{LC} - \frac{1}{4R^2 C^2}} \quad (8.3)$$

which shows that as $|R|$ decreases, the frequency decreases. However, if $|R| \gg \sqrt{L/C}$, as is the case in most RFIC oscillators, even during startup, this effect can be ignored. Also note that once steady state has been reached in a real oscillator, R approaches infinity and the oscillating frequency will approach

$$\omega_{\text{osc}} = \sqrt{\frac{1}{LC}} \quad (8.4)$$

The resulting waveform is shown in Figure 8.4. To form an oscillator, however, the effect of damping must be eliminated in order for the waveform to persist.

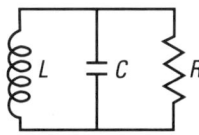

Figure 8.3 Parallel LC resonator.

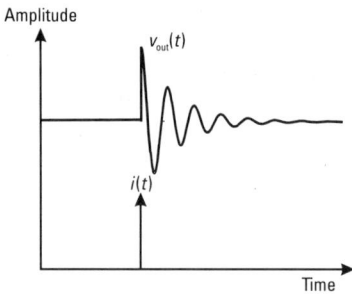

Figure 8.4 Damped LC resonator with current step applied.

8.4 Adding Negative Resistance Through Feedback to the Resonator

The resonator is only part of an oscillator. As can be seen from Figure 8.4, in any practical circuit, oscillations will die away unless feedback is added in order to sustain the oscillation. A feedback loop can be designed to generate a negative resistance as shown conceptually in Figure 8.5. If this parallel negative resistance [Figure 8.5(a)] is smaller than the positive parallel resistance in the circuit, then any noise will start an oscillation whose amplitude will grow with time. Similarly, in Figure 8.5(b), if the negative series resistance is larger than the positive resistive losses, then this circuit will also start to oscillate.

The oscillator can be seen as a linear feedback system, as shown in Figure 8.6. The oscillator is broken into two parts, which together describe the oscillator and the resonator.

At the input, the resonator is disturbed by an impulse which represents a broadband noise stimulus that starts up the oscillator. The impulse input results in an output that is detected by the amplifier. If the phase shift of the loop is correct and the gain around the loop is such that the pulse that the amplifier produces is equal in magnitude to the original pulse, then the pulse acts to maintain the oscillation amplitude with each cycle. This is a description of the Barkhausen criteria, which will now be described mathematically.

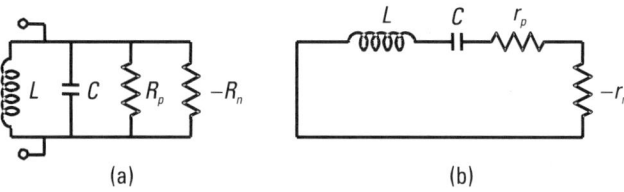

Figure 8.5 The addition of negative resistance to the circuit to overcome losses in (a) a parallel resonator and (b) a series resonator.

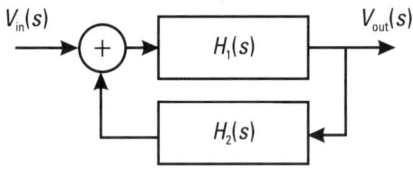

Figure 8.6 Linear model of an oscillator as a feedback control system.

The gain of the system in Figure 8.6 is given by

$$\frac{V_{\text{out}}(s)}{V_{\text{in}}(s)} = \frac{H_1(s)}{1 - H_1(s)H_2(s)} \tag{8.5}$$

We can see from the equation that if the denominator approaches zero, with finite $H_1(s)$, then the gain approaches infinity and we can get a large output voltage for an infinitesimally small input voltage. This is the condition for oscillation. By solving for this condition, we can determine the frequency of oscillation and the required gain to result in oscillation.

More formally, the system poles are defined by the denominator of (8.5). To find the poles of the closed-loop system, one can equate this expression to zero, as in

$$1 - H_1(s)H_2(s) = 0 \tag{8.6}$$

For sustained oscillation at constant amplitude, the poles must be on the $j\omega$ axis. To achieve this, we replace s with $j\omega$ and set the equation equal to zero.

For the open-loop analysis, rewrite the above expression as

$$H_1(j\omega)H_2(j\omega) = 1 \tag{8.7}$$

Since in general $H_1(j\omega)$ and $H_2(j\omega)$ are complex, this means that

$$|H_1(j\omega)||H_2(j\omega)| = 1 \tag{8.8}$$

and that

$$\angle H_1(j\omega)H_2(j\omega) = 2n\pi \tag{8.9}$$

where n is a positive integer.

These conditions for oscillation are known as the Barkhausen criterion, which states that for sustained oscillation at constant amplitude, the gain around the loop is 1 and the phase around the loop is 0 or some multiple of 2π. We note that $H_1 H_2$ is simply the product of all blocks around the loop and so can be seen as open-loop gain. Also, it can be noted that, in principle, it does not matter where one breaks the loop or which part is thought of as the feedback gain or which part is forward gain. For this reason, we have not specified what circuit components constitute H_1 and H_2, and, in fact, many different possibilities exist.

8.5 Popular Implementations of Feedback to the Resonator

Feedback (or negative resistance) is usually provided in one of three ways, as shown in Figure 8.7. (Note that other choices are possible.)

1. Using a tapped capacitor and amplifier to form a feedback loop. This is known as a Colpitts oscillator.
2. Using a tapped inductor and amplifier to form a feedback loop. This is known as a Hartley oscillator. Note this form of oscillator is not common in IC implementations.
3. Using two amplifiers (typically two transistors) in a positive feedback configuration. This is commonly known as the $-G_m$ oscillator.

According to the simple theory developed so far, if the overall resistance is negative, then the oscillation amplitude will continue to grow indefinitely. In a practical circuit, this is, of course, not possible. Current limiting, the power supply rails, or some nonlinearity in the device eventually limits the magnitude of the oscillation to some finite value, as shown in Figure 8.8. This reduces the

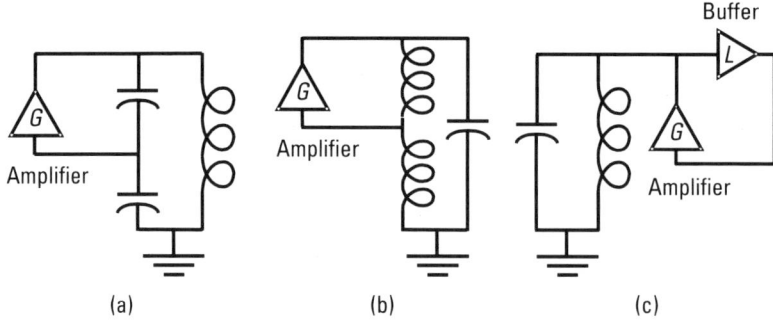

Figure 8.7 Resonators with feedback: (a) Colpitts oscillator; (b) Hartley oscillator; (c) $-G_m$ oscillator (biasing not shown).

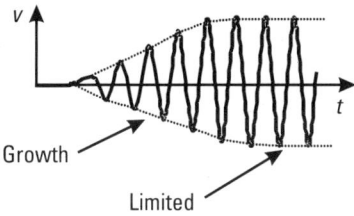

Figure 8.8 Waveform of an LC resonator with losses compensated. The oscillation grows until a practical constraint limits the amplitude.

effect of the negative resistance in the circuit until the losses are just canceled, which is equivalent to reducing the loop gain to 1.

8.6 Configuration of the Amplifier (Colpitts or $-G_m$)

The amplifier shown in Figure 8.7 is usually made using only one transistor in RF oscillators. The $-G_m$ oscillator (Figure 8.9) can be thought of as having either a common-collector amplifier made up of Q_2, where Q_1 forms the feedback, or a common-base amplifier consisting of Q_1, where Q_2 forms the feedback. Figure 8.9 may look a little unusual because the $-G_m$ oscillator is usually seen only in a differential form, in which case the two transistors are connected as a differential pair. The circuit is symmetrical when it is made differential (more on this in Section 8.10). However, the Colpitts and Hartley oscillators, each having only one transistor, can be made either common base or common collector. The common-emitter configuration is usually unsuitable because it requires large capacitors and RF chokes that are not usually available in a typical IC technology. The common-emitter configuration also suffers from the Miller effect because neither the collector nor the base is grounded. The

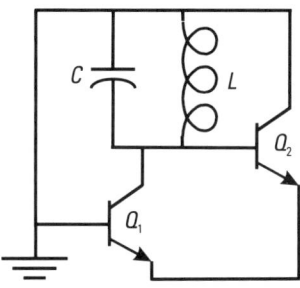

Figure 8.9 $-G_m$ oscillator (biasing not shown).

two favored choices (common base and common collector) are shown in Figure 8.10 as they would appear in the Colpitts oscillator.

8.7 Analysis of an Oscillator as a Feedback System

It can be instructive to apply the model of Figure 8.6 to the oscillator circuits discussed above. Expressions for H_1 and H_2 can be found and used in either an open-loop analysis or a closed-loop analysis. For the closed-loop analysis, the system's equations can also be determined, and then the poles of the system can be found. This is the approach we will take first. Later we will demonstrate the open-loop analysis technique. All of these techniques give us two basic pieces of information about the oscillator in question: (1) it allows us to determine the frequency of oscillation, and (2) it tells us the amount of gain required to start the oscillation.

8.7.1 Oscillator Closed-Loop Analysis

In this section, the common-base configuration of the Colpitts oscillator as shown in Figure 8.10 will be considered. The small-signal model of the oscillator is shown in Figure 8.11.

We start by writing down the closed-loop system equations by summing the currents at the collector (node v_c) and at the emitter of the transistor (node v_e). At the collector,

$$v_c \left(\frac{1}{R_p} + \frac{1}{sL} + sC_1 \right) - v_e (sC_1 + g_m) = 0 \qquad (8.10)$$

At the emitter we have

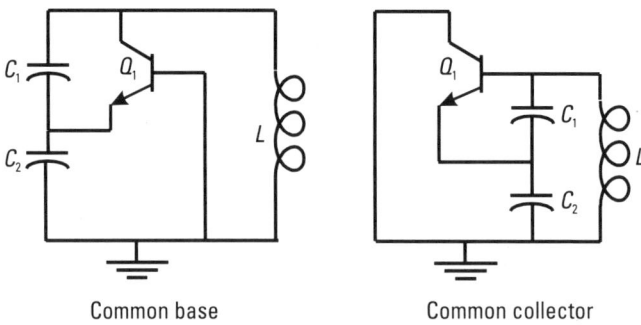

Common base   Common collector

Figure 8.10 Common-base and common-collector Colpitts oscillators (biasing not shown).

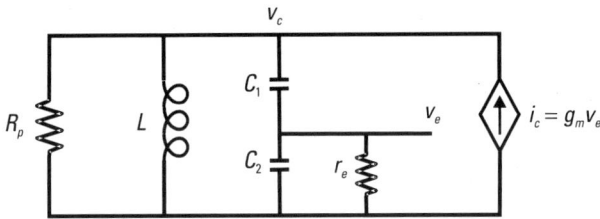

Figure 8.11 Closed-loop oscillator small-signal mode.

$$v_e \left(sC_1 + sC_2 + \frac{1}{r_e} \right) - v_c sC_1 = 0 \tag{8.11}$$

This can be solved in several ways; however, we will write it as a matrix expression:

$$[Y][v] = 0 \tag{8.12}$$

as in the following equation:

$$\begin{bmatrix} \frac{1}{R_p} + \frac{1}{sL} + sC_1 & -sC_1 - g_m \\ -sC_1 & sC_1 + sC_2 + \frac{1}{r_e} \end{bmatrix} \begin{bmatrix} v_c \\ v_e \end{bmatrix} = \begin{bmatrix} 0 \\ 0 \end{bmatrix} \tag{8.13}$$

The poles will be formed by the determinant of the matrix. To find the conditions for oscillation, we can set the determinant to zero and solve. The result is

$$\left(\frac{1}{R_p} + \frac{1}{sL} + sC_1 \right) \left(sC_1 + sC_2 + \frac{1}{r_e} \right) - sC_1 (sC_1 + g_m) = 0 \tag{8.14}$$

After multiplying out and collecting like terms, this results in

$$s^3 L C_1 C_2 + s^2 \left(\frac{L(C_1 + C_2)}{R_p} + \frac{LC_1}{r_e} - LC_1 g_m \right)$$

$$+ s \left(\frac{L}{R_p r_e} + C_1 + C_2 \right) + \frac{1}{r_e} = 0 \tag{8.15}$$

When the substitution is made that $s = j\omega$, even-order terms (s^2 and constant term) will be real, and odd-order terms (s^3 and s) will have a $j\omega$ in them. Thus, when the even-order terms are summed to zero, the result will be an expression for gain. When odd-order terms are summed to zero, the result will be an expression for the frequency. The result for the odd-order terms is

$$\omega = \sqrt{\left(\frac{C_1 + C_2}{C_1 C_2}\right)\frac{1}{L} + \frac{1}{r_e R_p C_1 C_2 L}} \qquad (8.16)$$

The first term can be seen to be ω_o, the resonant frequency of the resonator by itself. The second term can be simplified by noting that ω_o is determined by L resonating with the series combination of C_1 and C_2, as well as by noting that the Q of an inductor in parallel with a resistor is given by

$$Q_L = \frac{R_p}{\omega L} \qquad (8.17)$$

Then,

$$\omega = \sqrt{\omega_o^2 + \frac{\omega_o^2 L}{r_e R_p (C_1 + C_2)}}$$

$$= \omega_o \sqrt{1 + \frac{\omega L}{R_p} \cdot \frac{1}{\omega r_e (C_1 + C_2)}} \qquad (8.18)$$

$$= \omega_o \sqrt{1 + \frac{1}{Q_L \omega/\omega_c}}$$

where ω_c is the corner frequency of the highpass filter formed by the capacitive feedback divider.

Thus, if the inductor Q is high or if the operating frequency is well above the feedback corner frequency, then the oscillating frequency is given by ω_o. Otherwise, the frequency is increased by the amount shown. This effect will be revisited in Section 8.7.2 and Example 8.2.

The result for the even-order term in (8.15) is

$$g_m = \frac{\omega(C_1 + C_2)}{Q_L} \qquad (8.19)$$

Note that the approximation has been made that $r_e = 1/g_m$. Thus, this equation tells us what value of g_m (and corresponding value of r_e) will result in sustained oscillation at a constant amplitude. For a real oscillator, to overcome any additional losses not properly modeled and to guarantee startup and sustained oscillation at some nonzero amplitude, the g_m would have to be made larger than this value. How much excess g_m is used will affect the amplitude of oscillation. This is discussed further in Section 8.16.

8.7.2 Capacitor Ratios with Colpitts Oscillators

In this section, the role of the capacitive divider as it affects frequency of oscillation and feedback gain will be explored. It will be seen that this capacitor divider is responsible for isolating the loading of r_e on the resonant circuit and produces the frequency shift as mentioned above. The resonator circuit including the capacitive feedback divider is shown in Figure 8.12.

The capacitive feedback divider is made up of C_1, C_2, and r_e, and has the transfer function

$$\frac{v'_e}{v_c} = \frac{j\omega r_e C_1}{1 + j\omega r_e (C_1 + C_2)} = \left(\frac{C_1}{C_1 + C_2}\right)\left(\frac{j\frac{\omega}{\omega_c}}{1 + j\frac{\omega}{\omega_c}}\right) \quad (8.20)$$

This is a highpass filter with gain and phase as shown in Figure 8.13.

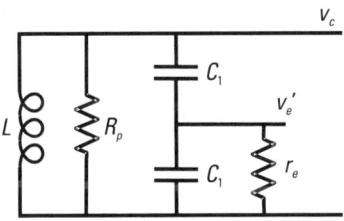

Figure 8.12 Z_{tank} using transformation of capacitive feedback divider.

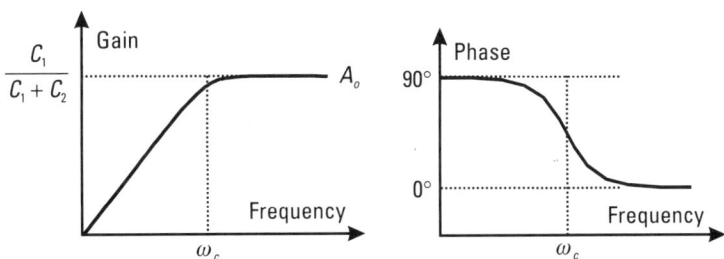

Figure 8.13 Plot of capacitive feedback frequency response.

The passband gain A_o is given by

$$A_o = \frac{C_1}{C_1 + C_2} \tag{8.21}$$

The corner frequency ω_c is given by

$$\omega_c = \frac{1}{r_e(C_1 + C_2)} \tag{8.22}$$

and the phase shift of the feedback network is

$$\phi = \frac{\pi}{2} - \tan^{-1}\left(\frac{\omega}{\omega_c}\right) \tag{8.23}$$

If the frequency of operation is well above the corner frequency ω_c, the gain is given by the capacitor ratio in (8.21) and the phase shift is zero. Under these conditions, the circuit can be simplified as described in the following paragraph. If this frequency condition is not met, there will be implications, which will be discussed later.

This high-pass filter also loads the resonator with r_e (the dynamic emitter resistance) of the transistor used in the feedback path. Fortunately, this resistance is transformed to a higher value through the capacitor divider ratio. This impedance transformation effectively prevents this typically low impedance from reducing the Q of the oscillator's LC resonator. The impedance transformation is discussed in Chapter 4 and is given by

$$r_{e,\text{tank}} = \left(1 + \frac{C_2}{C_1}\right)^2 r_e \tag{8.24}$$

for the Colpitts common-base oscillator, and

$$r_{e,\text{tank}} = \left(1 + \frac{C_1}{C_2}\right)^2 r_e \tag{8.25}$$

for the common-collector oscillator. The resulting transformed circuit as seen by the resonator is shown in Figure 8.14.

Therefore, in order to get the maximum effect of the impedance transformation, it is necessary to make C_2 large and C_1 small in the case of the common-base circuit and vice versa for the common-collector circuit. However, one must

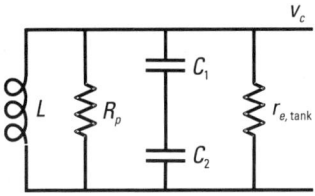

Figure 8.14 Z_{tank} using transformation of capacitive feedback divider.

keep in mind that the equivalent series capacitor nominally sets the resonance frequency according to

$$\omega = \frac{1}{\sqrt{LC_T}} = \sqrt{\frac{C_1 + C_2}{LC_1 C_2}} \qquad (8.26)$$

Example 8.1 Capacitor Ratio
A common-base Colpitts oscillator with a resonance at 1.125 GHz using an on-chip inductor is required. Explore the role of the capacitor ratio on the emitter resistance transformation, assuming that the largest available capacitor is 10 pF and the largest available inductor is 10 nH. Assume that the current through the transistor is set at 1 mA.

Solution
Resonance frequency is given by

$$\omega = \frac{1}{\sqrt{LC_T}} = \sqrt{\frac{C_1 + C_2}{LC_1 C_2}}$$

and $r_{e,\text{tank}}$ is given by

$$r_{e,\text{tank}} = \left(1 + \frac{C_2}{C_1}\right)^2 r_e$$

Large $r_{e,\text{tank}}$ is desired to reduce loss and minimize noise. To achieve this, it is advantageous for C_2 to be bigger than C_1. However, there will be a practical limit to the component values realizable on an integrated circuit. With 10 pF and 10 nH as the upper limits for capacitors and inductors on chip, Table 8.1 shows some of the possible combinations of L, C_1, and C_2 to achieve a frequency of 1.125 GHz.

Here it can be seen that a transformation of 25 is about as high as is possible at this frequency and with the specified component limits as shown in

Table 8.1
Inductor and Capacitor Values to Realize Oscillator at 1.125 GHz

L (nH)	C_T (pF)	C_1 (pF)	C_2 (pF)	Res. Freq. (GHz)	Impedance Transformation	$r_{e,tank}$
10	2	2.5	10	1.125	25	625
8	2.5	3.333	10	1.125	16	400
8	2.5	3.75	7.5	1.125	9	225
6.667	3	4.5	9	1.125	9	225

the first row. Note that L and C_2 are still on the high side, indicating that designing an oscillator at this frequency with an impedance transformation of 25 is quite challenging. If the transformation can be reduced to 16 or 9, then a number of other choices are possible, as shown in the table. Note that at 1 mA, the emitter resistance is about 25Ω. Multiplying by 9 or 25 results in 225Ω or 625Ω. For a typical 10-nH inductor at 1.125 GHz, the equivalent parallel resistance might be 300Ω for a Q of 4.243. Even with this low inductor Q, $r_{e,tank}$ degrades the Q significantly. In the best case with a transformation of 25, the Q is reduced to less than 3.

Example 8.2 Frequency Shift

If an oscillator is designed as in Example 8.1 with a 10-nH inductor with Q of 4.243, and it is operated at 2.21 times the corner frequency of the high-pass feedback network, then what is the expected frequency shift?

Solution
By (8.18),

$$\omega_{osc} = \omega_0 \sqrt{1 + \frac{1}{4.243 \cdot 2.21}} = 1.05\omega_0$$

so the frequency will be high by about 5%. This effect, due to the phase shift in the feedback path, is quite small and in practice can usually be neglected compared to the downward frequency shift due to parasitics and nonlinearities.

8.7.3 Oscillator Open-Loop Analysis

For this analysis, we redraw the Colpitts common-base oscillator with the loop broken at the emitter, as shown in Figure 8.15. Conceptually, one can imagine applying a small-signal voltage at v_e and measuring the loop gain at v'_e. Since

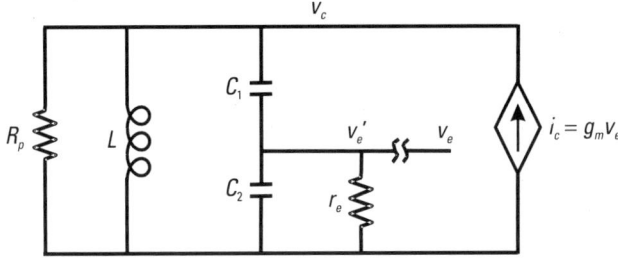

Figure 8.15 Feedback analysis of a Colpitts common-base oscillator.

v_c is the output of the oscillator, we can define forward gain as v_c/v_e and feedback gain as v_e'/v_c.

Forward Gain

The forward gain is

$$H_1(s) = \frac{v_c}{v_e} = g_m Z_{tank} \quad (8.27)$$

where Z_{tank} is defined in Figure 8.16 and has the following transfer function:

$$Z_{tank} = Z_L \| R_p \| Z_{FB} = \frac{1}{\frac{1}{j\omega L} + \frac{1}{R_p} + \frac{j\omega C_1 (1 + j\omega r_e C_2)}{1 + j\omega r_e (C_1 + C_2)}} \quad (8.28)$$

$$Z_{tank} = \frac{j\omega L R_p [1 + j\omega r_e (C_1 + C_2)]}{(R_p + j\omega L)[1 + j\omega r_e (C_1 + C_2)] + j\omega L R_p \cdot j\omega C_1 (1 + j\omega r_e C_2)} \quad (8.29)$$

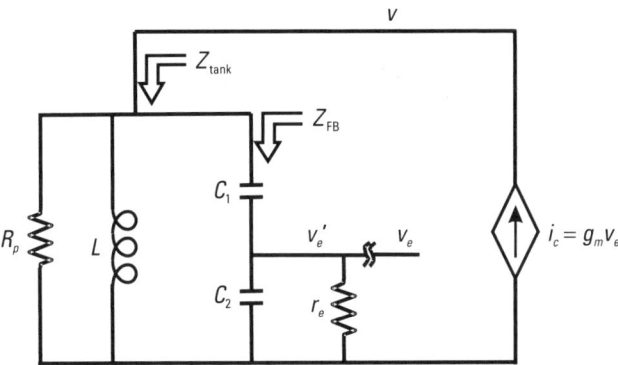

Figure 8.16 Definition of Z_{tank} and Z_{FB} in oscillator small-signal model.

Feedback Gain

The feedback circuit is just a highpass filter as described in the previous section and has the following transfer function:

$$H_2(j\omega) = \frac{v'_e}{v_c} = \frac{j\omega r_e C_1}{1 + j\omega r_e (C_1 + C_2)} \tag{8.30}$$

Loop Gain Expression

This can be solved as follows:

$$A = H_1 H_2 \tag{8.31}$$

$$= \frac{g_m \cdot j\omega r_e C_1 \cdot j\omega L R_p}{(R_p + j\omega L)[1 + j\omega r_e (C_1 + C_2)] + j\omega L R_p \cdot j\omega C_1 (1 + j\omega r_e C_2)}$$

Gathering terms:

$$A = H_1 H_2 = \frac{-\omega^2 \cdot g_m r_e C_1 L R_p}{B} \tag{8.32}$$

where $B = (-j\omega^3) r_e C_1 C_2 L R_p + (-\omega^2)[r_e(C_1 + C_2)L + R_p C_1 L] + j\omega[r_e R_p(C_1 + C_2) + L] + R_p$.

To determine oscillating conditions, (8.32) can be set equal to 1, and then the real and imaginary terms can be solved independently. The real part, which includes the even-order terms, is set equal to 1, and this sets the condition for gain. The result can be shown to be the same as for the closed-loop analysis, as done previously with final gain expression given by

$$g_m = \frac{\omega(C_1 + C_2)}{Q_L} \tag{8.33}$$

The imaginary part, which is defined by the odd-order terms, is also set equal to zero. This is equivalent to setting the phase equal to zero. The result, again, is equal to the previous derivation, with the result

$$\omega = \omega_o \sqrt{1 + \frac{1}{Q_L \omega/\omega_c}} \tag{8.34}$$

8.7.4 Simplified Loop Gain Estimates

To gain understanding, to explain this simple result, and to provide advice on how to do the design, in this section appropriate simplifications and approximations will be made by making use of the results shown in Section 8.7.2. As in the previous section, two expressions are written: one for the feedforward gain and one for the feedback gain.

If we assume we are operating above the capacitive feedback highpass corner frequency, then the feedback gain is given by

$$\frac{v_e}{v_c} = \frac{C_1}{C_1 + C_2} \tag{8.35}$$

Under these conditions, it can be seen that the capacitive voltage divider is a straight voltage divider with no phase shift involved. The loop gain can be seen to be

$$H_1 H_2 = \frac{g_m}{Y_{\text{tank}}} \frac{C_1}{C_1 + C_2} \tag{8.36}$$

$$= \left(\frac{C_1}{C_1 + C_2}\right) \cdot \frac{g_m}{\frac{1}{R_p} + \frac{1}{r_{e,\text{tank}}} + j\omega C_T - \frac{j}{\omega L}}$$

This can be set equal to 1 and solved for oscillating conditions. The imaginary terms cancel, resulting in the expected expression for resonant frequency:

$$\omega_o = \sqrt{\frac{1}{C_T L}} \tag{8.37}$$

The remaining real terms can be used to obtain an expression for the required g_m:

$$g_m = \left(\frac{1}{R_p} + \frac{1}{r_{e,\text{tank}}}\right) \cdot \left(\frac{C_1 + C_2}{C_1}\right) \tag{8.38}$$

where C_T, as before, is the series combination of C_1 and C_2. This final expression can be manipulated to show that it is equal to (8.19) and (8.33) and is here repeated:

$$g_m = \frac{\omega(C_1 + C_2)}{Q_L} \tag{8.39}$$

Here we can see that the transistor transconductance makes up for losses in the resistors R_p and $r_{e,\text{tank}}$. Since they are in parallel with the resonator, we would like to make them as large as possible to minimize the loss (and the noise). We get large R_p by having large inductor Q, and we get large $r_{e,\text{tank}}$ by using a large value of the capacitive transformer (by making C_2 bigger than C_1). Note that, as before, the value of g_m as specified in (8.38) or (8.39) is the value that makes loop gain equal to 1, which is the condition for marginal oscillation. To guarantee startup, loop gain is set greater than 1 or g_m is set greater than the value specified in the above equations.

Note in (8.39) that r_e seems to have disappeared; however, it was absorbed by assuming that $g_m = 1/r_e$.

8.8 Negative Resistance Generated by the Amplifier

In the next few sections, we will explicitly derive formulas for how much negative resistance is generated by each type of oscillator.

8.8.1 Negative Resistance of Colpitts Oscillator

In this section, an expression for the negative resistance of the oscillators will be derived. Consider first the common-base Colpitts configuration with the negative resistance portion of the circuit replaced by its small-signal model shown in Figure 8.17. Note that v'_π and the current source have both had their polarity reversed for convenience.

An equation can be written for v'_π in terms of the current flowing through this branch of the circuit.

$$i_i + g_m v'_\pi = j\omega C_2 v'_\pi + \frac{v'_\pi}{r_e} \qquad (8.40)$$

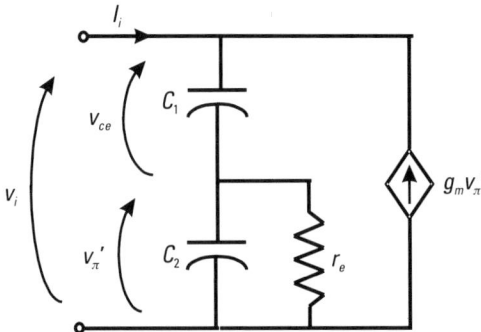

Figure 8.17 Small-signal model for the Colpitts common-base negative resistance cell.

This can be solved for v'_π noting that $g_m \approx 1/r_e$.

$$v'_\pi = \frac{i_i}{j\omega C_2} \tag{8.41}$$

Another equation can be written for v_{ce}.

$$v_{ce} = \frac{i_i + g_m v'_\pi}{j\omega C_1} \tag{8.42}$$

Substituting (8.41) into (8.42) gives

$$v_{ce} = \left(\frac{1}{j\omega C_1}\right)\left(i_i + \frac{g_m i_i}{j\omega C_2}\right) \tag{8.43}$$

Now using (8.41) and (8.43) and solving for $Z_i = v_i/i_i$ with some manipulation,

$$Z_i = \frac{v_i}{i_i} = \frac{v'_\pi + v_{ce}}{i_i} = \frac{1}{j\omega C_1} + \frac{1}{j\omega C_2} - \frac{g_m}{\omega^2 C_1 C_2} \tag{8.44}$$

this is just a negative resistor in series with the two capacitors. Thus, a necessary condition for oscillation in this oscillator is

$$r_s < \frac{g_m}{\omega^2 C_1 C_2} \tag{8.45}$$

where r_s is the equivalent series resistance on the resonator. It will be shown in Example 8.4 that the series negative resistance is maximized for a given fixed total series capacitance when $C_1 = C_2$. An identical expression to (8.45) can be derived for the Colpitts common-collector circuit.

8.8.2 Negative Resistance for Series and Parallel Circuits

Equation (8.44) shows the analysis results for the oscillator circuit shown in Figure 8.18 when analyzed as an equivalent series circuit of C_1, C_2, and R_{neg}. Since the resonance is actually a parallel one, the series components need to be converted back to parallel ones. However, if the equivalent Q of the RC circuit is high, the parallel capacitor C_p will be approximately equal to the series capacitor C_s, and the above analysis is valid. Even for low Q, these simple equations are useful for quick calculations.

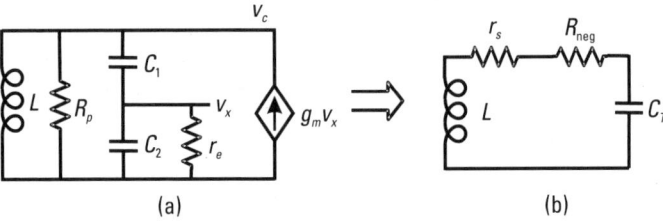

Figure 8.18 (a) Colpitts oscillator circuit; and (b) equivalent series model.

Example 8.3 Negative Resistance for Series and Parallel Circuits

Assume that, as before, $L = 10$ nH, $R_p = 300\Omega$, $C_1 = 2.5$ pF, $C_2 = 10$ pF, and the transistor is operating at 1 mA, or $r_e = 25\Omega$ and $g_m = 0.04$. Using negative resistance, determine the oscillator resonant frequency and apparent frequency shift.

Solution
As before, C_T is 2 pF and

$$\omega = \frac{1}{\sqrt{LC_T}} = \frac{1}{\sqrt{10 \text{ nH} \times 2 \text{ pF}}} = 7.07 \text{ Grad/sec}$$

or frequency $f_o = 1.1254$ GHz. The series negative resistance is equal to

$$r_s = -\frac{g_m}{\omega^2 C_1 C_2} = \frac{-0.04}{(7.07107 \text{ GHz})^2 \cdot 2.5 \text{ pF} \cdot 10 \text{ pF}} = -32.0\Omega$$

Then Q can be calculated from

$$Q = -\frac{1}{\omega r_s C_T} = \frac{-1}{7.07107 \text{ GHz} \cdot 32 \cdot 2 \text{ pF}} = -2.2097$$

$$r_{par} = r_s(1 + Q^2) = -32(1 + 2.2097^2) = -188\Omega$$

We note that the parallel negative resistance is smaller in magnitude than the original parallel resistance, indicating that the oscillator should start up successfully.

The above is sufficient for a hand calculation; however, to complete the example, the equivalent parallel capacitance can be determined to be

$$C_{par} = \frac{C_S}{1 + \frac{1}{Q^2}} = \frac{2 \text{ pF}}{1 + \frac{1}{2.2097^2}} = 1.66 \text{ pF}$$

This results in a new resonator resonant frequency of

$$\omega = \frac{1}{\sqrt{LC_{par}}} = \frac{1}{\sqrt{10 \text{ nF} \cdot 1.66 \text{ pF}}} = 7.76150 \text{ Grad/sec}$$

This is a frequency of 1.2353 GHz, which is close to a 10% change in frequency. The oscillating frequency is determined by resonance of the loop, which in this case results in a 5% change in frequency as seen in Example 5.2. This discrepancy, which can be verified by a simulation of the original circuit of Figure 8.18(a), is due to the phase shift in the nonideal capacitive feedback path. While calculating frequency shifts and explaining them is of interest to academics, it is suggested that for practical designs, the simple calculations be used since parasitics and nonlinear effects will cause a downward shift of frequency. Further refinement should come from a simulator.

8.8.3 Negative Resistance Analysis of $-G_m$ Oscillator

The analysis of the negative resistance amplifier, shown in Figure 8.9, and the more common differential form in Figure 8.22(c) is somewhat different. The small-signal equivalent model for this circuit is shown in Figure 8.19. Note that one transistor has had the normal convention reversed for V_π.

An expression for the current that flows into the circuit can be written as follows:

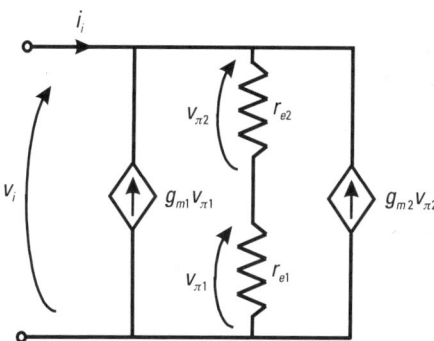

Figure 8.19 Small-signal equivalent model for the negative resistance cell in the negative resistance oscillator.

$$i_i = \frac{v_i}{r_{e1} + r_{e2}} - g_{m1} v_{\pi 1} - g_{m2} v_{\pi 2} \tag{8.46}$$

Now if it is assumed that both transistors are biased identically, then $g_{m1} = g_{m2}$, $r_{e1} = r_{e2}$, $v_{\pi 1} = v_{\pi 2}$, and the equation can be solved for $Z_i = v_i/i_i$.

$$Z_i = \frac{-2}{g_m} \tag{8.47}$$

Thus, in this circuit, a necessary condition for oscillation is that

$$g_m > \frac{2}{R_p} \tag{8.48}$$

where R_p is the equivalent parallel resistance of the resonator.

Example 8.4 Minimum Current for Oscillation

An oscillator is to oscillate at 3 GHz. Using a 5-nH inductor with $Q = 5$ and assuming no other loading on the resonator, determine the minimum current required to start the oscillations if a Colpitts oscillator is used or if a $-G_m$ oscillator is used.

Solution

Ignoring the effect of the losses on the frequency of oscillation, we can determine what total resonator capacitance is required.

$$C_{\text{total}} = \frac{1}{\omega_{\text{osc}}^2 L} = \frac{1}{(2\pi \cdot 3 \text{ GHz})^2 \, 5 \text{ nH}} = 562.9 \text{ fF}$$

The total capacitance is also given by

$$C_{\text{total}} = \frac{C_1 C_2}{C_1 + C_2}$$

Since C_{total} is fixed because we have chosen a frequency of oscillation, we can solve for C_2:

$$C_2 = \frac{C_1 C_{\text{total}}}{C_1 - C_{\text{total}}}$$

Now we can put this back into the negative resistance formula in (8.45):

$$r_{neg} = \frac{g_m}{\omega^2 C_1 C_2} = \frac{g_m}{\omega^2 C_1 C_{total}} - \frac{g_m}{\omega^2 C_1^2}$$

To find the minimum current, we find the maximum r_{neg} by taking the derivative with respect to C_1.

$$\frac{\partial r_{neg}}{\partial C_1} = \frac{-g_m}{\omega^2 C_1^2 C_{total}} + \frac{2 g_m}{\omega^2 C_1^3} = 0$$

This leads to

$$C_1 = 2 C_{total}$$

which means that the maximum obtainable negative resistance is achieved when the two capacitors are equal in value and twice the total capacitance. In this case, $C_1 = C_2 = 1.1258$ pF.

Now the loss in the resonator at 3 GHz is due to the finite Q of the inductor.

The series resistance of the inductor is

$$r_s = \frac{\omega L}{Q} = \frac{(2\pi \cdot 3 \text{ GHz}) 5 \text{ nH}}{5} = 18.85 \Omega$$

Therefore, $r_{neg} = r_s = 18.85 \Omega$. Noting that $g_m = I_c / v_T$,

$$I_c = \omega^2 C_1 C_2 v_T r_{neg}$$
$$= (2\pi \cdot 3 \text{ GHz})^2 (1.1258 \text{ pF})^2 (25 \text{ mV})(18.85 \Omega)$$
$$= 212.2 \ \mu\text{A}$$

In the case of the $-G_m$ oscillator there is no capacitor ratio to consider. The parallel resistance of the inductor is

$$R_p = \omega L Q = (2\pi \cdot 3 \text{ GHz}) 5 \text{ nH} (5) = 471.2 \Omega$$

Therefore $r_{neg} = R_p = 471.2 \Omega$. Noting again that $g_m = I_c / v_T$

$$I_c = \frac{2 v_T}{R_p} = \frac{2(25 \text{ mV})}{471.2 \Omega} = 106.1 \ \mu\text{A}$$

Thus, we can see from this example that a $-G_m$ oscillator can start with half as much collector current in each transistor as a Colpitts oscillator under the same loading conditions.

8.9 Comments on Oscillator Analysis

It has been shown that closed-loop analysis agrees exactly with the open-loop analysis. It can also be shown that analysis by negative resistance produces identical results. This analysis can be extended. For example, in a negative resistance oscillator, it is possible to determine if oscillations will be stable as shown by Kurokawa [1], with detailed analysis shown by [2]. However, what does it mean to have an exact analysis? Does this allow one to predict the frequency exactly? The answer is no. Even if one could take into account RF model complexities including parasitics, temperature, process, and voltage variations, the nonlinearities of an oscillator would still change the frequency. These nonlinearities are required to limit the amplitude of oscillation, so they are a built-in part of an oscillator. Fortunately, for a well-designed oscillator, the predicted results will give a reasonable estimate of the performance. Then, to refine the design, it is necessary to simulate the circuit.

Example 8.5 Oscillator Frequency Shifts and Open-Loop Gain

Explore the predicted frequency with the actual frequency of oscillation by doing open-loop and closed-loop simulation of an oscillator. Compare the results to the simple formula. This example can also be used to explore the amplitude of oscillation and its relationship to the open-loop gain.

Solution

For this example, the previously found capacitor and inductor values are used in the circuit shown in Figure 8.20.

Loop gain can be changed by adjusting g_m or the tank resistance R_p. Both will also affect frequency somewhat. R_p will affect ω_c through Q_L and g_m will affect ω_c indirectly, since $r_e = 1/g_m$. In this case, we varied both R_p and g_m. Results are plotted in Figure 8.21.

It can be seen from Figure 8.21(a) that the open-loop simulations consistently predict higher oscillating frequencies than the closed-loop simulations. Thus, nonlinear behavior results in the frequency being decreased. We note that the initial frequency estimate using the inductor and capacitor values and adding an estimate for the parasitic capacitance results in a good estimate of final closed-loop oscillating frequency. In fact, this estimate of frequency is better than the open-loop small-signal prediction of frequency. It can also be seen from Figure 8.21(b) that output signal amplitude is related to the open-loop gain, and as expected, as gain drops to 1 or less, the oscillations stop.

Voltage-Controlled Oscillators 269

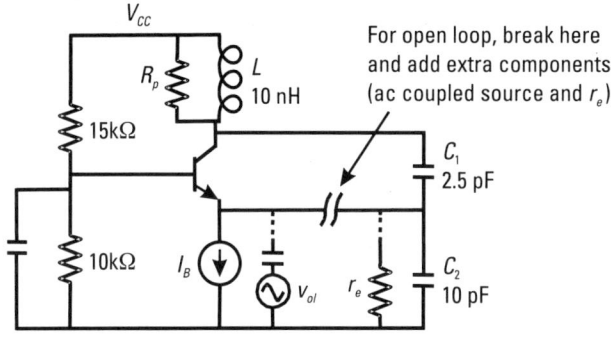

Figure 8.20 Circuit for oscillator simulations.

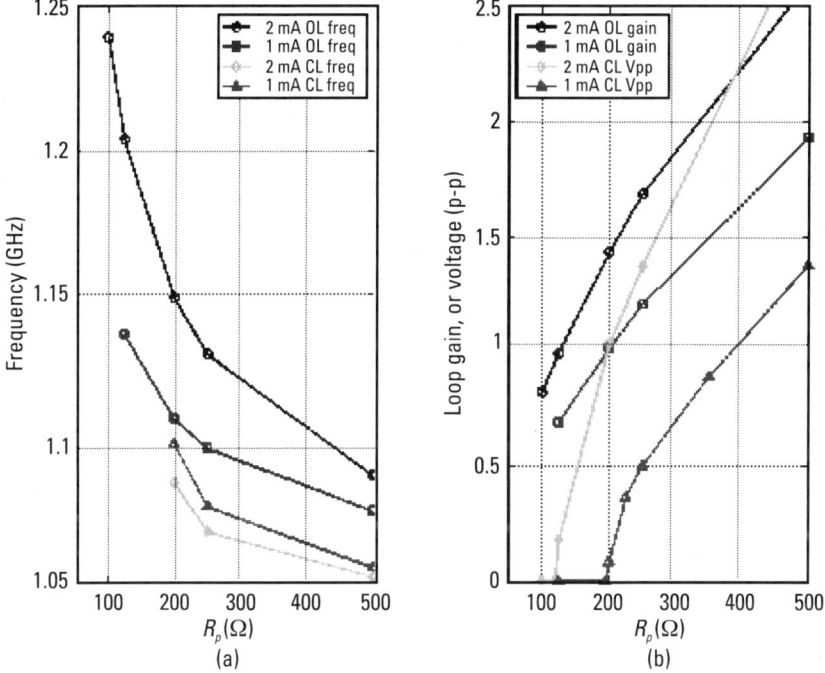

Figure 8.21 Plot of oscillator performance versus tank resistor: (a) open-loop and closed-loop frequency; and (b) loop gain and oscillation amplitude.

So how does one decide on the oscillator small-signal loop gain? In a typical RF integrated oscillator, a typical starting point is to choose a small-signal loop (voltage) gain of about 1.4 to 2 (or 3 to 6 dB); then the current is swept to determine the minimum phase noise. Alternatively, one might design for optimal output power; however, typically, output buffers are used to obtain the desired output power. In traditional negative resistance oscillators, analysis has shown that a small-signal open-loop voltage gain of 3 is optimal for output power [2]. Fortunately, this is close to the optimum value for phase noise performance.

8.10 Basic Differential Oscillator Topologies

The three main oscillators discussed so far can be made into differential circuits. The basic idea is to take two single-ended oscillators and place them back to back. The nodes in the single-ended circuits, which were previously connected to ground, in the differential circuit are tied together forming an axis of symmetry down the center of the circuit. The basic circuits with biasing are shown in Figure 8.22.

8.11 A Modified Common-Collector Colpitts Oscillator with Buffering

One problem with oscillators is that they must be buffered in order to drive a low impedance. Any load that is a significant fraction of the R_p of the oscillator would lower the output swing and increase the phase noise of the oscillator. It is common to buffer oscillators with a stage such as an emitter follower or emitter-coupled pair. These stages add complexity and require current. One design that gets around this problem is shown in Figure 8.23. Here, the common-collector oscillator is modified slightly by the addition of resistors placed in the collector [3, 4]. The output is then taken from the collector. Since this is a high-impedance node, the oscillator's resonator is isolated from the load without using any additional transistors or current. However, the addition of these resistors will also reduce the headroom available to the oscillator.

8.12 Several Refinements to the $-G_m$ Topology

Several refinements can be made to the $-G_m$ oscillator to improve its performance. In the version already presented in Figure 8.22, the transistors' bases and collectors are at the same dc voltage. Thus, the maximum voltage swing that can be obtained is about 0.8V. That is to say, the voltage on one side of the resonator drops about 0.4V while on the other side of the oscillator the voltage rises by about 0.4V. This means that the collector would be about 0.8V below the base and the transistor goes into saturation. In order to get larger swings out of this topology, we must decouple the base from the collector. One

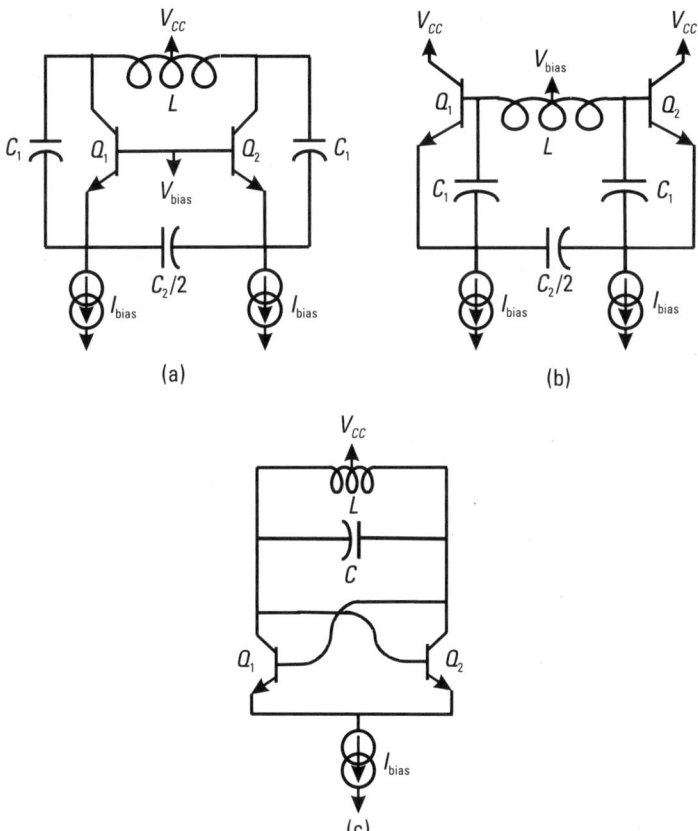

Figure 8.22 Basic differential oscillators: (a) Colpitts common base; (b) Colpitts common collector; and (c) $-G_m$ oscillator.

common way to do this is with capacitors. This improved oscillator is shown in Figure 8.24. The bases have to be biased separately now, of course. Typically, this is done by placing resistors in the bias line. These resistors have to be made large to prevent loss of signal at the base. However, these resistors can be a substantial source of noise.

Another variation on this topology is to use a transformer instead of capacitors to decouple the collectors from the bases, as shown in Figure 8.25 [5]. Since the bias can be applied through the center tap of the transformer, there is no longer a need for the RF blocking resistors in the bias line. Also, if a turns ratio of greater than unity is chosen, there is the added advantage that the swing on the base can be much smaller than the swing on the collector, helping to prevent transistor saturation.

Another modification that can be made to the $-G_m$ oscillator is to replace the high-impedance current source connected to the emitters of Q_1 and Q_2 in

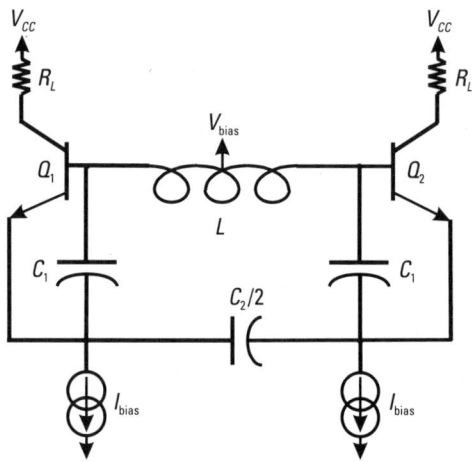

Figure 8.23 Modified Colpitts common-collector oscillator with self-buffering.

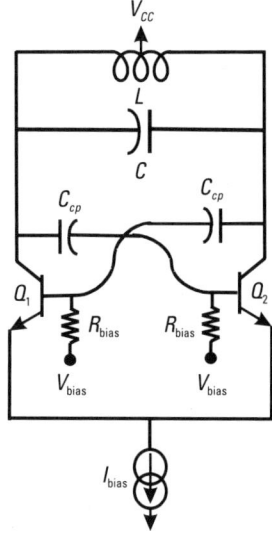

Figure 8.24 $-G_m$ oscillator with capacitive decoupling of the bases.

Figure 8.25 with a resistor. Since the resistor is not a high impedance source, the bias current will vary dynamically over the cycle of the oscillation. In fact, the current will be highest when the oscillator voltage is at its peaks and lowest during the zero crossings of the waveform. Since the oscillator is most sensitive to phase noise during the zero crossings, this version of the oscillator can often give very good phase noise performance. This oscillator is shown in Figure 8.26(a).

Voltage-Controlled Oscillators 273

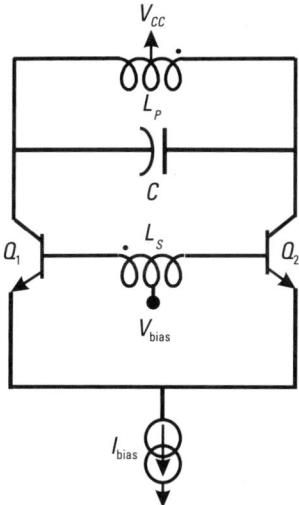

Figure 8.25 $-G_m$ oscillator with inductive decoupling of the bases.

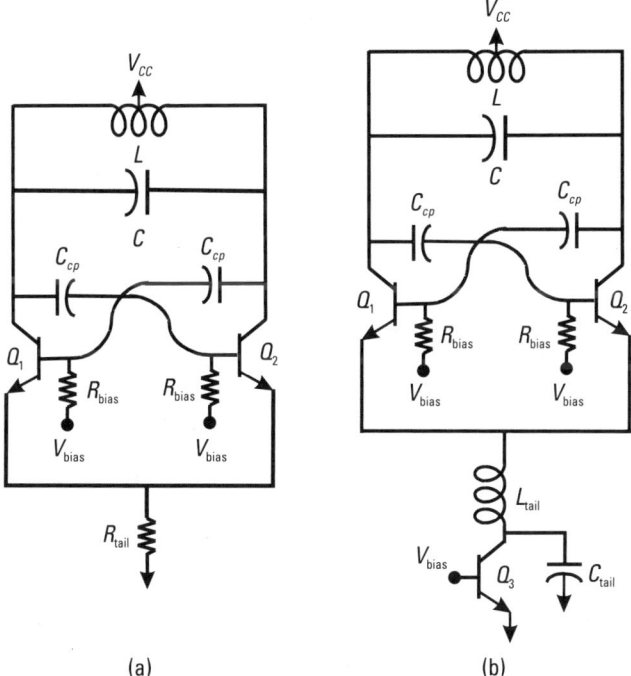

Figure 8.26 $-G_m$ oscillator with (a) resistive tail current source, and (b) current source noise filter.

A brief circuit description will now be provided. The circuit must be dc biased at some low current. As the oscillation begins, the voltage rises on one side of the resonator and one transistor starts to turn off while the other starts to turn on harder and draw more current. As the transistor draws more current, more current flows through R_{tail}, and thus the voltage across this resistor starts to rise. This acts to reduce the v_{BE} of the transistor, which acts as feedback to limit the current at the top and bottom of the swing. The collector waveforms are shown conceptually in Figure 8.27. Since the current is varying dynamically over a cycle, and since the resistor R_{tail} does not require as much headroom as a current source, this allows a larger oscillation amplitude for a given power supply.

An alternative to the resistor R_{tail} is to use a noise filter in the tail as shown in Figure 8.26(b) [6]. While the use of the inductor does require more chip area, its use can lead to a very low-noise bias, leading to low-phase-noise designs. Another advantage to using this noise filter is that before startup, the transistor Q_3 can be biased in saturation, because during startup the second harmonic will cause a dc bias shift at the collector of Q_3, pulling it out of saturation and into the active region. Also, since the second harmonic cannot pass through the inductor L_{tail}, there is no "ringing" at the collector of Q_3, further reducing its headroom requirement.

8.13 The Effect of Parasitics on the Frequency of Oscillation

The first task in designing an oscillator is to set the frequency of oscillation and hence set the value of the total inductance and capacitance in the circuit. To increase output swing, it is usually desirable to make the inductance as large as possible (this will also make the oscillator less sensitive to parasitic resistance). However, it should be noted that large monolithic inductors suffer from limited

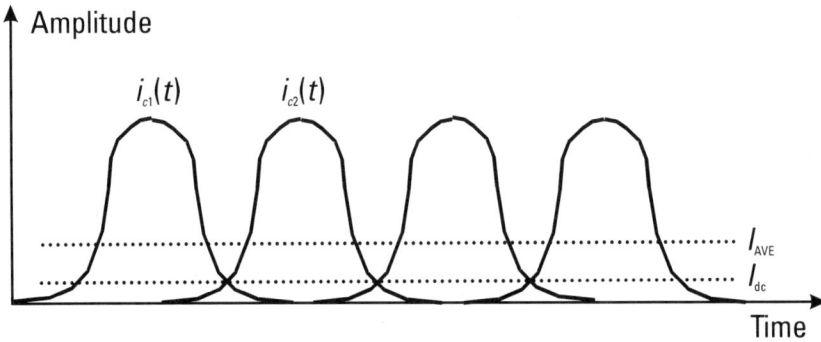

Figure 8.27 $-G_m$ oscillator with resistive tail collector currents.

Q. In addition, as the capacitors become smaller, their value will be more sensitive to parasitics. The frequency of oscillation for the Colpitts common-base oscillator, as shown in Figure 8.10(a), taking into account transistor parasitics, is given by

$$\omega_{osc} \approx \frac{1}{\sqrt{L\left(\dfrac{C_1 C_2 + C_1 C_\pi}{C_1 + C_2 + C_\pi} + C_\mu\right)}} \tag{8.49}$$

For the Colpitts common-collector oscillator, as shown in Figure 8.10(b), the frequency is given by

$$\omega_{osc} \approx \frac{1}{\sqrt{L\left(\dfrac{C_1 C_2 + C_2 C_\pi}{C_1 + C_2 + C_\pi} + C_\mu\right)}} \tag{8.50}$$

For the $-G_m$ oscillator, the frequency is given by

$$\omega_{osc} \approx \frac{1}{\sqrt{L\left(2C_\mu + \dfrac{C_\pi}{2} + C\right)}} \tag{8.51}$$

Note that in the case of the $-G_m$ oscillator, the parasitics tend to reduce the frequency of oscillation a bit more than with the Colpitts oscillator.

8.14 Large-Signal Nonlinearity in the Transistor

So far, the discussion of oscillators has assumed that the small-signal equivalent model for the transistor is valid. If this were true, then the oscillation amplitude would grow indefinitely, which is not the case. As the signal grows, nonlinearity will serve to reduce the negative resistance of the oscillator until it just cancels out the losses and the oscillation reaches some steady-state amplitude. The source of the nonlinearity is typically the transistor itself.

Usually the transistor is biased somewhere in the active region. At this operating point, the transistor will have a particular g_m. However, as the voltage swing starts to increase during startup, the instantaneous g_m will start to change over a complete cycle. The transistor may even start to enter the saturation region at one end of the swing and the cutoff region at the other end of the voltage swing. Which of these effects starts to happen first depends on the biasing of the transistor. Ultimately, a combination of all effects may be present.

Eventually, with increasing signal amplitude, the effective g_m will decrease to the point where it just compensates for the losses in the circuit and the amplitude of the oscillator will stabilize.

The saturation and cutoff linearity constraint will also put a practical limit on the maximum power that can be obtained from an oscillator. After reaching this limit, increasing the bias current will have very little effect on the output swing. Although increasing the current causes the small-signal g_m to rise, this just tends to "square up" the signal rather than to increase its amplitude.

Looking at the common-base or common-collector Colpitts oscillators as shown in Figure 8.10, it can be seen how this effect works on the circuit waveforms in Figures 8.28 and 8.29. In the case of the common-base circuit, when v_c is at the bottom of its swing, v_{ce} tends to be very small, causing the base collector junction to be forward biased. This also tends to make v_{be} quite large. These two conditions together cause the transistor to go into saturation. When v_c reaches the top of its swing, v_{be} gets very small and this drives the

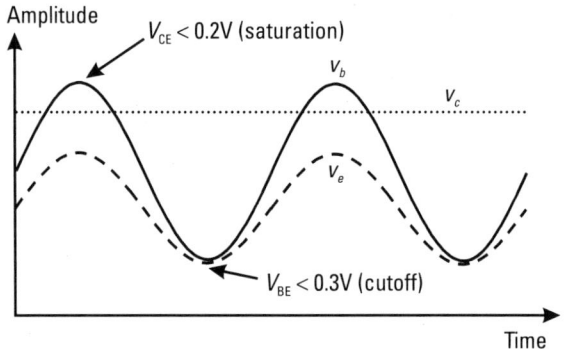

Figure 8.28 Waveforms for a common-collector oscillator that is heavily voltage-limited.

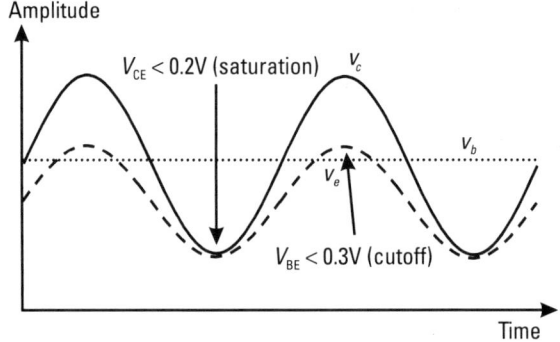

Figure 8.29 Waveforms for a common-base oscillator that is heavily voltage-limited.

oscillator into cutoff. A similar argument can be made for the common-collector circuit, except that it enters cutoff at the bottom of its swing and saturation at the top of its swing.

8.15 Bias Shifting During Startup

Once the oscillator starts to experience nonlinearity, harmonics start to appear. The even-ordered harmonics, if present, can cause shifts in bias conditions since they are not symmetric. They have no negative-going swing so they can change the average voltage or current at a node. Thus, they tend to raise the voltage at any node with signal swing on it, and after startup, bias conditions may shift significantly from what would be predicted by a purely dc analysis. For instance, the voltage at the emitter of the common-base or common-collector Colpitts oscillators will tend to rise. Another very good example of this is the $-G_m$ oscillator with resistive tail as shown in Figure 8.26. The node connected to R_{tail} is a virtual ground; however, there is strong second-harmonic content on this node that tends to raise the average voltage level and the current through the oscillator after startup.

8.16 Oscillator Amplitude

If the oscillator satisfies the conditions for oscillation, then oscillations will continue to grow until the transistor nonlinearities reduce the gain until the losses and the negative resistance are of equal value.

For a quantitative analysis of oscillation amplitude, we first start with a transistor being driven by a large sinusoidal voltage, as shown in Figure 8.30.

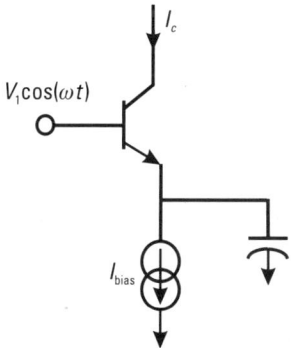

Figure 8.30 Transistor driven by a large sinusoidal voltage source.

Note that in a real oscillator, a sinusoid driving the base is a good approximation, provided the resonator has a reasonable Q. This results in all other frequency components being filtered out and the voltage (although not the current) is sinusoidal even in the presence of strong nonlinearity.

It is assumed that the transistor is being driven by a large voltage, so it will only be on for a very small part of the cycle, during which time it produces a large pulse of current. However, regardless of what the current waveform looks like, its average value over a cycle must still equal the bias current. Therefore,

$$\overline{i_c} = \frac{1}{T} \int_0^T i_c(t)\, dt = I_{\text{bias}} \tag{8.52}$$

The part of the current at the fundamental frequency of interest can be extracted by multiplying by a cosine at the fundamental and integrating.

$$i_{\text{fund}} = \frac{2}{T} \int_0^T i_c(t) \cos(\omega t)\, dt \tag{8.53}$$

This can be solved by assuming a waveform for $i_c(t)$. However, solving this equation can be avoided by noting that the current is only nonzero when the voltage is almost at its peak value. Therefore, the cosine can be approximated as unity and integration simplifies:

$$i_{\text{fund}} \approx \frac{2}{T} \int_0^T i_c(t)\, dt = 2I_{\text{bias}} \tag{8.54}$$

With this information, it is possible to define a large signal transconductance for the transistor given by

$$G_m = \frac{i_{\text{fund}}}{V_1} = \frac{2I_{\text{bias}}}{V_1} \tag{8.55}$$

Since $g_m = I_c/v_T$ and since G_m can never be larger than g_m, it becomes clear that this approximation is not valid if V_1 is less then $2v_T$.

We can now apply this to the case of the Colpitts common-collector oscillator as shown in Figure 8.10. We draw the simplified schematic replacing the transistor with the large-signal transconductance, as shown in Figure 8.31.

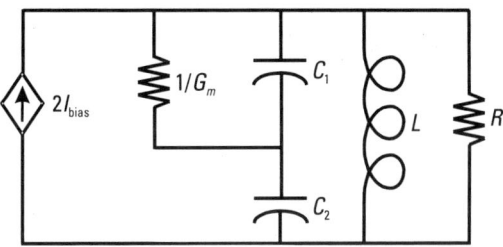

Figure 8.31 Colpitts common-collector oscillator with large-signal transconductance applied.

Note that we are using the T-model here for the transistor, so the current source is between collector and base.

We first note that the resonator voltage will be the bias current at the fundamental times the equivalent resonator resistance.

$$V_{tank} = 2I_{bias} R_{total} \tag{8.56}$$

This resistance will be made up of the equivalent loading of all losses in the oscillator and the loading of the transconductor on the resonator.

The transconductor presents the impedance

$$\frac{1}{G_m}\left(\frac{C_1 + C_2}{C_2}\right)^2 = \frac{1}{G_m n^2} \tag{8.57}$$

where n is the equivalent impedance transformation ratio.

This is in parallel with all other losses in the resonator R_p:

$$R_{total} = R_p \mathbin{//} \frac{1}{G_m n^2} = \frac{R_p}{1 + G_m n^2 R_p} \tag{8.58}$$

We can plug this back into the original expression:

$$V_{tank} = 2I_{bias} \frac{R_p}{1 + G_m n^2 R_p} \tag{8.59}$$

Now we also know that

$$G_m = \frac{2I_{bias}}{V_1} \tag{8.60}$$

and that

$$V_1 = \frac{C_2}{C_1 + C_2} V_{tank} = nV_{tank} \tag{8.61}$$

Therefore,

$$V_{tank} = 2I_{bias} \frac{R_p}{1 + \frac{2I_{bias}}{nV_{tank}} n^2 R_p} \tag{8.62}$$

$$V_{tank} = 2I_{bias} R_p \left(\frac{C_1}{C_1 + C_2} \right) \tag{8.63}$$

A very similar analysis can be carried out for the common-base Colpitts oscillator shown in Figure 8.10, and it will yield the result

$$V_{tank} = 2I_{bias} R_p \left(\frac{C_2}{C_1 + C_2} \right) \tag{8.64}$$

Note that it is often common practice to place some degeneration in the emitter of the transistor in a Colpitts design. This practice will tend to spread the pulses over a wider fraction of a cycle, reducing the accuracy of the above equation somewhat. However, it should still be a useful estimate of oscillation amplitude.

The application of the theory to the $-G_m$ oscillator is slightly more complicated. We start by breaking the loop and applying a voltage to the bases of the transistors and looking at the collector currents that result.

We can see from Figure 8.32 that this is just a differential pair with a large voltage applied across the input. The resulting formula for such a configuration has already been seen in Chapter 6 while doing a linearity analysis of a differential pair. The output currents are given by

$$i_c(\theta) = \frac{I_{bias}}{1 + e^{\frac{V_{tank}}{v_T} \cos(\theta)}} \tag{8.65}$$

Similar to the previous analysis, we find the fundamental component of the current by solving the following integral:

$$i_{fund} = \frac{2}{\pi} \int_{\theta=0}^{\theta=\pi} \frac{I_{bias}}{1 + e^{\frac{V_{tank}}{v_T} \cos(\theta)}} \cos(\theta) \, d\theta \tag{8.66}$$

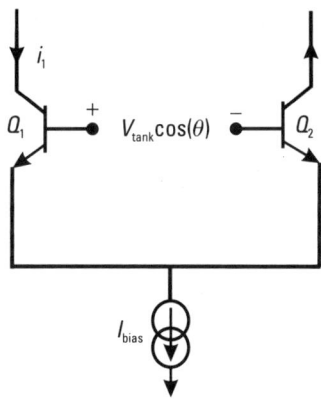

Figure 8.32 Short tail pair with a large sinusoidal voltage applied to the base.

It can be shown, provided $V_{tank}/v_T > 8$ (which is reasonable for most practical oscillator amplitudes), that

$$\frac{i_{fund}}{I_{tank}} = \frac{2}{\pi} \tag{8.67}$$

Thus, if the parallel resistance of the resonator is R_p, the peak voltage developed across the resonator differentially will be given by

$$V_{tank} = i_{fund} R_p = 2 \cdot \left(\frac{2}{\pi} I_{tank}\right) \cdot R_p = \frac{2}{\pi} I_{tank} R_p \tag{8.68}$$

Note that there are two currents of the type described in (8.67), one for each transistor. However, the current flows down through the supply, so only develops a voltage across half the parallel resonator resistance.

The case where the current source in Figure 8.32 is made from a resistor is slightly more complicated. The total current flowing through the oscillator will change over a cycle (with minimum current at the zero crossings and maximum current at the voltage peaks). Equation (8.65) can be modified to take into account the resistor.

$$i_c(\theta) \approx \frac{V_B + \left|\frac{V_{tank}\cos(\theta)}{2}\right| - V_{BEQ}}{R_{bias}} = \frac{I_{BQ} + \frac{|V_{tank}\cos(\theta)|}{2R_{bias}}}{1 + e^{\frac{V_{tank}}{v_T}\cos(\theta)}} \tag{8.69}$$

where R_{bias} is the tail resistor that is serving as a current source, I_{BQ} is the quiescent bias current before oscillations begin, and V_{BEQ} is the quiescent base emitter voltage of Q_1 or Q_2. It is assumed that V_{BEQ} is constant in this expression, which is a good approximation for the half cycle when the transistor is on. In the other half cycle, the denominator will force the expression to zero for any reasonable value of V_{tank} ($V_{tank} \gg v_T$), so the approximation will not seriously affect the shape of the resulting waveform.

This expression can be used to find the average dc operating current in the circuit (which will be higher than the quiescent current). This current also depends on the final amplitude of the VCO.

$$I_{AVE} = \frac{1}{2\pi} \int_0^{2\pi} \frac{I_{BQ} + \frac{|V_{tank} \cos(\theta)|}{2R_{bias}}}{1 + e^{\frac{V_{tank}}{v_T} \cos(\theta)}} d\theta \approx I_{BQ} + \frac{V_{tank}}{\pi R_{bias}} \quad (8.70)$$

The fundamental component of the current can also be extracted from (8.69) as before.

$$i_{fund} = \frac{2}{\pi} \int_0^{\pi} \left[\frac{I_{BQ} + \frac{|V_{tank} \cos(\theta)|}{2R_{bias}}}{1 + e^{\frac{V_{tank}}{v_T} \cos(\theta)}} \right] \cos(\theta) \, d\theta \approx \frac{2}{\pi} I_{BQ} + \frac{V_{tank}}{4R_{bias}}$$

$$(8.71)$$

This allows us to determine the ratio of current at the fundamental to average current:

$$k = \frac{i_{fund}}{I_{AVE}} = \frac{\frac{2}{\pi} I_{BQ} + \frac{V_{tank}}{4R_{bias}}}{I_{BQ} + \frac{V_{tank}}{\pi R_{bias}}} \quad (8.72)$$

In the limit of large and small V_{tank}, it can be seen that k is bounded by

$$\frac{2}{\pi} \leq k \leq \frac{\pi}{4} \quad (8.73)$$

Therefore, the oscillation amplitude can once again be given in terms of dc current as

$$V_{\text{tank}} = i_{\text{fund}} R_p = k I_{\text{AVE}} R_p \tag{8.74}$$

Thus, equations to predict oscillation amplitude have now been derived. By comparing (8.73) and (8.74) to (8.68), it can be seen that a given amount of dc current will lead to more current at the fundamental frequency in the case of the resistive tail as opposed to the current tail.

8.17 Phase Noise

A major challenge in most oscillator designs is to meet the phase noise requirements of the system. An ideal oscillator has a frequency response that is a simple impulse at the frequency of oscillation. However, real oscillators exhibit "skirts" caused by instantaneous jitter in the phase of the waveform. Noise that causes variations in the phase of the signal (distinct from noise that causes fluctuations in the amplitude of the signal) is referred to as phase noise. The waveform of a real oscillator can be written as

$$V_{\text{osc}} = A \cos[\omega_o t + \phi_n(t)] \tag{8.75}$$

where $\phi_n(t)$ is the phase noise of the oscillator. Here amplitude noise is ignored because it is usually of little importance in most system specifications. Because of amplitude limiting in integrated oscillators, typically AM noise is lower than FM noise. There are several major sources of phase noise in an oscillator, and they will be discussed next.

8.17.1 Linear or Additive Phase Noise and Leeson's Formula

In order to derive a formula for phase noise in an oscillator, we will start with the feedback model of an oscillator as shown in Figure 8.33 [7].

From control theory, it is known that

$$\frac{N_{\text{out}}(s)}{N_{\text{in}}(s)} = \frac{H_1(s)}{1 - H(s)} \tag{8.76}$$

where $H(s) = H_1(s) H_2(s)$. $H(s)$ can be written as a truncated Taylor series:

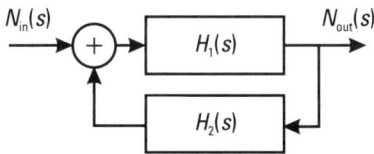

Figure 8.33 Feedback model of an oscillator used for phase-noise modeling.

$$H(j\omega) \approx H(j\omega_o) + \Delta\omega \frac{dH}{d\omega} \tag{8.77}$$

Since the conditions of stable oscillation must be satisfied, $H(j\omega_o) = 1$. We let $H_1(j\omega_o) = H_1$, where H_1 is a constant determined by circuit parameters. Now (8.76) can be rewritten using (8.77) as

$$\frac{N_{out}(s)}{N_{in}(s)} = \frac{H_1}{-\Delta\omega \frac{dH}{d\omega}} \tag{8.78}$$

Noise power is of interest here, so

$$\left|\frac{N_{out}(s)}{N_{in}(s)}\right|^2 = \frac{|H_1|^2}{(\Delta\omega)^2 \left|\frac{dH}{d\omega}\right|^2} \tag{8.79}$$

This equation can now be rewritten using $H(\omega) = |H|e^{j\phi}$ and the product rule

$$\frac{dH}{d\omega} = \frac{d|H|}{d\omega} e^{j\phi} + |H| j e^{j\phi} \frac{d\phi}{d\omega} \tag{8.80}$$

noting that the two terms on the right are orthogonal:

$$\left|\frac{dH}{d\omega}\right|^2 = \left|\frac{d|H|}{d\omega}\right|^2 + |H|^2 \left|\frac{d\phi}{d\omega}\right|^2 \tag{8.81}$$

At resonance, the phase changes much faster than magnitude, and $|H| \approx 1$ near resonance. Thus, the second term on the right is dominant, and this equation reduces to

$$\left|\frac{dH}{d\omega}\right|^2 = \left|\frac{d\phi}{d\omega}\right|^2 \tag{8.82}$$

Now substituting (8.82) back into (8.79),

$$\left|\frac{N_{out}(s)}{N_{in}(s)}\right|^2 = \frac{|H_1|^2}{(\Delta\omega)^2 \left|\frac{d\phi}{d\omega}\right|^2} \tag{8.83}$$

This can be rewritten again with the help of the definition of Q given in Chapter 4:

$$\left|\frac{N_{out}(s)}{N_{in}(s)}\right|^2 = \frac{|H_1|^2 \omega_o^2}{4Q^2(\Delta\omega)^2} \tag{8.84}$$

In the special case for which the feedback path is unity, then $H_1 = H$, and since $|H| = 1$ near resonance it reduces to

$$\left|\frac{N_{out}(s)}{N_{in}(s)}\right|^2 = \frac{\omega_o^2}{4Q^2(\Delta\omega)^2} \tag{8.85}$$

Equation (8.85) forms the noise shaping function for the oscillator. In other words, for a given noise power generated by the transistor amplifier part of the oscillator, this equation describes the output noise around the tone.

Phase noise is usually quoted as an absolute noise referenced to the carrier power, so (8.85) should be rewritten to give phase noise as

$$\text{PN} = \frac{|N_{out}(s)|^2}{2P_S} = \left(\frac{|H_1|\omega_o}{(2Q\Delta\omega)}\right)^2 \left(\frac{|N_{in}(s)|^2}{2P_S}\right) \tag{8.86}$$

where P_S is the signal power of the carrier and noting that phase noise is only half the noise present. The other half is amplitude noise, which is of less interest. Also, in this approximation, conversion of amplitude noise to phase noise (also called AM to PM conversion) is ignored. This formula is known as Leeson's equation [8].

The one question that remains here is, What exactly is N_{in}? If the transistor and bias were assumed to be noiseless, then the only noise present would be due to the resonator losses. Since the total resonator losses are due to its finite resistance, which has an available noise power of kT, then

$$|N_{in}(s)|^2 = kT \tag{8.87}$$

The transistors and the bias will add noise to this minimum. Note that since this is not a simple amplifier with a clearly defined input and output, it would not be appropriate to define the transistor in terms of a simple noise figure. Considering the bias noise in the case of the $-G_m$ oscillator, as shown in Figure 8.22(c), noise will come from the current source when the transistors Q_1 and Q_2 are switched. If ρ is the fraction of a cycle for which the transistors are completely switched, i_{nt} is the noise current injected into the oscillator from the biasing network during this time. During transitions, the transistors act like an amplifier, and thus collector shot noise i_{cn} from the resonator transistors usually dominates the noise during this time. The total input noise becomes

$$|N_{in}(s)|^2 \approx kT + \frac{i_{nt}^2 R_p}{2}\rho + i_{cn}^2 R_p(1-\rho) \tag{8.88}$$

where R_p is the equivalent parallel resistance of the tank. Thus, we can define an excess noise factor for the oscillator as excess noise injected by noise sources other than the losses in the tank:

$$F = 1 + \frac{i_{nt}^2 R_p}{2kT}\rho + \frac{i_{cn}^2 R_p(1-\rho)}{kT} \tag{8.89}$$

Note that as the Q of the tank increases, R_p increases and noise has more gain to the output; therefore, F is increased. Thus, while (8.85) shows a decrease in phase noise with an increase in Q, this is somewhat offset by the increase in F. If noise from the bias i_{cn} is filtered and if fast switching is employed, it is possible to achieve a noise factor close to unity.

Now (8.86) can be rewritten as

$$\text{PN} = \left(\frac{|H_1|\omega_o}{(2Q\Delta\omega)}\right)^2 \left(\frac{FkT}{2P_S}\right) \tag{8.90}$$

Note that in this derivation, it has been assumed that flicker noise is insignificant at the frequencies of interest. This may not always be the case, especially in CMOS designs. If ω_c represents the flicker noise corner where flicker noise and thermal noise are equal in importance, then (8.90) can be rewritten as

$$\text{PN} = \left(\frac{|H_1|\omega_o}{(2Q\Delta\omega)}\right)^2 \left(\frac{FkT}{2P_S}\right)\left(1 + \frac{\omega_c}{\Delta\omega}\right) \tag{8.91}$$

It can be noted that (8.91) predicts that noise will roll off at slopes of -30 or -20 dB/decade depending on whether flicker noise is important. However, in real life, at high frequency offsets there will be a thermal noise floor. A typical plot of phase noise versus offset frequency is shown in Figure 8.34.

It is important to make a few notes here about the interpretation of this formula. Note that in the derivation of this formula, it has been assumed that

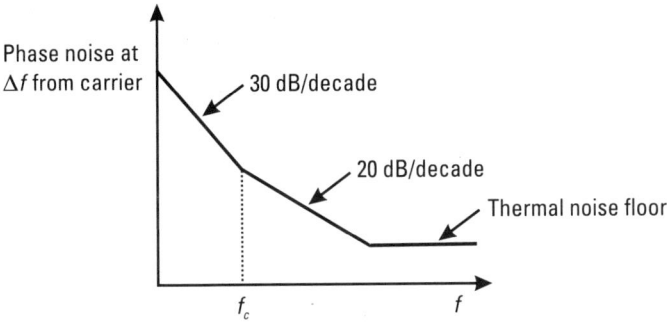

Figure 8.34 Phase noise versus frequency.

the noise N_{in} is injected into the resonator. Thus, $|H_1| = |H| = 1$ at the top of the resonator. However, at other points in the loop, the signal level is not the same as at the resonator. For instance, in the case of the common-base Colpitts oscillator, when looking at the midpoint between the two capacitors, then the signal is reduced by $C_2/(C_1 + C_2)$. A common mistake is to assume that, in this circuit, the phase noise at this point would be intrinsically worse than at the top of the resonator because the output power is lower by a factor $[C_2/(C_1 + C_2)]^2$. However, the noise is reduced by the same amount, leaving the phase noise at the same level at both points in the feedback loop. Note that this is only true for offset frequencies for which the noise is higher than the thermal noise floor.

Example 8.6 Phase Noise Limits

A sales representative for Simply Fabless Semiconductor Inc. has told a potential customer that Simply Fabless can deliver a 5-GHz receiver including an on-chip *phase-locked loop* (PLL). The VCO in the part is to run off a 1.8V supply, consume no more than 1 mW of power, and deliver a phase noise performance of −105 dBc/Hz at 100-kHz offset. It has fallen on the shoulders of engineering to design this part. It is known that, in the technology to be used, the best inductor Q is 15 for a 3-nH device. Assume that capacitors or varactors will have a Q of 50. What is the likelihood that engineering will be able to deliver the part with the required performance to the customer?

Solution

We will assume a $-G_m$ topology for this design and start with the assumption that the inductor and capacitive resistance are the only load on the device (we will ignore all other losses).

The $r_{p/L}$ of the inductor is

$$r_{p/L} = \omega L Q = 2\pi \cdot 5 \text{ GHz} \cdot 3 \text{ nH} \cdot 15 = 1,413.7 \Omega$$

The capacitance in the design will be

$$C_{total} = \frac{1}{\omega_{osc}^2 L} = \frac{1}{(2\pi \cdot 5 \text{ GHz})^2 \, 3 \text{ nH}} = 337.7 \text{ fF}$$

Thus, the parallel resistance due to the capacitor will be

$$r_{p/C} = \frac{Q}{\omega C_{total}} = \frac{50}{(2\pi \cdot 5 \text{ GHz}) \cdot 337.7 \text{ fF}} = 4{,}712.9 \, \Omega$$

Thus, the equivalent parallel resistance of the resonator is $1{,}087.5 \, \Omega$.

With a supply of 1.8V and a power consumption of 1 mW, the maximum current that the circuit can draw is 555.5 μA.

The peak voltage swing in the oscillator will be

$$V_{tank} = \frac{2}{\pi} I_{bias} R_p = \frac{2}{\pi} (555.5 \ \mu\text{A})(1087.5 \, \Omega) = 0.384 \text{V}$$

This means that the oscillator will have an RF output power of

$$P = \frac{V_{tank}^2}{2R_p} = \frac{(0.384 \text{V})^2}{2(1{,}087.5 \, \Omega)} = 67.8 \ \mu\text{W}$$

The Q of the oscillator will be

$$Q = R_P \sqrt{\frac{C_{total}}{L}} = 1{,}087.5 \, \Omega \sqrt{\frac{337.7 \text{ fF}}{3 \text{ nH}}} = 11.53$$

If we now assume that all low-frequency upconverted noise is small and further assume that active devices add no noise to the circuit and therefore $F = 1$, we can now estimate the phase noise.

$$\text{PN} = \left[\frac{A\omega_o}{(2Q\Delta\omega)}\right]^2 \left(\frac{FkT}{2P_S}\right)$$

$$= \left[\frac{1.12(2\pi \cdot 5 \text{ GHz})}{2(11.53)(2\pi \cdot 100 \text{ kHz})}\right]^2 \left(\frac{(1)(1.38 \times 10^{-23} \text{ J/K})(298\text{K})}{2(67.8 \ \mu\text{W})}\right)$$

$$= 1.79 \cdot 10^{-10}$$

This is −97.5 dBc/Hz at 100-kHz offset, which is 7.5 dB below the promised performance. Thus, the specifications given to the customer are most likely very difficult (people claiming that anything is impossible are often interrupted by those doing it), given the constraints. This is a prime example of one of the most important principles in engineering. If the sales department is running open loop, then the system is probably unstable and you may be headed for the rails [9].

Example 8.7 Choosing Inductor Size

Big inductors, small inductors, blue inductors, red inductors? What kind is best? Assuming a constant bias current and noise figure for the amplifier, and further assuming a constant Q for all sizes of inductance, determine the trend for phase noise in a $-G_m$ oscillator relative to inductance size. Assume the inductor is the only loss in the resonator.

Solution

Since the Q of the inductor is constant regardless of inductor size and it is the only loss in the resonator, then the Q of the resonator will be constant.

The parallel resistance of the resonator will be given by

$$R_p = Q_{\text{ind}} \omega_o L$$

For low values of inductor, R_p will be small. We can assume that the oscillation amplitude is proportional to

$$V_{\text{tank}} \propto R_p$$

We are only interested in trends here, so constants are not important. Thus, the power in the resonator is given by

$$P_S \propto \frac{V_{\text{tank}}^2}{R_P} = \frac{(R_p)^2}{R_P} = R_p$$

Now phase noise is

$$\text{PN} = \left[\frac{\omega_o}{(2Q\Delta\omega)}\right]^2 \left(\frac{FkT}{2P_S}\right)$$

Q is a constant, and we assume a constant frequency and noise figure. The only thing that changes is the output power.

Thus,

$$\text{PN} \propto \frac{1}{P_S} \propto \frac{1}{L}$$

Thus, as L increases, the phase noise decreases as shown in Figure 8.35.

At some point the inductor will be made so large that increasing it further will no longer make the signal swing any bigger. At this point,

$$P_S \propto \frac{V_{tank}^2}{R_p} \propto \frac{1}{R_p}$$

Again everything else is constant except for the power term, so

$$\text{PN} \propto \frac{1}{P_S} \propto L$$

Thus, once the amplitude has reached its maximum, making the inductor any bigger will tend to increase the phase noise.

These two curves will intersect at this point. Therefore, we can draw the trend lines as seen in Figure 8.35.

So far, the discussion has been of oscillators that have no tuning scheme. However, most practical designs incorporate some method to change the frequency of the oscillator. In these oscillators, the output frequency is proportional to the voltage on a control terminal:

$$\omega_{osc} = \omega_o + K_{VCO} V_{cont} \tag{8.92}$$

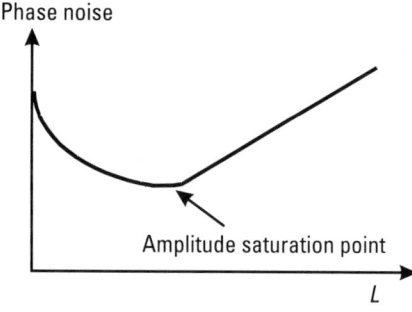

Figure 8.35 Phase noise versus tank inductance.

where K_{VCO} is the gain of the VCO and V_{cont} is the voltage on the control line. If it is assumed that V_{cont} is a low-frequency sine wave of amplitude V_m, and using the narrow-band FM approximation, the resulting output voltage is

$$v_{out}(t) = A\cos(\omega_o t) + \frac{AV_m K_{VCO}}{2\Delta\omega}[\cos(\omega_o + \Delta\omega)t - \cos(\omega_o - \Delta\omega)t] \quad (8.93)$$

where A is the carrier power and $\Delta\omega$ is the frequency of the controlling signal. Thus, if it is assumed that the sine wave is a noise source, then the noise power present at $\pm\Delta\omega$ is given by

$$\text{Noise} = \left(\frac{AV_m K_{VCO}}{2\Delta\omega}\right)^2 \quad (8.94)$$

This can be converted into phase noise by dividing by the signal power:

$$\text{PN} = \left(\frac{V_m K_{VCO}}{2\Delta\omega}\right)^2 \quad (8.95)$$

8.17.2 Some Additional Notes About Low-Frequency Noise

From the preceding analysis, it is easy to see how one might estimate the effect of low-frequency noise on the phase noise of the oscillator. Using a simple small-signal noise analysis, one can find out how much noise is present at the varactor terminals. Then, knowing the K_{VCO}, the amount of phase noise can be estimated.

However, this is not necessarily the whole story. Noise on any terminal, which controls the amplitude of the oscillation, can lead to fluctuations in the amplitude. These fluctuations, if they occur at low frequencies, are just like noise and can actually dominate the noise content in some cases. However, a small-signal analysis will not reveal this.

Example 8.8 Control Line Noise Problems

A VCO designer has designed a VCO to operate between 5.7 and 6.2 GHz, and the tuning voltage is set to give this range as it is tuned between 1.5V and 2.5V. The design has been simulated to have a phase noise of −105 dBc/Hz at a 100-kHz offset. The design has been given to the synthesizer designers who wish to place it in a loop. The loop will have an off-chip RC filter and the tuning line of the VCO will be brought out to a pin. The synthesizer team decides to use a pad with an *electrostatic discharge* (ESD) strategy that makes use of a 300-Ω series resistor. What is the likely impact of this ESD strategy on this design?

Solution
First, we estimate the gain of the oscillator:

$$K_{VCO} = \frac{6.2 \text{ GHz} - 5.7 \text{ GHz}}{2.5\text{V} - 1.5\text{V}} = 500 \text{ MHz/V}$$

This is a high-gain VCO. It should also be noted that this is a very crude estimate of the gain, as the varactors will be very nonlinear. Thus, in some regions, the gain could be as much as twice this value.

Next we determine how much noise voltage is produced by this resistor:

$$v_n = \sqrt{4kTr} = \sqrt{4(1.38 \times 10^{-23} \text{ J/K})(298\text{K})(300\Omega)} = 2.22 \text{ nV/}\sqrt{\text{Hz}}$$

We are concerned with how much noise ends up on the varactor terminals at 100 kHz. Note that it is at 100 kHz, *not* 6 GHz ± 100 kHz. At this frequency, any varactor is likely to be a pretty good open circuit. Thus, all the noise voltage is applied directly to the varactor terminals and is transformed into phase noise.

$$\text{PN} = \left(\frac{V_m K_{VCO}}{2\Delta\omega}\right)^2 = \left(\frac{(2.22 \text{ nV/}\sqrt{\text{Hz}})(500 \text{ MHz/volt})}{2(100 \text{ kHz})}\right)^2 = 3.08 \times 10^{-11}$$

This is roughly −105.1 dBc/Hz at 100-kHz offset. Given that originally the VCO had a phase noise of −105 dBc/Hz at a 100-kHz offset and we have now doubled the noise present, the design will lose 3 dB and give a performance of −102 dBc/Hz at 100-kHz offset. This means that the VCO will no longer meet specifications. This illustrates the importance of keeping the control line noise as low as possible. It is also easy to see that good-intentioned colleagues can usually be counted on to compromise your design.

8.17.3 Nonlinear Noise

A third type of noise in oscillators is due to the nonlinearity in the transistor mixing noise with other frequencies. For instance, referring to Figure 8.36, assume that there is a noise at some frequency f_n. This noise will get mixed with the oscillation tone f_o to the other sideband at $2f_o - f_n$. This is the only term that falls close to the carrier. The other terms fall out of band and are therefore of much less interest.

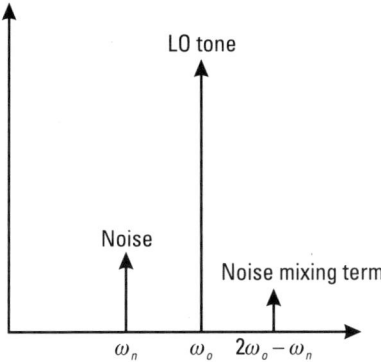

Figure 8.36 Conceptual figure to show the effect of nonlinear mixing.

The magnitude of this noise can be estimated with the following analysis. The analysis begins by considering a transistor being driven by a large sinusoidal voltage, as shown in Figure 8.37, and a small noise source. It is assumed that the transistor can be described by the following power series:

$$i_C \approx I_C \left[1 + \frac{v_i}{v_T} + \frac{1}{2}\left(\frac{v_i}{v_T}\right)^2 + \frac{1}{6}\left(\frac{v_i}{v_T}\right)^3 \cdots + \frac{1}{n!}\left(\frac{v_i}{v_T}\right)^n \right]$$
$$= k_o + k_1 v_i + k_2 v_i^2 + \ldots + k_n v_i^n \tag{8.96}$$

Note that truncation after only a few terms is not possible due to the fact that the oscillation tone is much greater than v_T. Now let us assume that the input is given by

$$v_i = v_o \cos(\omega_o t) + v_n \cos(\omega_n t) \tag{8.97}$$

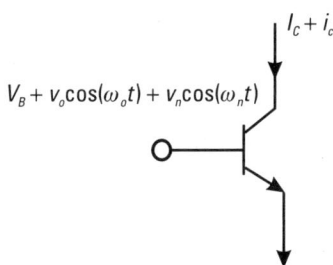

Figure 8.37 A transistor driven by a large sinusoid in the presence of noise.

where v_o is the fundamental tone in the oscillator and v_n is some small noise source at some frequency ω_n. Substituting (8.97) into (8.96), the components at frequency ω_n can be extracted and are given by

$$i_{\omega_n} \approx k_1 v_n + \frac{6}{4}k_3 v_n v_o^2 + \frac{30}{16}k_5 v_n v_o^4 + \frac{140}{64}k_7 v_n v_o^6 + \frac{630}{256}k_9 v_n v_o^8$$
$$+ \frac{2,772}{1,024}k_{11} v_n v_o^{10} + \ldots \quad (8.98)$$

assuming that $v_o \gg v_n$. Note that the constants can be derived as a series or computed with the aid of a software package. The third-order intermodulation term can likewise be extracted from (8.96) and (8.97) and is given by

$$i_{2\omega_o - \omega_n} \approx \frac{3}{4}k_3 v_n v_o^2 + \frac{20}{16}k_5 v_n v_o^4 + \frac{105}{64}k_7 v_n v_o^6 + \frac{504}{256}k_9 v_n v_o^8$$
$$+ \frac{2,310}{1,024}k_9 v_n v_o^{10} + \ldots \quad (8.99)$$

Note that as the number of terms gets large, the ratio of the ith term of (8.98) and the ith term of (8.99) approaches 1, so that

$$0 < \frac{i_{2\omega_o - \omega_n}}{i_{\omega_n}} < 1 \quad (8.100)$$

For most practical oscillation amplitudes, the ratio in (8.100) will be approximately 1. Leeson's formula can provide the amplitude of the linear noise, which is present at frequency ω_n. If it is further assumed that there is equal linear noise content at both ω_n and $2\omega_o - \omega_n$, then this excess noise is added on top of what was already accounted for in the linear analysis. Since these noise sources are uncorrelated, the powers, rather than voltages, must be added and this means that about 3 dB of noise is added to what is predicted by the linear analysis. Thus, the noise content at an offset frequency is

$$\text{PN} = \left[\frac{A\omega_o}{(2Q\Delta\omega)}\right]^2 \left(\frac{FkT}{2P_S}\right) \quad (8.101)$$

where Q is the quality factor of the resonator, P_S is the power of the oscillator, F is the noise figure of the transistor used in the resonator, k is Boltzmann's constant, T is the temperature of operation, ω_o is the frequency of operation, $\Delta\omega$ is the frequency offset from the carrier, and A' takes into account the nonlinear noise and is approximately $\sqrt{2}$. Note that flicker noise and the thermal noise floor have not been included in this equation but are straightforward to add (see discussion about Figure 8.34).

The term A added to Leeson's formula is usually referred to in the literature as the excess small-signal gain. However, A has been shown, for most operating conditions, to be equal to 3 dB, independent of coefficients in the power series used to describe the nonlinearity, the magnitude of the noise present, the amplitude of oscillation, or the excess small-signal gain in the oscillator.

8.18 Making the Oscillator Tunable

Varactors in a bipolar process can be realized using either the base-collector or the base-emitter junctions or else using a MOS varactor in BiCMOS processes [10, 11]. However, when using any of these varactors, there is also a parasitic diode between one side and the substrate. Unlike the base-collector or base-emitter junctions, which have a high Q, this parasitic junction has a low Q due to the low doping of the substrate. This makes it desirable to remove it from the circuit. Placing the varactors in the circuit such that the side with the parasitic diodes tied together at the axis of symmetry can do this.

Example 8.9 VCO Varactor Placement
Make a differential common-base oscillator tunable. Use base-collector junctions as varactors and choose an appropriate place to include them in the circuit.

Solution
The most logical place for the varactors is shown in Figure 8.38. This gives a tuning voltage between power and ground and prevents the parasitic substrate diodes from affecting the circuit.

As can be seen from the last section, low-frequency noise can be very important in the design of VCOs. Thus, the designer should be very careful how the varactors are placed in the circuit. Take, for instance, the $-G_m$ oscillator circuit shown in Figure 8.39. Suppose that a low-frequency noise current was injected into the resonator either from the transistors Q_1 and Q_2 or from the current source at the top of the resonator. Note that at low frequencies the inductors behave like short circuits. This current will see an impedance equal to the output impedance of the current source in parallel with the transistor loading (two forward-biased diodes in parallel). This would be given by

$$R_{\text{Load}} = r_{\text{cur}} // r_{e1} // r_{e2} \approx \frac{r_e}{2} \tag{8.102}$$

where r_{cur} is the output impedance of the current source and the transistors are assumed to be identical.

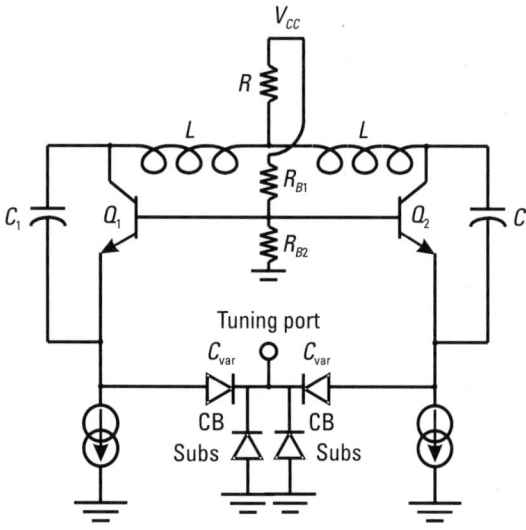

Figure 8.38 A Colpitts common-base design including output buffers and correct varactor placement.

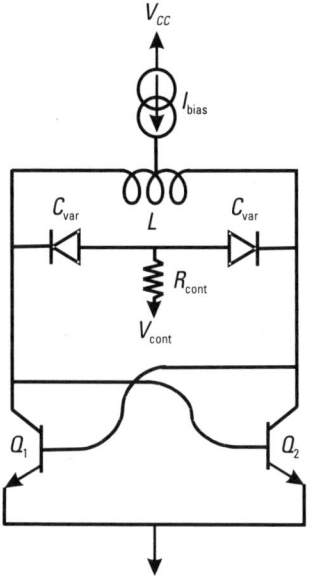

Figure 8.39 An oscillator topology sensitive to low-frequency noise.

This impedance given by (8.102) could easily be in the tens of ohms. Compare this to the circuit shown in Figure 8.24. In the case of Figure 8.24, noise currents have only the dc series resistance of the inductor coil over which to develop a voltage. Thus, if low-frequency noise starts to dominate in the design, the topology of Figure 8.24 would obviously be the better choice.

There are some circumstances where a topology, like the one shown in Figure 8.39, is unavoidable. An example is when using a MOS varactor. This varactor typically requires both positive and negative voltages. When using such a varactor, there is no choice but to place it in the circuit without the terminals connected to V_{CC}.

Example 8.10 Colpitts VCO Design

Consider the Colpitts VCO shown in Figure 8.40. We want to design a VCO to run between 2.4 and 2.5 GHz. The supply will be 3.3V. A differential octagonal inductor has been previously optimized and characterized to have a Q of 14 at 2.5 GHz and an inductance of 4 nH. Base-collector varactors are available in the process and have a Q of 30 at the bottom of their tuning range (excluding the parasitic diode) at 2.5 GHz. They have a $C_{max}:C_{min}$ ratio of 2:1.

Design
The first step is to determine how much capacitance will be needed. We can find the mean total capacitance including parasitics:

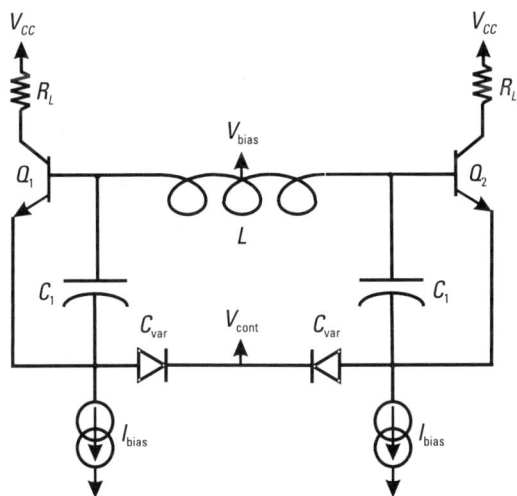

Figure 8.40 A Colpitts common-collector design example.

$$C_{total} = \frac{1}{\omega_{osc}^2 L} = \frac{1}{(2\pi \cdot 2.45 \text{ GHz})^2 4 \text{ nH}} = 1.05 \text{ pF}$$

Now the capacitance must be broken between C_1 and C_{var}. Since the r_e of the transistors will load the resonator, we would like to transform this resistance to a higher value. This can be done to greatest advantage if C_1 is made larger than C_{var}. However, with more imbalances in the capacitor sizes, the negative resistance they will generate is lower, and as a result, the oscillator is less efficient. These tradeoffs must be carefully considered, preferably with the aid of a simulator. Since we have to start somewhere, let us choose $C_{var} = 0.8 C_1$. We can now work out the values for C_1 and C_{var}:

$$2C_{total} = \frac{C_1 C_{var}}{C_{var} + C_{var}} = \frac{0.8 C_1^2}{1.8 C_1} \Rightarrow C_1 = 4.5 C_{total} = 4.72 \text{ pF}$$

and therefore $C_{var} = 3.78$ pF nominally, but we still have to work out its minimum and maximum values.

$$C_{var\,max} = \frac{2 C_1 C_{total\,max}}{C_1 - 2 C_{total\,max}} = 4.12 \text{ pF}$$

$$C_{var\,min} = \frac{2 C_1 C_{total\,min}}{C_1 - 2 C_{total\,min}} = 3.55 \text{ pF}$$

This is a ratio of only 1.16:1, which means that we can make this capacitor with a smaller varactor in parallel with some fixed capacitance.

This leads to two equations with two unknowns:

$$C_{Fixed} + C_{var\,min} = 3.55 \text{ pF}$$

$$C_{Fixed} + 2 C_{var\,min} = 4.12 \text{ pF}$$

solving these two equations, we find that we can make C_{var} with a 2.98-pF fixed capacitor and a varactor with a C_{max} of 1.14 pF.

With the frequency of oscillation set and the capacitors sized, the next thing to consider is the large signal behavior of the oscillator. We will assume that the VCO will drive a 50Ω load for the purposes of this example. Thus, the best choice for the R_L resistors would be 50Ω.

Now we must set the output swing. We would like to make the oscillator swing at the bases of Q_1 and Q_2 as large as possible so that we get good power and therefore help to minimize phase noise. (Note that in a real design there may be many other tradeoffs such as power consumption.) The voltage can swing downwards until the current sources start to give out. Most current sources in a bipolar technology (depending on how they are built) are happy until they hit about 0.5V or less.

Let us choose a swing of about 2 · 0.8V = 1.6V swing total.

If we bias the base at 1.65V and it swings down 0.8V from there, then it will sit at about 0.85V. The V_{BE} at this point will be small, <0.5V (the transistors are off except at the peak of the swing), so the current sources should avoid saturation. As another check, the emitter swing will only be 0.55 · 0.8V = 0.44V. So if they are biased at 0.8V, then they should swing between 0.8 and about 0.35 just at the bottom of the swing, which should be fine. The collector is biased at about 100 mV below supply. The peak current in the oscillator will be quite high; the base will swing as high as 2.45V. Thus, the collector can swing as low as 2V without serious problems. The only way that the collector could swing down to 2V is if the current had a peak value of 26 mA or about 15 times the quiescent value. This is not likely, so we are safe.

Now we need to set the current level to get this voltage swing.

The inductor has a Q of 14 at 2.5 GHz. Therefore, its parallel resistance is

$$R_L = Q_{ind}\omega_o L = 14 \cdot (2 \cdot \pi \cdot 2.5 \text{ GHz}) 4 \text{ nH} = 880 \Omega$$

The varactor has a Q of 30 in the C_{max} condition. This corresponds to a series resistance of

$$r_s = \frac{1}{Q_{var}\omega_o C_{var}} = \frac{1}{(2 \cdot \pi \cdot 2.5 \text{ GHz}) \cdot 30 \cdot 1.14 \text{ pF}} = 1.86 \Omega$$

In the middle of the tuning range, the varactors have a capacitance of 0.855 pF. If we assume that the series resistance of these varactors remains constant, then the equivalent parallel resistance is

$$R_{cap} = \frac{|Z_{var}|}{R_s} \cdot \frac{1}{\omega_o C_{var}}$$

$$= \frac{1}{1.86 \Omega (2 \cdot \pi \cdot 2.45 \text{ GHz}) \cdot 0.855 \text{ pF}} \cdot \frac{1}{(2 \cdot \pi \cdot 2.45 \text{ GHz}) \cdot 0.855 \text{ pF}}$$

$$= 3.1 \text{ k}\Omega$$

Now this resistance must be converted into the equivalent resistance at the bases of Q_1 and Q_2 (note that C_2 is made up of 0.855 pF of varactor capacitance and 2.98 pF of fixed capacitance at the middle of the tuning range):

$$R'_{cap} = 2\left(1 + \frac{C_2}{C_1}\right)^2 R_{cap} = 2\left(1 + \frac{3.83 \text{ pF}}{4.72 \text{ pF}}\right)^2 3.1 \text{ k}\Omega = 20.3 \text{ k}\Omega$$

Thus, the total parallel resonator resistance is $R_p = 843\Omega$.

$$V_{tank} \approx 2I_{bias}\left(\frac{C_1}{C_1 + C_2}\right) R_p$$

$$I_{bias} = \frac{V_{tank}}{2R_p}\left(\frac{C_1 + C_2}{C_1}\right) = \frac{1.6\text{V}}{2(843\Omega)}\left(\frac{4.72 \text{ pF} + 3.83 \text{ pF}}{4.72 \text{ pF}}\right) = 1.71 \text{ mA}$$

The next thing to do is to size the transistors used in the tank. Transistors were chosen to be 25 μm.

We can also estimate the phase noise of this oscillator.

The r_e of the transistor at this bias is

$$r_e = \frac{v_T}{I_C} = \frac{25 \text{ mV}}{1.69 \text{ mA}} = 14.8\Omega$$

Since this value will seriously affect our estimate, we also take into account 5Ω for parasitic emitter resistance (this value can be determined with a dc simulation).

$$R_{r_e tank} = 2\left(1 + \frac{C_1}{C_2}\right)^2 r_e = 2\left(1 + \frac{4.72 \text{ pF}}{3.78 \text{ pF}}\right)^2 19.8\Omega = 200\Omega$$

This can be added to the existing losses of the overall resonator resistance of 843Ω to give 162Ω.

Now we can compute the Q:

$$Q = R_{tank}\sqrt{\frac{C_{total}}{L}} = 162\Omega\sqrt{\frac{1.18 \text{ pF}}{4 \text{ nH}}} = 2.78$$

We need to estimate the available power:

$$P_S = \frac{V_{tank}^2}{2R_{tank}} = \frac{(1.6\text{V})^2}{2(162\Omega)} = 7.9 \text{ mW}$$

We can now estimate the phase noise of the oscillator. We will assume that the phase noise due to K_{VCO} is not important.

We will assume a noise factor of 1 for the transistor.

$$\text{PN}(f_m) = 10 \log\left\{\left[\frac{Af_o}{(2Qf_m)}\right]^2 \left(\frac{FkT}{2P_S}\right)\right\} = 10 \log\left(\frac{0.098 \text{ Hz}^2}{f_m^2}\right)$$

The oscillator was simulated and the following results will now be shown.

Transistor voltage waveforms can be seen in Figure 8.41. Note that the transistor safely avoids saturating. Also note that emitter voltage stays above 0.5V, leaving the current source enough room to operate.

The collector current waveform is shown in Figure 8.42. Note how nonlinear the waveform is. The transistor is on for only a short portion of the cycle.

The differential collector output voltage is shown in Figure 8.43. Notice all the harmonics present in this waveform. Given the collector currents, this is to be expected.

The differential resonator voltage is shown in Figure 8.44. It is much more sinusoidal, as it is filtered by the LC resonator. The peak voltage is around 1.8V, which is only slightly higher than the simple estimate (1.6V) we did earlier.

The tuning voltage versus frequency for the VCO is shown in Figure 8.45. Note that we are only getting about 60 MHz of tuning range and that the frequency has dropped about 120 MHz compared to the design calculations.

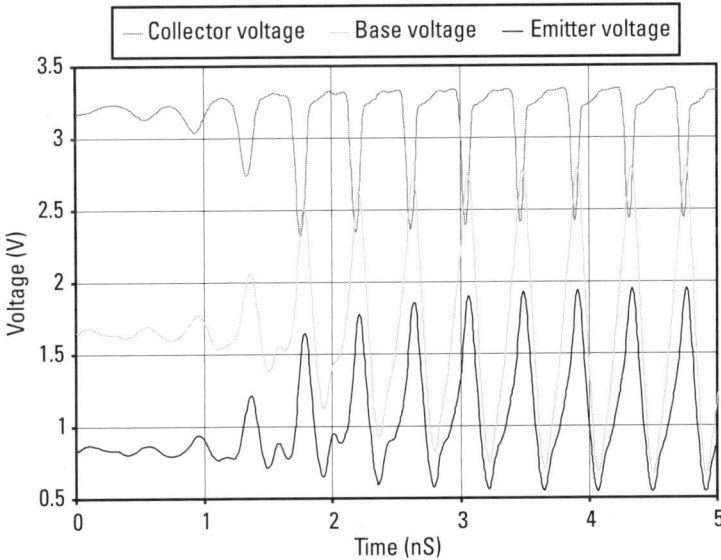

Figure 8.41 Base, collector, and emitter voltages.

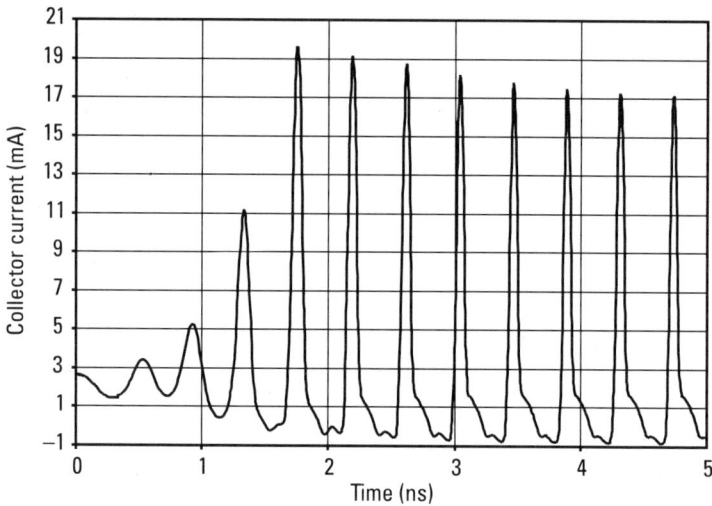

Figure 8.42 Collector current.

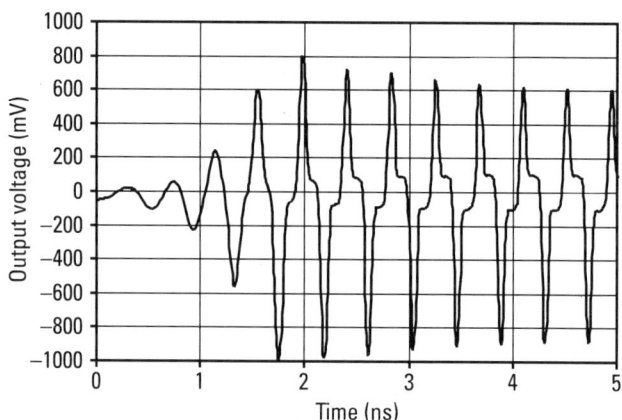

Figure 8.43 Differential output voltage waveform.

This clearly shows the effect of transistor parasitics. This design will have to be tweaked in simulation in order to make its frequency accurately match what was asked for at the start. This is best done with a simulator, carefully taking into account the effect of all stray capacitance. The results of the phase noise simulation and calculation are shown in Figure 8.46.

8.19 VCO Automatic-Amplitude Control Circuits

The purpose of adding *automatic-amplitude control* (AAC) to a VCO design is to create a VCO with good phase noise and very robust performance over

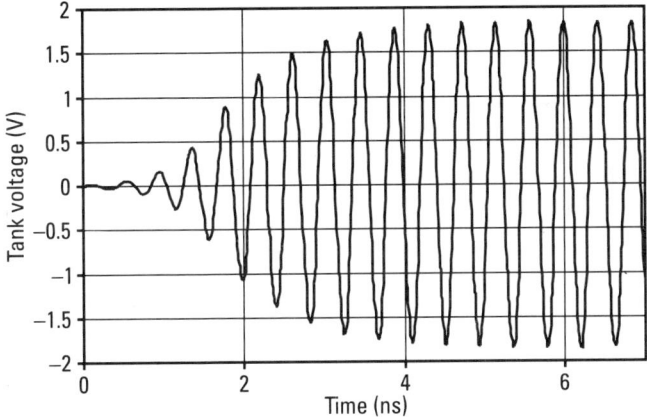

Figure 8.44 Differential resonator voltage.

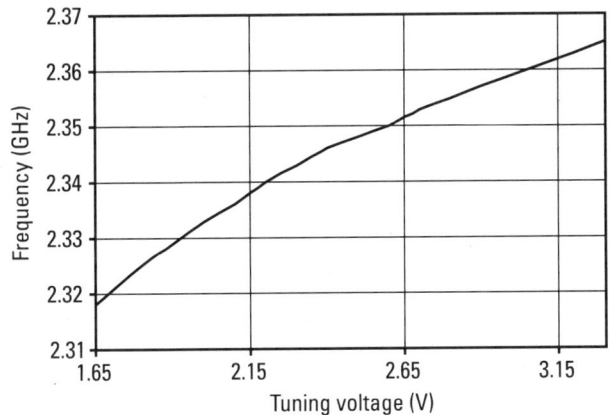

Figure 8.45 Frequency versus varactor tuning voltage.

process, temperature, and frequency variations [12]. A VCO schematic with simple additional feedback circuitry is shown in Figure 8.47.

The current in the VCO is set by transistor Q_6, which acts as a current source. Transistors Q_3 and Q_4 are used to limit the swing of the oscillator to $\pm 2V_{BE}$. Once the oscillation gets this large, these transistors start to turn on briefly at the top and bottom of the oscillator's swing, loading the resonator with their dynamic emitter resistance. This will effectively de-Q the circuit and prevent the signal from growing any larger. This will prevent transistors Q_1 and Q_2 from entering saturation. However, if transistors Q_3 and Q_4 have to be heavily turned on to limit the swing, they will also start to affect the phase noise performance of the circuit.

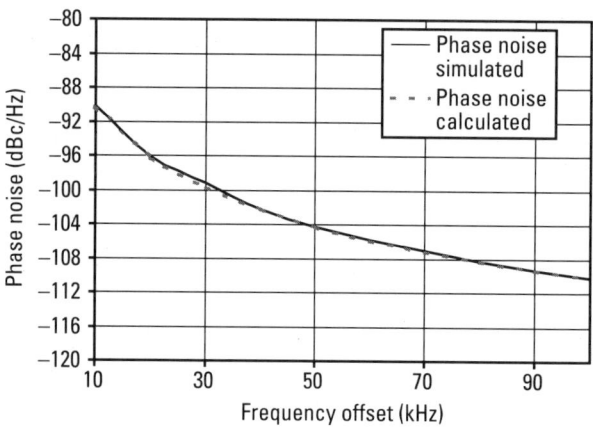

Figure 8.46 Simulated and calculated phase noise.

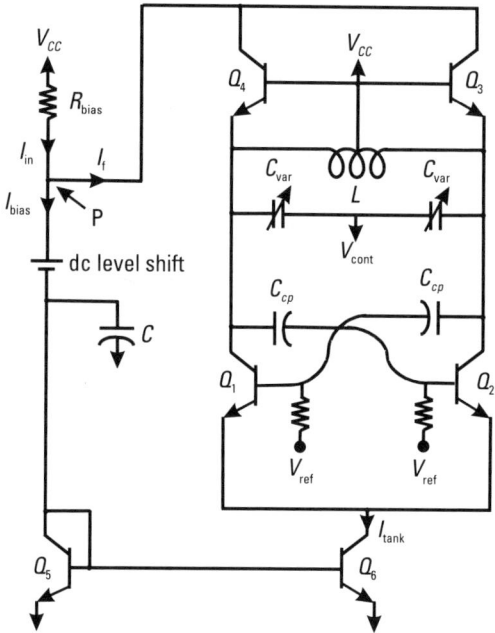

Figure 8.47 VCO topology with feedback in the bias to control the amplitude.

Transistors Q_3 and Q_4 limit the amplitude of the oscillation directly, but are also the basis for the feedback loop that is the second mechanism used to make sure that the VCO is operating at an optimal level. Once these transistors start to turn on, they start to draw current I_f. Their collectors are connected back to the resistor R_{bias}. Q_3 and Q_4 then steal current away from the bias, causing the current in Q_6 to be reduced. This in turn reduces the current in the VCO. Since the VCO amplitude is related to its current, the amplitude of the VCO is thus reduced until transistors Q_3 and Q_4 just barely turn on. This ensures that the VCO always draws just enough current to turn on these transistors and no more, although the reference current through R_{bias} may vary for any number of reasons. The reference current through R_{bias} must, therefore, be set higher than the optimum, as the loop can only work to reduce the current through the oscillator, but can never make it higher. Put another way, unless transistors Q_3 and Q_4 turn on, the loop has zero gain.

The loop can be drawn conceptually as shown in Figure 8.48. The point P shown in Figure 8.47 acts as a summing node for the three currents I_{in}, I_{bias}, and I_f. The current mirror amplifies this current and produces the resonator current, which is taken by the VCO core and produces an output voltage proportional to the input resonator current. The limiting transistors at the top of the resonator take the VCO amplitude and convert it into a current that is fed back to the input of the loop.

The transfer function for the various blocks around the loop can now be derived. In the current mirror, a capacitor C has been placed in the circuit to limit the frequency response of the circuit. It creates a dominant pole in the system (this helps to control the effect of parasitic poles on the system) and limits the frequency response of the loop. The transfer function for this part of the loop is given by

$$A_1(s) = \frac{I_{tank}(s)}{I_{bias}(s)} \approx \frac{\frac{g_{m6}}{C}}{s + \frac{r_{\pi 5} + r_{\pi 6} + g_{m5} r_{\pi 5} r_{\pi 6}}{r_{\pi 5} r_{\pi 6} C}} \quad (8.103)$$

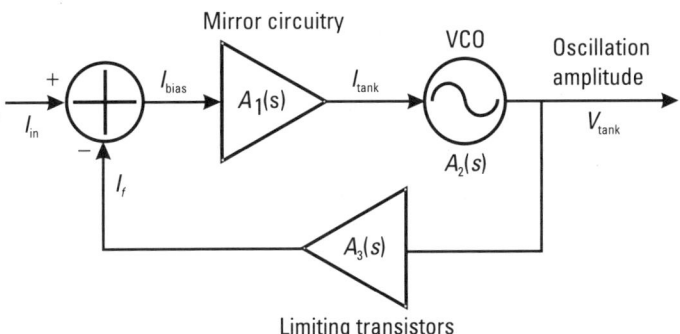

Figure 8.48 Conceptual drawing of the AGC feedback loop.

This equation has a dominant pole at

$$P_1 \approx \frac{g_{m5}}{C} \tag{8.104}$$

The behavior of the oscillator must also be determined in so far as it affects the behavior of the loop. We have already shown that the VCO amplitude can be approximated by (8.68). This formula has obvious limitations in describing certain aspects of oscillator performance. Specifically, for large amplitudes, the oscillation amplitude will cease to grow with increasing current and for low current the VCO will not start. More importantly, this expression also fails to capture the frequency response of the oscillator amplitude.

For the purposes of this analysis, the oscillator resonator is treated as a resonator with a pulse of current applied to it by transistors Q_1 and Q_2 each half cycle. From this simple model, the transient behavior of the circuit can be determined. The resonator forms a time constant $R_p C_{var}$ that is equivalent to a pole in the response of the oscillation amplitude versus bias current. This pole can be used to give frequency dependence to (8.68).

$$A_2(s) = \frac{V_{tank}(s)}{I_{tank}(s)} = \frac{2}{\pi C_{var}} \left(\frac{1}{s + \frac{1}{R_p C_{var}}} \right) \tag{8.105}$$

This pole can also be written in terms of the Q and frequency of oscillation:

$$P_2 = \frac{1}{C_{var} R_p} = \frac{\omega_{osc}}{2Q} \tag{8.106}$$

It is interesting to note that a resonator with higher Q will respond slower and therefore have a lower frequency pole than a low-Q oscillator. This makes intuitive sense, since it is up to the losses in the resonator to cause a change in amplitude.

The last part of the loop consists of the limiting transistors. This is the hardest part of the loop to characterize because by their very nature, the limiters are very nonlinear. The transistor base is essentially grounded, while the emitter is attached to the resonator of the oscillator. In the scheme that has been shown, the base is connected to a voltage higher than the resonator voltage. For any reasonable applied resonator voltage, the current will form narrow pulses with large peak amplitude. Thus, this current will have strong harmonic content. This harmonic content will lead to a nonzero dc current, which is the property of interest. Finding it requires solving the following integral.

$$I_{C\_AVE} = \frac{1}{2\pi} \int_0^{2\pi} I_S e^{\frac{V_{tank}}{2v_T}\sin(\theta)} d\theta = I_S I_0\left(\frac{V_{tank}}{2v_T}\right) \quad (8.107)$$

where $I_0(x)$ is a modified Bessel function of the first kind of order zero, and I_S is the saturation current. For fairly large $V_{tank}/2v_T$, there is an approximate solution:

$$I_{C\_AVE} \approx \frac{I_S \cdot e^{\frac{V_{tank}}{2v_T}}}{\sqrt{2\pi \frac{V_{tank}}{2v_T}}} \quad (8.108)$$

Now we can write the gain of this part of the loop, which is a nonlinear function of V_{tank} (all other parts of the loop were described by linear functions):

$$A_3(s) = 2\frac{\partial I_{C\_AVE}}{\partial V_{tank}} = \frac{I_S \cdot e^{\frac{V_{tank}}{2v_T}}}{\sqrt{\pi v_T V_{tank}}} - \frac{I_S \cdot e^{\frac{V_{tank}}{2v_T}}}{\sqrt{\frac{\pi}{V_T}} V_{tank}^{3/2}} \quad (8.109)$$

Here it is assumed that this part of the loop has poles at a significantly higher frequency than the one in the VCO and the one in the mirror.

These equations can be used to design the loop and demonstrate the stability of this circuit. The capacitor C_2 is placed in the circuit to create a dominant and controllable pole P_1 significantly below the other pole P_2. Generally, the frequency of P_2 and gain of $A_2(s)$ are set by the oscillator requirements and are not adjustable. For more stability, the loop gain can be adjusted by either changing the gain in $A_1(s)$ (by adjusting the ratio of the current mirror) or by adjusting the gain of $A_3(s)$ (this can be done by changing the size of the limiting transistors Q_3 and Q_4). Reducing the gain of the loop is a less desirable alternative than adjusting P_1, because as the loop gain is reduced, its ability to settle to an exact final value is reduced.

Example 8.11 The Design of a VCO AAC Loop

Design an AAC loop for the VCO schematic shown in Figure 8.47. Use $L = 5$ nH ($Q = 10$), $V_{CC} = 5$V. The AAC loop should be set up so that the dc gain around the loop is 40 dB to ensure that the final dc current through the oscillator is set accurately. The loop is to have 0-dB gain at 1 GHz to ensure that parasitic phase shift has minimal impact on the stability of the design. Also, find the phase margin of the loop. Assume that $J_s = 1 \times 10^{-18}$ A/μm and $\beta = 100$ in this technology. Use no more than a gain of 1:10 in the current mirror.

Solution

We can first compute R_p in the usual way, assuming that the inductor is the sole loss in the system. It is easy to see that at 2 GHz this will be 628.3 Ω. Next, we also need to know the value of the two capacitors in the resonator. Noting that there are two capacitors,

$$C_{var} = \frac{2}{\omega_{osc}^2 L} = \frac{1}{(2\pi \cdot 2 \text{ GHz})^2 5 \text{ nH}} = 2.53 \text{ pF}$$

Now since the transistor limiters will turn on for a voltage of about 0.85V in a modern bipolar technology, we can assume that the VCO amplitude will end up being close to this value. Therefore, we can compute the final value for the resonator current from (8.68).

$$I_{tank} = \frac{V_{tank}}{0.635 R_p} = \frac{1.7 \text{V}}{(0.635) 628.3 \Omega} = 4.26 \text{ mA}$$

We will choose to use the 10:1 ratio for the mirror in order to get the best dc gain of 10 from this stage. Thus, the current through the mirror transistor will be 426 μA. We can now compute the values of transconductance and resistance in the model:

$$g_{m5} = \frac{I_{bias}}{v_T} = \frac{426 \text{ μA}}{25 \text{ mV}} = 17.04 \frac{\text{mA}}{\text{V}}$$

$$g_{m6} = \frac{I_{bias}}{v_T} = \frac{4.26 \text{ mA}}{25 \text{ mV}} = 170.4 \frac{\text{mA}}{\text{V}}$$

$$r_{\pi 5} = \frac{\beta}{g_{m4}} = \frac{100}{17.04 \text{ mS}} = 5.87 \text{ k}\Omega$$

$$r_{\pi 6} = \frac{\beta}{g_{m4}} = \frac{100}{170.4 \text{ mS}} = 587 \Omega$$

Once we have chosen the gain for the current mirror, we also know the gain from the resonator:

$$A_2(s) = \frac{V_{tank}(0)}{I_{tank}(0)} = \frac{\frac{2}{\pi} \frac{1}{C_{var}}}{s + \frac{1}{R_p C_{var}}} = \frac{2}{\pi} R_p = 399 \frac{\text{V}}{\text{A}}$$

and since the gain from the mirror will be 10, we can now find the gain required from the limiters.

$$\text{Loop gain} = 40 \text{ dB} = 20 \log(10 \cdot 399 \cdot A_2)$$

$$A_2 = 25.1 \frac{\text{mA}}{\text{V}}$$

Thus, from (8.109) we can size the limiting transistors to give the required gain:

$$A_2 = \frac{I_S e^{\frac{V_{\text{tank}}}{2v_T}}}{\sqrt{\pi v_T V_{\text{tank}}}} - \frac{I_S e^{\frac{V_{\text{tank}}}{2v_T}}}{\sqrt{\frac{\pi}{v_T} V_{\text{tank}}^{3/2}}}$$

$$I_S = A_2 \left[\frac{e^{\frac{V_{\text{tank}}}{2v_T}}}{\sqrt{\pi v_T V_{\text{tank}}}} - \frac{e^{\frac{V_{\text{tank}}}{2v_T}}}{\sqrt{\frac{\pi}{v_T} V_{\text{tank}}^{3/2}}} \right]^{-1}$$

$$= 2.51 \frac{\text{mA}}{\text{V}} \left[\frac{e^{\frac{1.7\text{V}}{2(25 \text{ mV})}}}{\sqrt{\pi (25 \text{ mV})(1.7\text{V})}} - \frac{e^{\frac{1.7\text{V}}{2(25 \text{ mV})}}}{\sqrt{\frac{\pi}{25 \text{ mV}} 1.7\text{V}^{3/2}}} \right]^{-1}$$

$$= 1.59 \times 10^{-17} \text{ A}$$

This sets the size of these transistors to be 15.9 μm, which is a reasonable size and will likely not load the resonator very much.

Now we can find the pole in the oscillator. It is located at

$$f_2 = \frac{\omega_2}{2\pi} = \frac{1}{2\pi R_p C_{\text{var}}} = \frac{1}{2\pi(628.3\Omega)(2.53 \text{ pF})} = 100.1 \text{ MHz}$$

This means that at 1 GHz, this pole will provide 20 dB of attenuation from the dc value. Thus, the other pole will likewise need to provide 20 dB of attenuation and must also be located at 100 MHz. Since this will lead to both poles being at the same frequency, this design will end up with poor phase margin. We will finish the example as is and fix the design in the next example.

The other pole is located at

$$f_1 = \frac{\omega_1}{2\pi} = \frac{g_{m5}}{2\pi C} \Rightarrow C = \frac{g_{m5}}{2\pi f_1} = \frac{(10.34 \text{ mS})}{2\pi(100 \text{ MHz})} = 16.45 \text{ pF}$$

This is a large but not unreasonable MIM cap, but it could also be made out of a poly cap. In order to get the phase margin, it is probably easiest to plug the formulas into your favorite math program. You can produce a plot like the one shown in Figure 8.49. The graph shows that this design has about 12° of phase margin. This is not great. If there is only a little excess phase shift in the loop from something that we have not considered, then this design could very well be unstable. This could be a bad thing. In order to get more phase margin, the poles could be separated or the loop gain could be reduced. Note that the reduction of the loop gain will affect its ability to accurately set the conditions of the oscillator.

Example 8.12 Improvements to an AAC Loop

Make some suggestions about how to improve stability and simulate a preliminary design.

Solution

Let us assume that the inductance is still fixed; thus, the pole in the oscillator cannot be adjusted. One obvious thing we can do to make things better is to reduce the loop gain; however, let us start by separating the poles. If we can make the limiting transistors three times smaller at 5.3 μm, but increase the current mirror ratio from 10:1 to 30:1, the gain will remain the same. This will reduce the current in Q_5, which will mean that g_{m5} goes down by a factor of 3. Since the pole is g_{m5}/C, this will move the pole frequency to 33 MHz. We could separate the poles further by increasing C. Note that instead of reducing the pole frequency, this pole could be moved to a much higher

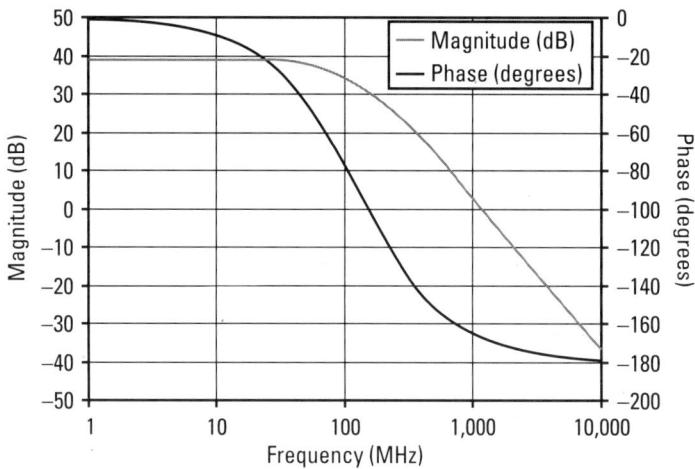

Figure 8.49 Gain and phase response of the AAC loop.

frequency (say, 3 GHz), but then the second harmonic would not be attenuated by the loop, which could lead to other problems.

Note that the gain and amount of pole separation are still not enough to make the design practical. If we were designing a product, we would continue to refine these values until we are sure that the feedback loop will not oscillate under any conditions.

We again employ a math tool to plot the frequency response of the equations.

From the plot in Figure 8.50 we can see that there is now about 16° of phase margin, which is still not great, but safer than the original design. Instead of trying to refine the design, let us simply perform some initial simulations to demonstrate the operation of the circuit.

The supply was chosen to be 5V, the base was biased at 2.5V, and the reference current was chosen as 300 μA. The results are plotted in Figure 8.51, which shows the base and collector waveforms of Q_1 or Q_2. Note that the collector voltage is always higher than the base voltage, ensuring that the transistor never saturates.

Figure 8.52 shows the current in transistor Q_6. Note that we have started the reference current at almost 8 mA. Once the loop begins to operate, it brings this current back to a value necessary to give the designed amplitude. This is about 4.5 mA. Note that this is higher than the estimated current. This is because the resonator voltage is higher than we estimated at the start of the example, and thus more current is required. Figure 8.53 shows the response of the first design biased under similar conditions. Note that with the poles at the same frequency, the response is much less damped. One would expect that in

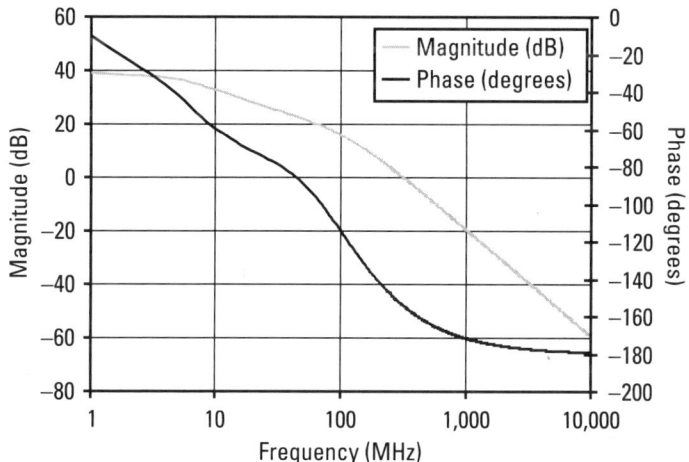

Figure 8.50 Gain and phase response of the improved AAC loop.

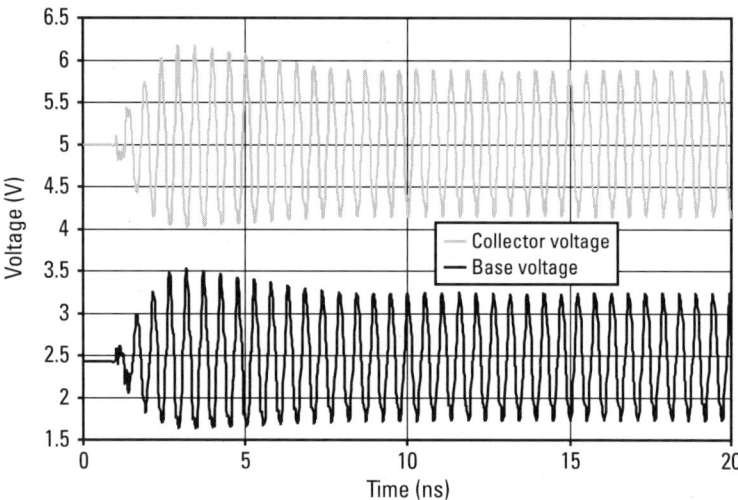

Figure 8.51 Base and collector voltages of the VCO with AAC.

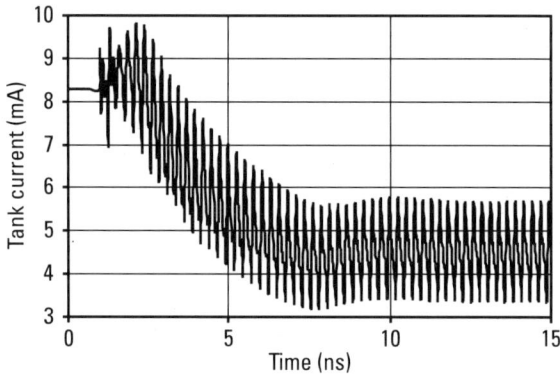

Figure 8.52 Resonator current for the VCO with AAC.

the case for which the poles are widely spaced, the response would look much more first order.

Figure 8.54 shows the differential resonator voltage. Note that it grows and then settles back to a more reasonable value. This peak voltage is slightly higher than that designed for, primarily due to finite loop gain and because we were estimating where the limiting transistors would need to operate. The 1.7V estimate is quite a rough one. Nevertheless, it served its purpose.

Figure 8.55 is a plot of the current through one of the limiting transistors. This also shows the transient response of the system.

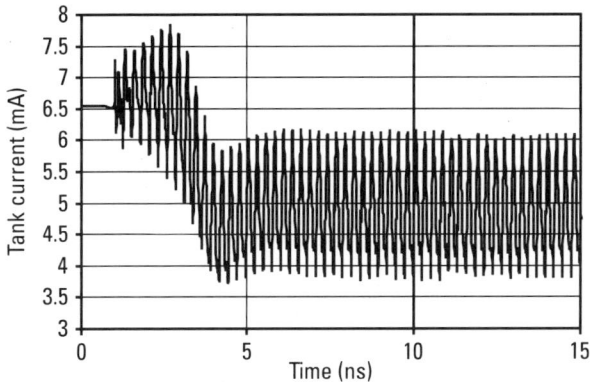

Figure 8.53 Resonator current for the VCO with ACC (reduced phase margin).

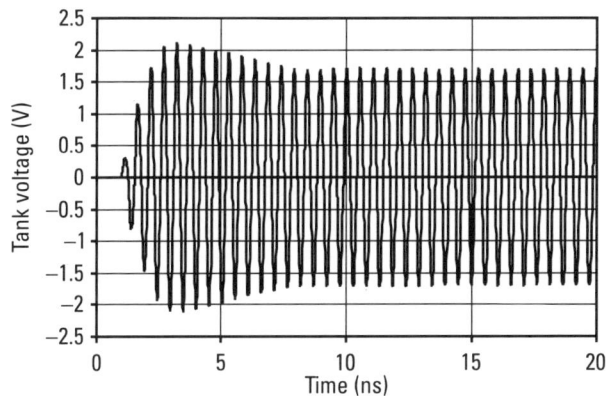

Figure 8.54 Resonator voltage for the VCO with AAC loop.

8.20 Other Oscillators

Although we have stressed LC-based VCOs in this chapter as the most common RF oscillators due to their excellent phase noise, there are many other ways to build a circuit that generates harmonic waveforms. One example is a voltage-controlled, emitter-coupled multivibrator oscillator as shown in Figure 8.56.

Q_1 and Q_2 alternately turn on, and I_1 goes through the capacitor C, causing a ramp of voltage until the other transistor is turned on. First note that Q_5 and Q_6 are always on, each sinking current I_1. Also note that Q_3 and Q_4 are always on due to I_x and I_y, and thus v_{b2} is always one diode drop less than v_{c1}, and v_{b1} is always one diode drop lower than v_{c2}. In the waveforms shown in Figure 8.56, we have assumed that a diode or base-emitter junction, when forward biased, has a voltage drop of $V_{BE(ON)} = 0.8V$. Thus, if Q_1 switches

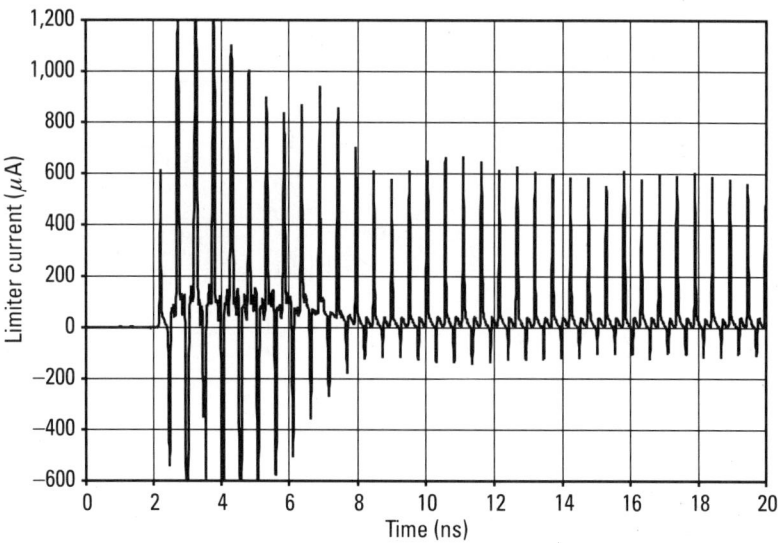

Figure 8.55 Limiting transistor current.

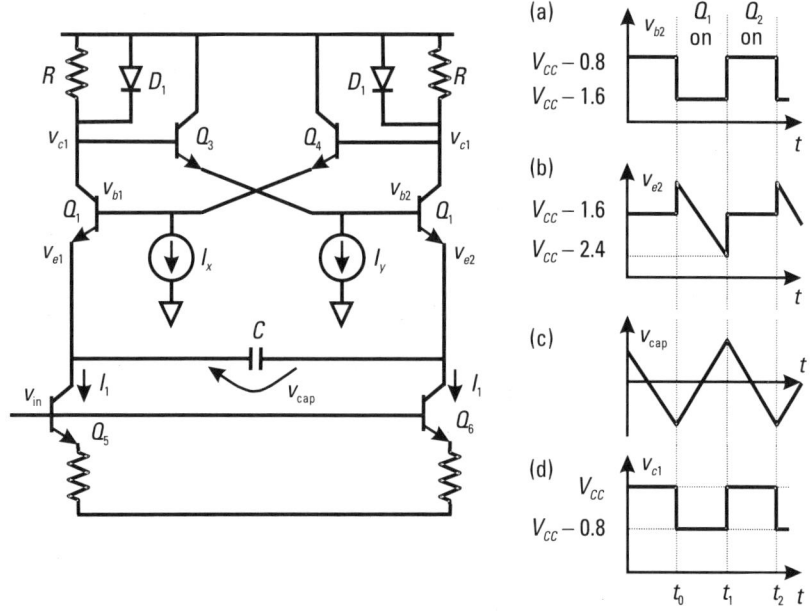

Figure 8.56 Multivibrator VCO with waveforms showing operation: (a) on the base of Q_2; (b) the emitter of Q_2; (c) across the capacitor C; and (d) the collector of Q_1.

from off to on, v_{c1} switches from V_{CC} to one diode drop below V_{CC}, and similarly, as Q_2 switches off and on, v_{c2} switches between V_{CC} and one diode drop below V_{CC}. We note that R is large enough for the diodes to clamp the voltage at v_{c1} or v_{c2}.

Between t_0 and t_1, D_1 and Q_1 are on and D_2 and Q_2 are off. Thus, current flows through D_1, Q_1, Q_5, C, and Q_6. The important thing to note is that current I_1 flows through C and a ramp of voltage occurs as shown in Figure 8.56(c). We also note that node v_{e1} is clamped to two diode drops below V_{CC} by Q_4 and Q_1. Thus, the voltage ramp translates into a negative voltage ramp at v_{e2}. As soon as this voltage reaches three diode drops below V_{CC}, Q_2 turns on, since v_{b2} is held at two diode drops below V_{CC} by D_1 and Q_3. Q_2 turning on also turns on D_2 and turns off Q_1 and D_1, and the voltage on v_{c1} goes back up close to V_{CC}. The rise in v_{c1} is a positive feedback that helps to turn on Q_2 and also serves to raise the voltage v_{e2} by one diode drop to a clamped value of two diode drops below V_{CC}, as seen in Figure 8.56(b). During the time from t_1 to t_2, Q_2 and D_2 are on, while Q_1 and D_1 are off, so there is the opposite ramp across the capacitor [Figure 8.56(d)] and there is a negative ramp of v_{e1}.

The ramp slope is determined by the current I_1 and the capacitor values by

$$\frac{I_1}{C} = \frac{\Delta v_{\text{cap}}}{\Delta t} \qquad (8.110)$$

By noting that during each cycle, a ramp of amplitude $2V_{\text{BE(ON)}}$ occurs twice, the frequency can be found as

$$f = \frac{1}{T} = \frac{I_1}{4V_{\text{BE(ON)}}C} \qquad (8.111)$$

Thus, it can be seen that the frequency can be changed by modifying the bias current. Since this can be changed over a broad range, this oscillator can have a very wide tuning range. The oscillating frequency is limited by the minimum reasonable capacitor size and the speed of switching between saturation and cutoff. To achieve high speed in digital circuits, typically emitter-coupled logic or current-mode logic is used and the need for such switching is avoided.

Another common oscillator topology is the ring oscillator. This is a very simple oscillator made up of an odd number of inverters with the output fed back to the input as shown in Figure 8.57.

When power is applied to this circuit stage, assume the input to 1 is low and the output capacitance C_1 is charged up. When the next stage input sees

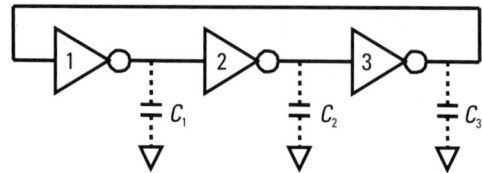

Figure 8.57 A simple ring oscillator.

a high, it will discharge C_2. When the third stage sees a low, it charges C_3. This will in turn discharge stage 1.

Thus, f is related to I/C, where I is the charging or discharging current and C is the capacitance size.

Thus, frequency can be controlled by changing I. This circuit can operate to high frequency if simple inverters are used. It can be made with CMOS or bipolar transistors. However, ring oscillators, not having inductors, are usually thought to be noisier than LC oscillators, but this depends on the design and the technology.

References

[1] Kurokawa, K., "Some Basic Characteristics of Broadband Negative Resistance Oscillator Circuits," *The Bell System Technical J.*, July 1969.

[2] Gonzalez, G., *Microwave Transistor Amplifiers Analysis and Design*, 2nd ed., Upper Saddle River, NJ: Prentice-Hall, 1997.

[3] Voinigescu, S. P., D. Marchesan, and M. A. Copeland, "A Family of Monolithic Inductor-Varactor SiGe-HBT VCOs for 20 GHz to 30 GHz LMDS and Fiber-Optic Receiver Applications," *Proc. RFIC Symposium*, June 2000, pp. 173–176.

[4] Dauphinee, L., M. Copeland, and P. Schvan, "A Balanced 1.5 GHz Voltage Controlled Oscillator with an Integrated LC Resonator," *International Solid-State Circuits Conference*, 1997, pp. 390–391.

[5] Zannoth, M., et al., "A Fully Integrated VCO at 2 GHz," *IEEE J. Solid-State Circuits*, Vol. 33, Dec. 1998, pp. 1987–1991.

[6] Hegazi, E., H. Sjoland, and A. Abidi, "A Filtering Technique to Lower LC Oscillator Phase Noise," *IEEE J. Solid-State Circuits*, Vol. 36, Dec. 2001, pp. 1921–1930.

[7] Razavi, B., "A Study of Phase Noise in CMOS Oscillators," *IEEE J. Solid-State Circuits*, Vol. 31, March 1996, pp. 331–343.

[8] Leeson, D. B., "A Simple Model of Feedback Oscillator Noise Spectrum," *Proc. IEEE*, Feb. 1966, pp. 329–330.

[9] Adams, S., *The Dilbert Principle*, New York: HarperCollins, 1996.

[10] Porret, A. S., et al., "Design of High-Q Varactors for Low Power Wireless Applications Using a Standard CMOS Process," *IEEE J. Solid-State Circuits,* Vol. 35, March 2000, pp. 337–345.

[11] Svelto, F., S. Deantoni, and R. Castello, "A 1.3 GHz Low-Phase Noise Fully Tunable CMOS LC VCO," *IEEE J. Solid-State Circuits,* Vol. 35, March 2000, pp. 356–361.

[12] Margarit, M., et al., "A Low-Noise, Low-Power VCO with Automatic Amplitude Control for Wireless Applications," *IEEE J. Solid-State Circuits,* Vol. 34, June 1999, pp. 761–771.

Selected Bibliography

Chen, W., and J. Wu, "A 2-V 2-GHz BJT Variable Frequency Oscillator," *IEEE J. Solid-State Circuits,* Vol. 33, Sept. 1998, pp. 1406–1410.

Craninckx, J., and M. S. J. Steyaert, "A 1.8-GHz Low-Phase-Noise CMOS VCO Using Optimized Hollow Spiral Inductors," *IEEE J. Solid-State Circuits,* Vol. 32, May 1997, pp. 736–744.

Craninckx, J., and M. S. J. Steyaert, "A Fully Integrated CMOS DCS-1800 Frequency Synthesizer," *IEEE J. Solid-State Circuits,* Vol. 33, Dec. 1998, pp. 2054–2065.

Craninckx, J., and M. S. J. Steyaert, "A 1.8-GHz CMOS Low-Phase-Noise Voltage-Controlled Oscillator with Prescaler," *IEEE J. Solid-State Circuits,* Vol. 30, Dec. 1995, pp. 1474–1482.

Hajimiri, A., and T. H. Lee, "A General Theory of Phase Noise in Electrical Oscillators," *IEEE J. Solid-State Circuits,* Vol. 33, June 1999, pp. 179–194.

Jansen, B., K. Negus, and D. Lee, "Silicon Bipolar VCO Family for 1.1 to 2.2 GHz with Fully-Integrated Tank and Tuning Circuits," *Proc. International Solid-State Circuits Conference,* pp. 392–393.

Niknejad, A. M., J. L. Tham, and R. G. Meyer, "Fully-Integrated Low Phase Noise Bipolar Differential VCOs at 2.9 and 4.4 GHz," *Proc. European Solid-State Circuits Conference,* 1999, pp. 198–201.

Razavi, B., "A 1.8 GHz CMOS Voltage-Controlled Oscillator," *Proc. International Solid-State Circuits Conference,* pp. 388–389.

Rogers, J. W. M., J. A. Macedo, and C. Plett, "The Effect of Varactor Non-Linearity on the Phase Noise of Completely Integrated VCOs," *IEEE J. Solid-State Circuits,* Vol. 35, Sept. 2000, pp. 1360–1367.

Soyuer, M., et al., "A 2.4-GHz Silicon Bipolar Oscillator with Integrated Resonator," *IEEE J. Solid-State Circuits,* Vol. 31, Feb. 1996, pp. 268–270.

Svelto, F., S. Deantoni, and R. Castello, "A 1.3 GHz Low-Phase Noise Fully Tunable CMOS LC VCO," *IEEE J. Solid-State Circuits,* Vol. 35, March 2000, pp. 356–361.

Soyuer, M., et al., "An 11-GHz 3-V SiGe Voltage Controlled Oscillator with Integrated Resonator," *IEEE J. Solid-State Circuits,* Vol. 32, Dec. 1997, pp. 1451–1454.

9

High-Frequency Filter Circuits

9.1 Introduction

The need for filters in a radio has already been discussed extensively in Chapter 2. They must be included in a radio to prevent blockers from overloading the receiver and to provide image rejection. Traditionally, these filters have used off-chip passive filters, for example, LC circuits or surface acoustic wave devices; however, these off-chip filters can add to the overall cost of a radio substantially, so recently there has been much interest in trying to move these filters on chip. On-chip filters do come with many challenges, however. Often, to make them practical, they must have active circuitry to enhance the Q. Thus, on-chip filters require power and have potential problems with linearity. Due to process variations, they will also require some form of feedback in many cases to set their frequency of operation and to set the Q precisely. In addition, if care is not taken, filters with active Q enhancement can become unstable due to their high Q.

At the time of this writing, on-chip filters are still not very common, but more and more researchers are starting to investigate all aspects of these circuits, and it will only be a matter of time before these filters start to appear in products, thus removing one more barrier to higher levels of integration. We will focus on image reject filters (to replace or complement the image reject mixers discussed in Chapter 7); however, filters can be used to solve other problems as well, such as harmonic filtering or blocker rejection. While the filters presented here are only second order, it is understood that more complex filters may be needed for a specific application.

9.2 Second-Order Filters

All filters that we will consider in this chapter can be composed of cascaded second-order filters. As will be shown, these filters can be implemented using LC resonators and active Q enhancement circuitry. It has already been shown in Chapter 4 that a second-order bandpass filter has a transfer function of the form

$$T(s) = \frac{A_o s}{s^2 + s\mathrm{BW} + \omega_o^2} = \frac{A_o s}{s^2 + \frac{\omega_o}{Q}s + \omega_o^2} \quad (9.1)$$

where BW is the 3-dB bandwidth of the filter, ω_o is the center frequency, and Q is the Q of the filter. This transfer function has two zeros. One zero is at dc and the other is at a frequency of infinity. If the zeros are instead placed on the $j\omega$ axis, then a bandstop or notch filter is created. A notch filter has a transfer function of the form

$$T(s) = \frac{s^2 + \omega_z^2}{s^2 + \frac{\omega_o}{Q}s + \omega_o^2} \quad (9.2)$$

The addition of zeros in the transfer function is very useful, as it allows us to place a notch at the image frequency and get very high image rejection from a low-order filter without adjusting the filter's bandwidth. A plot of this transfer function is shown in Figure 9.1.

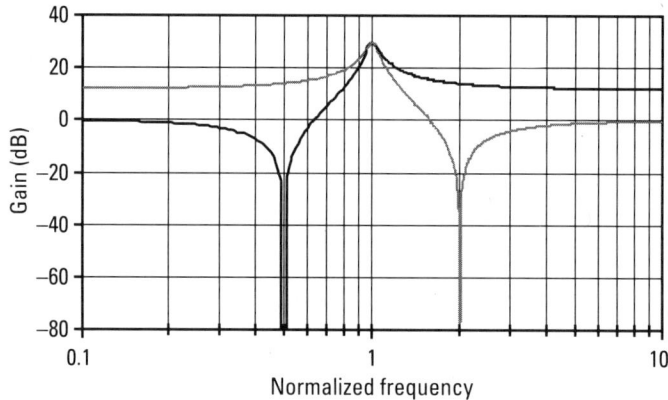

Figure 9.1 Response of two second-order filters with notches.

9.3 Integrated RF Filters

RF filters are everywhere in RFICs, although typically they do not have very high Qs. For example, most RF LNAs will have an LC tank as their load, as shown in Figure 9.2. This is an example of a low-Q bandpass filter. As will be seen, it is also possible to design an LNA as a bandstop filter.

9.3.1 A Simple Bandpass LC Filter

The gain of the LNA in Figure 9.2 is given by

$$\frac{v_{out}}{v_{in}} = \frac{-g_m}{sC + \frac{1}{sL} + \frac{1}{R}} = \frac{-\frac{sg_m}{C}}{s^2 + s\frac{1}{RC} + \frac{1}{LC}} \quad (9.3)$$

At the resonant frequency of the LC circuit, the gain will simply be given by $-g_m R$, as we might expect. So the question then is: How good can a simple tuned LNA be as a filter?

Example 9.1 How Good a Filter Is a Simple Tuned LNA?

Assume that an LNA has a center frequency of 1.9 GHz and the IF of the receiver is at 300 MHz. The image frequency will be at 2.5 GHz. If it is further assumed that the LC resonator has a Q of 5 (limited by the inductor), then what is the resulting 3-dB bandwidth? What is the image rejection?

Solution
The 3-dB bandwidth is ω_o/Q, or in this case about 400 MHz. At the image, the gain can be calculated by considering the equivalent lowpass filter (which

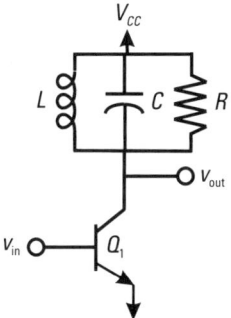

Figure 9.2 Simple RF LNA with turned load and no input matching.

would be first order in this case) with a single-sided bandwidth of 200 MHz. Then the image is 600 MHz from the center, or at three times the corner frequency; therefore, the gain is 1/3 or −9.5 dB. Equivalently, the number of decades (log 3 = 0.477) could also be calculated and then multiplied by −20 dB/decade (−20 · 0.477 = −9.54).

For good image rejection, a higher Q is needed or a higher order filter must be employed. Note that for higher Q, the bandwidth gets narrower, requiring the filter to be tuned to compensate for mismatch and process variation.

This filter will provide virtually no channel selection either. In comparison to a typical wireless standard, in which a channel width is usually no more than a few hundred kilohertz, or at most a few megahertz, and the whole frequency spectrum available may be no more than 200 MHz, the bandwidth of this filter is very wide by comparison. To make matters even worse, if a filter of this type were actually to have such a narrow bandwidth, its gain would be huge.

Example 9.2 Determining Required Bandwidth

Determine the bandwidth required in the last example to get 60 dB of attenuation at the image frequency.

Solution

Since in this case the image is 600 MHz away from the passband, a first-order low-pass equivalent corner frequency at about 600 kHz is needed to get 3 decades of attenuation for the image at 600-MHz offset. The resulting bandwidth is 1.2 MHz at a center frequency of 1.9 GHz and this implies a Q of 1,583 or about 0.06% accuracy. Such accuracy is not possible without some tuning scheme, since the actual accuracy due to process variation will be more like 20%.

9.3.2 A Simple Bandstop Filter

Placing a parallel LC resonator in the collector of an LNA makes a crude bandpass filter, but bandstop filters can be made as well. For example, a cascode LNA with a series resonator attached to it as shown in Figure 9.3 forms a bandstop filter [1].

In this case, the current that is produced at the collector of Q_1 is split between Q_2 and the series resonant circuit made up of L, C, and R. Thus, it can be shown that the gain in this circuit is given by

$$A_v = -g_m R_L \left(\frac{s^2 + \frac{R}{L}s + \frac{1}{LC}}{s^2 + \frac{R + \frac{1}{g_m}}{L}s + \frac{1}{LC}} \right) \quad (9.4)$$

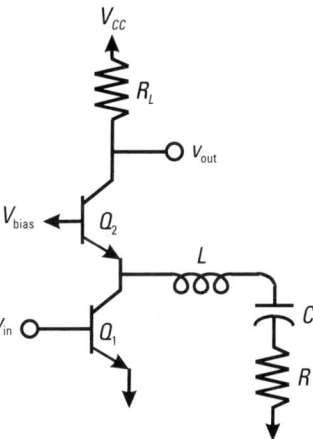

Figure 9.3 An LNA circuit with series resonator added to provide bandstop filtering.

At the resonance frequency, the gain of this circuit will be given by

$$A_v \bigg|_{\omega = \frac{1}{\sqrt{LC}}} = -g_m R_L \frac{R}{R + \frac{1}{g_m}} \quad (9.5)$$

Thus, provided that $R \ll 1/g_m$, the gain at the center frequency becomes much less than 1 and a notch appears in the gain. This is because the transfer function in (9.4) has a complex conjugate pair of zeros close to the $j\omega$ axis. Note that if $R = 0$, the circuit will have a gain of zero at the resonant frequency. Another way of thinking about this is that the LC resonator acts like a short circuit at the resonant frequency. Thus, at this frequency, almost all of the current will flow into the resonator rather than into the cascode transistor, so the gain is reduced.

9.3.3 An Alternative Bandstop Filter

An alternative circuit that also realizes a notch in the transfer function of an LNA is shown in Figure 9.4. In the approach shown in this figure, the conventional topology for a cascode LNA is modified by replacing the inductor in the emitter with a resonator. At the resonance frequency, a high impedance is presented to the emitter of the driving transistor. A high impedance here will mean that the driver will have a very low gain (ideally a zero gain) at this frequency. Thus, the LNA will reject the signals at this frequency. In a typical application, the notch would be adjusted to occur at the image frequency. Below the resonance frequency, the resonator will look inductive and will have an

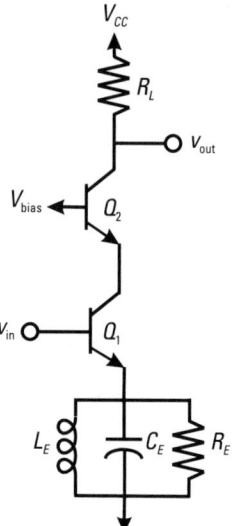

Figure 9.4 Alternative way of providing notch filtering to an LNA.

impedance close to that of the actual inductor placed in the circuit. Thus, in the passband the LNA will still look like an LNA with inductive degeneration.

The transfer function can be determined using the simplified small-signal model for the circuit shown in Figure 9.5. The output is modeled as a simple voltage-controlled current source driving the resistor at the collector of the cascode transistor. For the purposes of this analysis, the cascode is assumed to have a current gain of 1 with no phase shift.

From this diagram, it is easy to see that the input impedance of the amplifier is given by

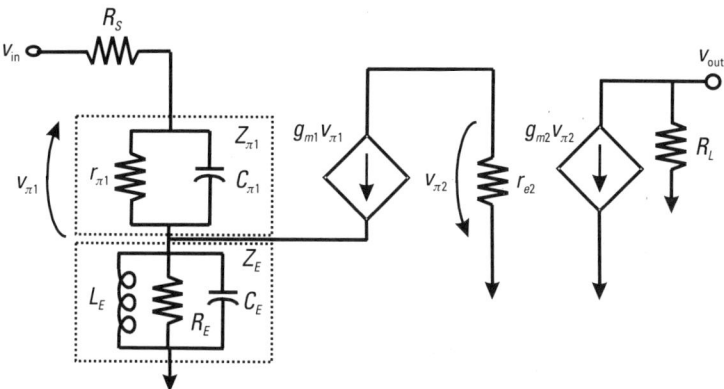

Figure 9.5 Simple model of the LNA for analysis.

$$Z_{\text{in}} = Z_{\pi 1} + Z_E(1 + g_{m1}Z_{\pi 1}) \tag{9.6}$$

If $Z_{\pi 1}$ is assumed to be capacitive, then this expression can be rewritten as

$$Z_{\text{in}} = \frac{\left(\dfrac{1}{C_{\pi 1}} + \dfrac{1}{C_E}\right)s^2 + \left(\dfrac{\dfrac{1}{R_E} + g_{m1}}{C_{\pi 1}C_E}\right)s + \dfrac{1}{L_E C_E C_{\pi 1}}}{s\left(s^2 + \dfrac{s}{C_E R_E} + \dfrac{1}{L_E C_E}\right)} \tag{9.7}$$

This formula will simplify to (6.61) (without L_b present) if $C_E = 0$ and $R_E = \infty$. Thus, as shown in Chapter 6, it can be seen that inductive degeneration can generate positive resistance looking into the base. Similarly, capacitive degeneration generates negative resistance looking into the base. One of the major reasons why capacitive degeneration of amplifiers is not commonly used is because this negative resistance can lead to instability. Fortunately, for this filter circuit, negative resistance is only generated above the notch frequency. Above the resonant frequency of the resonator in the emitter, the overall resistance will be negative over a narrow frequency band if the inductance and capacitance ratio is chosen properly in this circuit. Since the input will resonate in the passband frequencies below where the input resistance goes negative, the circuit can be designed to be stable. Nevertheless, a more rigorous analysis follows.

The transfer function $T(s)$ for this circuit can be derived using Figure 9.5. With only minimal effort, it can be shown that

$$T(s) = \frac{v_{\text{out}}(s)}{v_{\text{in}}(s)} = \frac{-g_m Z_\pi R_L}{Z_{\text{in}} + R_S} \tag{9.8}$$

Substituting (9.7) into (9.8) and after much manipulation,

$$T(s) = \frac{\dfrac{-g_{m1}R_L}{C_\pi R_S}\left(s^2 + \dfrac{s}{C_E R_E} + \dfrac{1}{L_E C_E}\right)}{As^3 + Bs + C} \tag{9.9}$$

where $A = \dfrac{1}{C_E R_E} + \dfrac{1}{R_S}\left(\dfrac{1}{C_{\pi 1}} + \dfrac{1}{C_E}\right)$, $B = \dfrac{\dfrac{1}{R_E} + g_{m1}}{C_E C_{\pi 1} R_S} + \dfrac{1}{L_E C_E}$, and $C = \dfrac{1}{L_E C_E C_{\pi 1} R_S}$.

This transfer function again has a pair of complex conjugate zeros. Thus, the gain will have a notch in it at the resonant frequency of the LC resonator in the emitter. As mentioned previously, the notch is due to the large impedance at this frequency on the emitter at the resonance frequency, and thus the gain will be quite low.

Examining (9.9), stability can be more rigorously ascertained. In order to do this, the pole locations for (9.9) must be determined. This is a third-order system with three poles. Instead of finding the poles analytically, they are plotted with the assistance of a math program in Figure 9.6. It can be seen that even as R_E approaches infinity (as is the intention in this application), the poles will remain in the left half plane. In fact, from the plot it is apparent that the circuit losses would have to be highly overcompensated to drive them into the right half plane (an equation for this will follow shortly). By overcompensated we mean that the negative resistance generated by the active circuitry is more than is required to make the net resistance of the resonator infinity. Thus, there is typically a good margin of safety. Note that this circuit will generate negative resistance looking into the base of Q_1, so stability must be considered carefully, especially if the source impedance is complex.

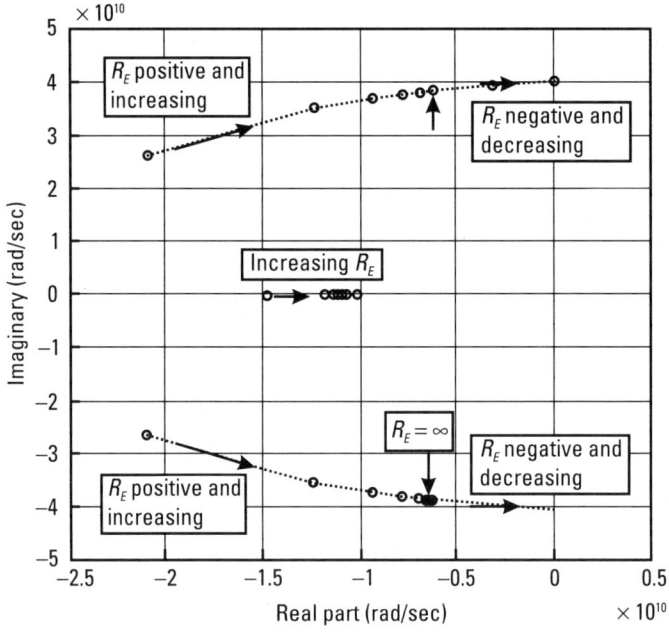

Figure 9.6 Plot of the poles of the filter showing potential instability.

9.4 Achieving Filters with Higher Q

So far, we have considered a few simple second-order filters that can be placed around amplifiers to provide frequency selectivity. Since on-chip inductors do not have a high Q, the bandwidth of the passband filter is not very high. Likewise, with the bandstop filters, the amount of attenuation at the notch frequency will be limited by the Q of the inductors used in building the filter. To increase the Q, it is necessary to build some active circuitry around the LC resonators. This can be done by creating negative resistance to offset the losses in the inductors and other circuits. In Chapter 8, a few very common circuits for doing this were studied. There are many possible ways to generate negative resistance, and in the next few sections, a few of them will be considered.

9.4.1 Differential Bandpass LNA with Q-Tuned Load Resonator

The first example of a circuit that will be considered is the differential version of the previously shown bandpass circuit with the addition of a $-G_m$ cell, as shown in Figure 9.7. Note that to complete this design, some form of emitter degeneration would usually be used in Q_1 and Q_2 and input matching would be needed.

The $-G_m$ cell generates a negative resistance of $-2/g_m$ (or $-4v_T/I_{sharp}$), which is placed in parallel with R, the equivalent total losses in the tank. If this parallel combination is used instead of R in (9.3), the gain becomes

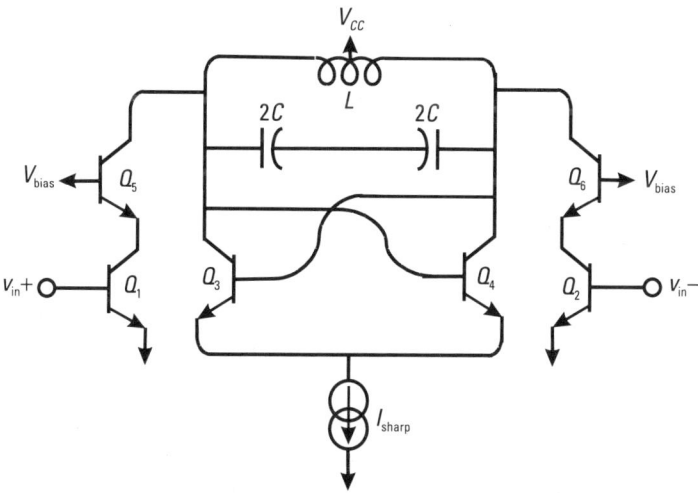

Figure 9.7 An LNA circuit with Q-enhanced resonator.

$$\frac{v_{out}}{v_{in}} = \frac{-\frac{sg_{m1}}{C}}{s^2 + s\frac{2 - Rg_m}{2RC} + \frac{1}{LC}} = \frac{-\frac{sg_{m1}}{C}}{s^2 + s\frac{4v_T - RI_{sharp}}{4v_T RC} + \frac{1}{LC}} \quad (9.10)$$

Note that this resonator should never be tuned to have an infinite Q or else it will have its poles on the $j\omega$ axis and will be on the verge of becoming an oscillator. The gain at the center frequency is also proportional to Q, and will start to rise as the Q is increased, as shown in Figure 9.8. Thus, as the Q is increased, the linearity will start to degrade. Note that with positive feedback, it is possible to set the Q very high; however, loading the circuit will load the Q, so a buffer will be required or I_{sharp} will have to be adjusted for the additional resistive loading.

The bandwidth of this circuit is then given by

$$BW = \frac{4v_T - RI_{sharp}}{4v_T RC} \quad (9.11)$$

Example 9.3 Effect of Process Tolerance on Bandwidth

Assume that the circuit of Figure 9.7 is constructed. What current I_{sharp} is required to give a bandwidth of 1.2 MHz at 2 GHz? Assume that the total parallel resonator loss is $R = 500\Omega$, that $C = 2$ pF, and $L = 3.17$ nH. If the current can vary by 10% due to process variation, what is the effect on the bandwidth?

Solution

We can find the current required by manipulation of (9.11).

$$I_{sharp} = \frac{4v_T - BW \cdot 4v_T RC}{R}$$

$$= \frac{4(25 \text{ mV}) - (2\pi \cdot 1.2 \text{ MHz}) \cdot 4(25\text{mV})(500\Omega)(2 \text{ pF})}{500\Omega}$$

$$= 198.5 \ \mu A$$

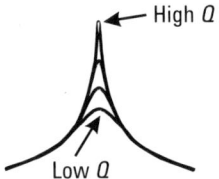

Figure 9.8 Increasing the Q of the bandpass filter also increases the gain.

If the current is decreased by 10% then I_{sharp} will be 178.7 μA. This will mean that the bandwidth of the circuit will grow to be 106.75 MHz or roughly two orders of magnitude. If the current increases by 10%, then it will have a value of 218.4 μA. Blindly plugging into (9.11) will generate a negative bandwidth, but this means that the circuit has gone unstable and will be oscillating. The moral of this little tale is that if you want a high-Q resonator, then you had better have tight control over the controlling current.

9.4.2 A Bandstop Filter with Colpitts-Style Negative Resistance

The bandstop filters can also be improved with the addition of active circuitry. For example, the circuit of Figure 9.3 can be modified to have a Colpitts-style negative resistance generating circuit, as shown in Figure 9.9.

This Colpitts circuit generates a negative resistance of $g_{m3}/\omega^2 C_1 C_2$. Thus, if the total series loss is given by R, then the gain of this amplifier is given by

$$A_v = -g_{m1} R_L \left(\frac{s^2 + \dfrac{R - \dfrac{g_{m3}}{\omega^2 C_1 C_2}}{L} s + \dfrac{C_1 + C_2}{L C_1 C_2}}{s^2 + \dfrac{R - \dfrac{g_{m3}}{\omega^2 C_1 C_2} + \dfrac{1}{g_{m2}}}{L} s + \dfrac{C_1 + C_2}{L C_1 C_2}} \right) \quad (9.12)$$

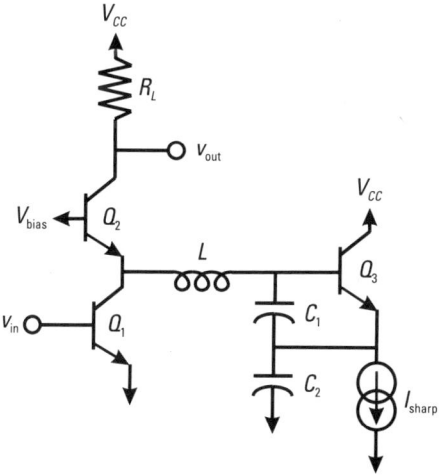

Figure 9.9 Bandstop filter with single-ended Colpitts-style Q enhancement.

In this case, the gain at the resonance frequency will go to zero when the zeros are on the $j\omega$ axis. This will happen when all the circuit losses are perfectly canceled and the LC resonator makes a perfect short circuit to ground. In this case,

$$R = \frac{g_{m3}}{\omega^2 C_1 C_2} = \frac{I_{sharp\_opt}}{v_T \omega^2 C_1 C_2} \qquad (9.13)$$

Therefore, the current to give the deepest notch is

$$I_{sharp\_opt} = R v_T \omega^2 C_1 C_2 \qquad (9.14)$$

Note that even in the case with the zeros on the $j\omega$ axis, the poles of the system are still safely in the left half plane. For the system to be unstable, the negative resistance generated by the Colpitts stage needs to be equal to

$$\frac{g_{m3}}{\omega^2 C_1 C_2} = R + \frac{1}{g_{m2}} \qquad (9.15)$$

or the current to create an unstable notch filter would be

$$I_{sharp\_osc} = \left(R + \frac{1}{g_{m2}}\right) \omega^2 C_1 C_2 v_T \qquad (9.16)$$

By taking the ratio of these two currents, it is possible to see that the current I_{sharp_osc}, which generates oscillations, is always bigger than the current I_{sharp_opt}, which produces the perfect notch. The ratio of these two currents is given by

$$\frac{I_{sharp\_osc}}{I_{sharp\_opt}} = \frac{R + \frac{1}{g_{m2}}}{R} = 1 + \frac{1}{g_{m2} R} \qquad (9.17)$$

Thus, there is a safety margin, as the ratio is always bigger than 1. However, g_{m2} should not be allowed to become too large, as this results in less separation between the notch current and the oscillation current and in higher possibility for instability.

Conceptually, what happens is that even though the path looking into the series resonator is a perfect short circuit (or even has a small negative resistance), the circuit is still loaded with the emitter resistance of Q_2 that serves to damp the circuit which therefore cannot oscillate, as shown in Figure 9.10.

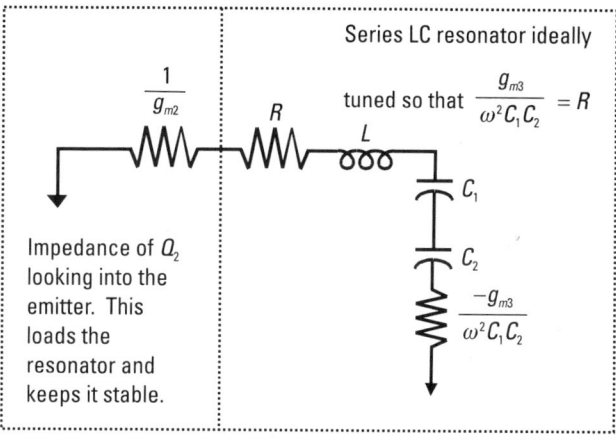

Figure 9.10 How the series notch filter remains stable even when all losses are cancelled.

9.4.3 Bandstop Filter with Transformer-Coupled $-G_m$ Negative Resistance

The parallel resonator-style bandstop filter can also have negative resistance placed around it (shown in Figure 9.11). The filter has a $-G_m$ negative resistance

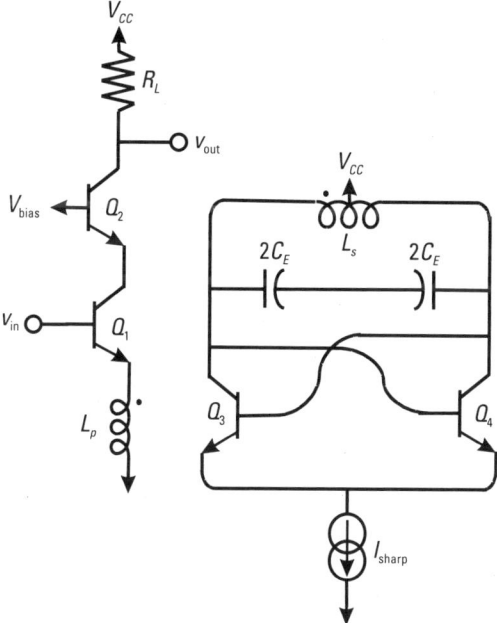

Figure 9.11 Bandstop filter with a transformer-coupled $-G_m$ negative resistance.

circuit coupled into the resonator with the aid of an on-chip transformer. The $-G_m$ cell generates a negative resistance of $-2/g_m$, which is then transformed through the transformer to the emitter of the transistor Q_1. If the transformer is assumed to have an inductance ratio of 1 for simplicity (note that nonunity inductance ratios can result in linearity, stability, and noise advantages, but a unity ratio will keep the math simpler), then (9.9) can be modified by first noting that the total resistance associated with the resonator will now be

$$R_{Total} = -\frac{2}{g_m} \,//\, R_E = \frac{2R_E}{2 - g_m R_E} \tag{9.18}$$

Note that R_E is assumed to be the total positive resistance associated with the resonator. Now (9.9) becomes

$$T(s) = \frac{\dfrac{-g_{m1} R_L}{C_{\pi 1} \dfrac{2R_E}{2 - g_m R_E}} \left(s^2 + \dfrac{s}{C_E \dfrac{2R_E}{2 - g_m R_E}} + \dfrac{1}{L_E C_E} \right)}{s^3 + Ds^2 + Es + F} \tag{9.19}$$

where $D = \dfrac{2 - g_m R_E}{2 C_E R_E} + \dfrac{1}{R_S}\left(\dfrac{1}{C_{\pi 1}} + \dfrac{1}{C_E}\right)$, $E = \dfrac{\dfrac{2 - g_m R_E}{2 R_E} + g_{m1}}{C_E C_{\pi 1} R_S} + \dfrac{1}{L_E C_E}$,

and $F = \dfrac{1}{L_E C_E C_{\pi 1} R_S}$.

The stability analysis of this circuit has already been considered in Section 9.3.3. Though reassuring, a plot of the poles fails to provide much design insight into the problem of making the filter stable. In this section, a simpler interpretation of stability will be presented. The mechanism for damping the oscillation must come from either the source or load impedance. In the previous circuit, the cascode transistor provided damping for the resonator. In this circuit, the source impedance must damp the filter, as shown in Figure 9.12; however, this analysis proceeds much like the one in the previous section.

For perfect notching, the negative resistance must equal the tank losses R_E, so,

$$\frac{2}{g_m} = R_E \tag{9.20}$$

Thus, the optimal current must be

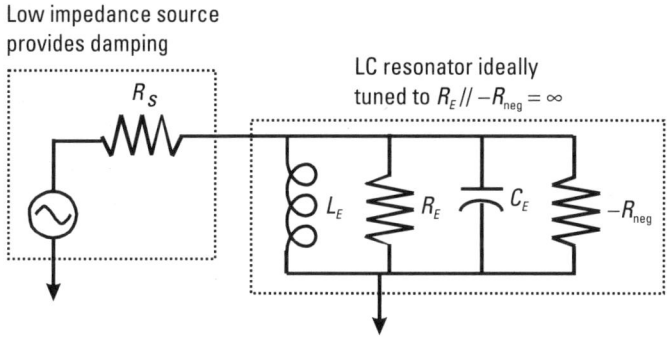

Figure 9.12 How the filter is damped without diminishing image rejection.

$$I_{\text{sharp\_opt}} = \frac{2v_T}{R_E} \qquad (9.21)$$

To start an oscillation,

$$\frac{2}{g_m} = R_E \;//\; R_S \qquad (9.22)$$

Therefore, for oscillations to start, a current of

$$I_{\text{sharp\_osc}} = \frac{2v_T(R_E + R_S)}{R_E R_S} \qquad (9.23)$$

is required. The ratio of these two currents is then given by

$$\frac{I_{\text{sharp\_osc}}}{I_{\text{sharp\_opt}}} = 1 + \frac{R_E}{R_S} \qquad (9.24)$$

Thus, if the source resistance is smaller than the tank resistance R_E, then the tank can be safely tuned to provide infinite Q and still have ample damping. However, if the tank resistance is smaller than the source resistance, then even a small error in tuning the tank to infinite Q could result in oscillation, but this is usually not the case.

9.5 Some Simple Image Rejection Formulas

Image rejection with passband filters has already been discussed. The notch filter's natural application is image rejection as well, so it is useful to develop some simple formulas for the image rejection they provide. For the circuit of Figure 9.9, the gain in the stop band G_{SB} is given by

$$G_{SB} = -g_{m1} R_L \left(\frac{R - \frac{g_{m3}}{\omega^2 C_1 C_2}}{R - \frac{g_{m3}}{\omega^2 C_1 C_2} + \frac{1}{g_{m2}}} \right) \quad (9.25)$$

$$= -g_{m1} R_L \left(\frac{R - \frac{I_{sharp}}{\omega^2 C_1 C_2 v_T}}{R - \frac{I_{sharp}}{\omega^2 C_1 C_2 v_T} + \frac{1}{g_{m2}}} \right)$$

and the gain in the passband G_{PB} is approximately

$$G_{PB} = -g_{m1} R_L \quad (9.26)$$

It is assumed that the passband is sufficiently far away that the series resonator is essentially an open circuit there. Therefore, the *image rejection* (IR) provided by this filter is

$$IR = 20 \log \left| \frac{G_{PB}}{G_{SB}} \right| = 20 \log \left| \frac{R - \frac{I_{sharp}}{\omega^2 C_1 C_2 v_T} + \frac{1}{g_{m2}}}{R - \frac{I_{sharp}}{\omega^2 C_1 C_2 v_T}} \right| \quad (9.27)$$

If I_{sharp} can be tuned to perfectly cancel the loss in the resonator, this circuit can have perfect image rejection. In practice, the accuracy to which I_{sharp} can be set will limit the image rejection available from this circuit.

Example 9.4 Effect of Process Tolerance on Image Rejection

Assume that the current I_{sharp} in the circuit of Figure 9.9 can be set to an accuracy of 1%. Determine the image rejection that can be expected from this circuit. Assume that Q_1 and Q_2 are biased at 3 mA and that the loss in the resonator is 5Ω.

Solution

In this case, we assume that I_{sharp} will take on a value of

$$I_{sharp} = \omega^2 C_1 C_2 v_T R \Delta$$

where Δ is the process tolerance of I_{sharp}, which can have value in this case of anywhere between $1.01 > \Delta > 0.99$. At a bias current of 3 mA, g_{m2} will be 120 mA/V; thus, the image rejection is given by

$$\text{IR} = 20 \log \left| \frac{R(1-\Delta) + \dfrac{1}{g_{m2}}}{R(1-\Delta)} \right|$$

Plugging in numbers at both extremes gives an image rejection of 44.5 dB. Thus, the circuit will give good image rejection provided that tolerances can be small. This is a big improvement over the previous design that had stability problems as well as tolerance problems, but more is still needed to make this practical.

Formulas for image rejection for the circuit of Figure 9.11 can also be developed. Noting that well below resonance an LC tank will have an impedance given roughly by the reactance of the inductor, and well above resonance it will have an impedance roughly that of its capacitor, we can develop the following equations. First, the gain in the passband G_{PB} of the filter is given by

$$G_{\text{PB}} = \frac{R_L}{Z_E} = \frac{R_L}{\omega_{\text{PB}} L_E} \tag{9.28}$$

since the resonator in the emitter is below resonance there.

The gain in the stop band G_{SB} is given by

$$G_{\text{SB}} = \frac{Z_L}{Z_E} = \frac{R_L}{R_{\text{Total}}} \tag{9.29}$$

where the resonator in the emitter is now resonating with total resistance R_{Total} (which is made up of resonator losses R_E and negative resistance generated by active circuitry).

Thus, making use of (9.18), (9.28), and (9.29), the image rejection IR can be approximated as

$$\text{IR} = 20 \log \left| \frac{R_L}{\omega_{\text{PB}} L_E} \cdot \frac{R_{\text{Total}}}{R_L} \right| = 20 \log \left| \frac{2 R_E}{(2 - g_m R_E) \omega_{\text{PB}} L_E} \right| \tag{9.30}$$

As before, this will be limited by process tolerance.

9.6 Linearity of the Negative Resistance Circuits

All receive-path circuits, including the filter resonator, have to process signals. If a very large signal is present on the resonator, it will cease to work correctly. Such large signals will tend to change the effective g_m of the transistors in the resonator, such as Q_3 and Q_4 of Figure 9.11, and therefore the negative resistance. Thus, as signals get larger, we can expect degradation of image rejection.

To determine the maximum signal size the circuit can handle, we need to do a large signal analysis. If a transistor is driven with a voltage source v_{in} and has no degeneration, then the output current can be expressed as an infinite series:

$$i_c = I_C \left[1 + \frac{v_{in}}{v_T} + \frac{1}{2}\left(\frac{v_{in}}{v_T}\right)^2 + \frac{1}{6}\left(\frac{v_{in}}{v_T}\right)^3 + \cdots \right] \qquad (9.31)$$

We now find the g_m of this circuit without making a small-signal assumption:

$$g_m = \frac{di_c}{dv_{in}} = I_C \left[\frac{1}{v_T} + \frac{v_{in}}{v_T^2} + \frac{1}{2}\frac{v_{in}^2}{v_T^3} + \cdots \right] = g_{mss}\left[1 + \frac{v_{in}}{v_T} + \frac{1}{2}\frac{v_{in}^2}{v_T^2} + \cdots \right] \qquad (9.32)$$

If v_{in} remains relatively small, this takes on the small-signal value g_{mss} of I_C/v_T. However, as the signal grows, this value changes. Thus, in the case of the Colpitts-style resonator as shown in Figure 9.9, the amount of negative resistance generated R_{neg} relative to the small-signal negative resistance R_{negss} is

$$\frac{R_{neg}}{R_{negss}} = \frac{\frac{-g_m}{\omega^2 C_1 C_2}}{\frac{-g_{mss}}{\omega^2 C_1 C_2}} = \left[1 + \frac{v_{in}}{v_T} + \frac{1}{2}\frac{v_{in}^2}{v_T^2} + \cdots \right] \qquad (9.33)$$

The current flowing into the resonator will cause a voltage of $v_{be3} = i_{in}/sC_1$ to be developed across the base emitter (assuming that the impedance of C_1 is much lower than the transistor). As this voltage approaches v_T, the operating point of this transistor will start to shift and its effectiveness will degrade. Therefore, an input current of

$$i_{in\_max} = v_T s C_1 \qquad (9.34)$$

can be tolerated. If the Colpitts negative resistance circuit is sized so that C_1 is large and C_2 small, then the linearity of this circuit will improve.

In the case of the $-G_m$ resonator as shown in either Figure 9.7 or 9.11, the amount of negative resistance generated is

$$\frac{R_{neg}}{R_{negss}} = \frac{\frac{-2}{g_m}}{\frac{-2}{g_{mss}}} = \left[1 + \frac{v_{in}}{v_T} + \frac{1}{2}\frac{v_{in}^2}{v_T^2} + \cdots\right]^{-1} \quad (9.35)$$

The voltage across the resonator is twice that across either Q_3 or Q_4, so ($v_{res} = 2v_{in}$):

$$\frac{R_{neg}}{R_{negss}} = \frac{\frac{-2}{g_m}}{\frac{-2}{g_{mss}}} = \left[1 + \frac{v_{res}}{2v_T} + \frac{1}{2}\frac{v_{res}^2}{4v_T^2} + \cdots\right]^{-1} \quad (9.36)$$

Therefore, as a voltage of about $2v_T$ is applied to the resonator, its effectiveness will degrade. Note that if resistors were added to the circuit to degenerate the transistors Q_3 and Q_4, then signal size could be increased and the linearity would improve. Alternatively, if a transformer with turns ratio greater than 1 is employed, then the voltage on the resonator could be stepped down, reducing the required signal handling requirement of these transistors.

9.7 Noise Added Due to the Filter Circuitry

As has already been discussed, there are three major sources of noise in an LNA: base shot noise, collector shot noise, and base resistance of the driver transistor. In a filter, there is the additional noise due to the active Q enhancement circuitry.

Noise due to the base resistance of Q_1 is in series with the input voltage, so it sees the full amplifier gain. The output noise due to base resistance is given by

$$v_{no,r_b} \approx \sqrt{4kTr_b} \cdot \frac{R_L}{Z_E} \quad (9.37)$$

It is assumed that Z_E is the impedance in the emitter and the LNA, and the load at the collector is again assumed to be R_L.

Collector shot noise is in parallel to collector signal current and is directly sent to the output load resistor.

$$v_{no,I_C} \approx \sqrt{2qI_C} R_L \qquad (9.38)$$

Base shot noise can be converted to input voltage by considering the impedance Z_{eq} it sees. If Z_{eq} is the source impedance plus any matching circuitry in parallel with the transistor input impedance, then

$$v_{no,I_B} \approx \sqrt{\frac{2qI_C}{\beta}} Z_{eq} \cdot \frac{R_L}{Z_E} \qquad (9.39)$$

Note that if the circuit is differential, then each of the above three equations must be multiplied by root 2.

These noise sources will be present in any tuned LNA, such as the ones studied in Chapter 6, and now to these noise sources the noise due to resonators must be added. This must be considered on a case-by-case basis. For example, in the case of the circuit shown in Figure 9.7, if we assume that the noise produced by the resonator is dominated by the collector shot noise, then the output noise current is given by

$$I_{out\_-Gm} \approx \sqrt{\left(\sqrt{2q\frac{I_{sharp}}{2}}\right)^2 + \left(\sqrt{2q\frac{I_{sharp}}{2}}\right)^2} = \sqrt{2qI_{sharp}} \qquad (9.40)$$

This noise current is then developed into a voltage across the load resistance; thus,

$$v_{no,I_{sharp}} \approx \sqrt{2qI_{sharp}} \cdot R_L \qquad (9.41)$$

Therefore, the noise present at the output produced by the circuit relative to the noise that would have been present without the filter is

$$\frac{N_{added\_filter}}{N_{added\_LNA}} \approx$$

$$\frac{2qI_{sharp} \cdot (R_L)^2 + \frac{4qI_C}{\beta} Z_{eq}^2 \cdot \left(\frac{R_L}{Z_E}\right)^2 + 4qI_C R_L^2 + 8kTr_b \cdot \left(\frac{R_L}{Z_E}\right)^2}{\frac{4qI_C}{\beta} Z_{eq}^2 \cdot \left(\frac{R_L}{Z_E}\right)^2 + 4qI_C R_L^2 + 8kTr_b \cdot \left(\frac{R_L}{Z_E}\right)^2} \qquad (9.42)$$

This is the same as for an LNA except with an additional term due to I_{sharp}, and thus the LNA built with the filter can never be as quiet as a true

LNA. Note that in the case of the bandpass filter, since the gain is very high, this noise current could produce a large output voltage. This same noise current in the case of the notch filters will create a large noise voltage on the notch resonator due to its high Q. However, the output voltage will be much lower due to the presence of a lower impedance.

Note once again that if a transformer were added to this circuit, then the current could be stepped down to reduce the impact on the noise figure of the circuit. The circuit that was considered in Figure 9.11 also uses a $-G_m$ circuit, but it is not differential, so with only minor modifications, (9.42) becomes

$$\frac{N_{added\_filter}}{N_{added\_LNA}} \approx \frac{2qI_{sharp} \cdot (R_L)^2 + \frac{2qI_C}{\beta} Z_{eq}^2 \cdot \left(\frac{R_L}{Z_E}\right)^2 + 2qI_C R_L^2 + 4kTr_b \cdot \left(\frac{R_L}{Z_E}\right)^2}{\frac{2qI_C}{\beta} Z_{eq}^2 \cdot \left(\frac{R_L}{Z_E}\right)^2 + 2qI_C R_L^2 + 4kTr_b \cdot \left(\frac{R_L}{Z_E}\right)^2}$$

(9.43)

It should be noted that the collector shot noise current from the $-G_m$ circuit also ends up developing a voltage at the collector in the same way as in the previous circuit. The Colpitts circuit is harder to analyze analytically, so rather than generating long and tedious equations, it is better simply to simulate it. Essentially, an extra noise current will be added between the driver transistor and the cascode, which will impact the noise figure of the circuit, just as in the other two circuits.

9.8 Automatic Q Tuning

The notch filters discussed here cannot be considered much more than a curiosity unless they can be tuned automatically on chip. To get a deep notch, the current through the resonator must be set precisely so that the losses are perfectly canceled. Too much or too little and the image rejection will suffer. Thus, some form of feedback must be added to the circuit to make it practical.

As a starting point, consider an oscillator with no loading, as shown in Figure 9.13. The oscillator will form the basis of the Q tuning circuit for the filter. A simple current mirror sets the current through the oscillator with a resistor as the reference. If the resonator current is set above that necessary for perfect cancellation of the losses, the circuit will oscillate.

Now consider the circuit of Figure 9.13, modified to include two sensing transistors Q_7 and Q_8, as shown in Figure 9.14. They are biased at a very low

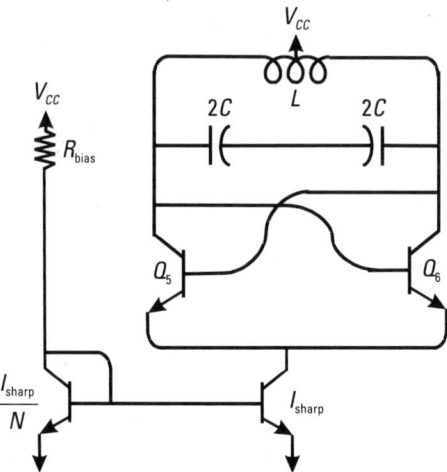

Figure 9.13 A resonator without damping, which oscillates if there is enough current.

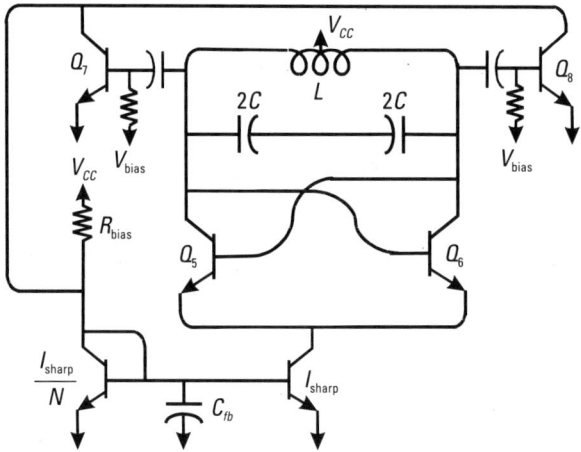

Figure 9.14 A resonator with feedback to control the current.

quiescent point so that they are operating almost in Class B mode. Note that this is essentially the same circuit as the automatic amplitude control for the VCO previously discussed. The only difference is that here the transistors are biased so that they turn on as soon as there is any swing, rather than waiting until it reaches some fraction of a V_{BE}. When they turn on, they provide a full wave rectified current around the loop to the current mirror. This feedback loop controls the current so that the oscillator just starts, which means the negative g_m circuitry perfectly cancels the resonator and circuit losses. As before, a capacitor C_{fb} is included for stability. This circuit becomes the master reference for the filter.

Now the reference is also used to control the current flowing to the notch filter itself (the slave), as shown in Figure 9.15. Here the application is shown with the third type of filter, but it would work equally well with the other notch filter (an example of such a circuit is shown in Figure 9.16). If the components used in building the oscillator are the same as those used in the filter, then the point at which the oscillator is perfectly tuned will also be the perfect Q tuning point for the filter, so the filter now has an automatic Q tuning mechanism. However, if the master oscillator and the slave filter have

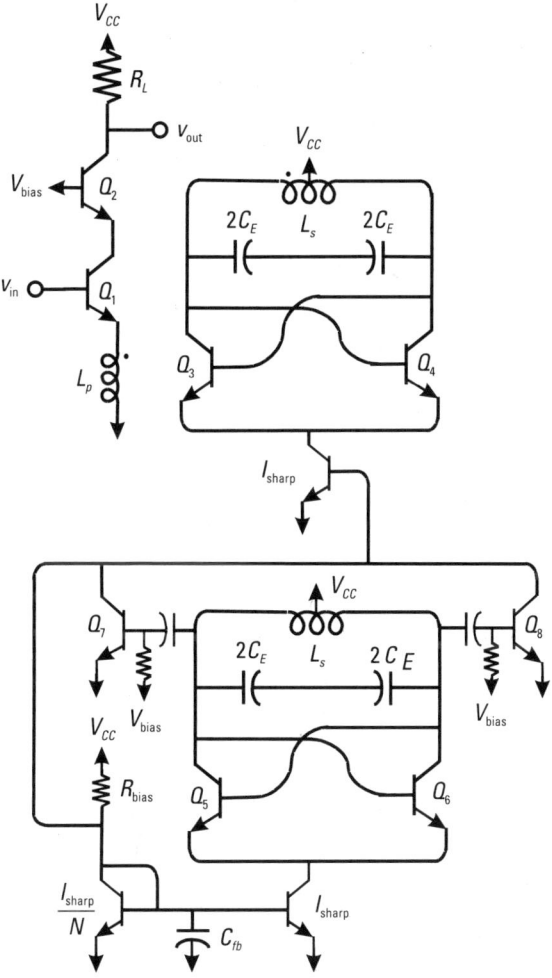

Figure 9.15 The oscillator as a master to the filter slave.

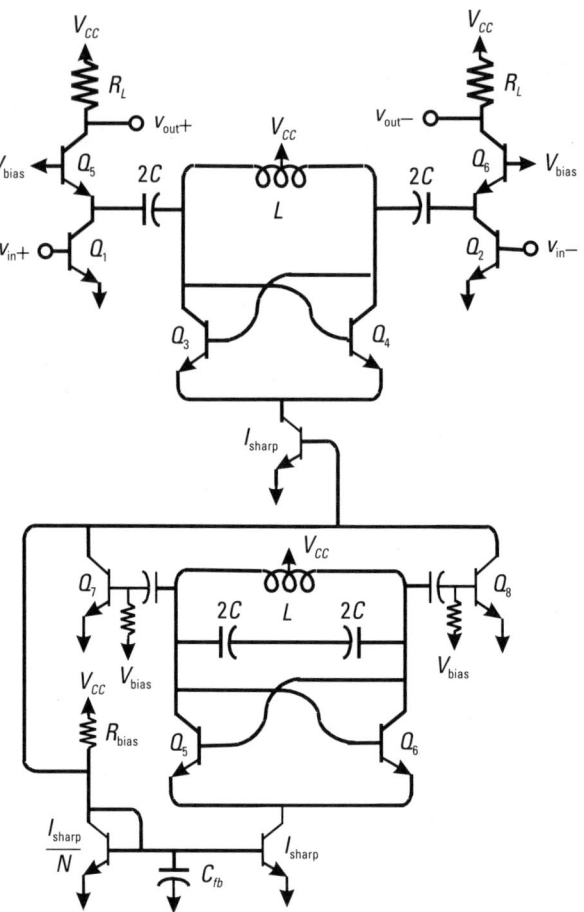

Figure 9.16 An alternative notch filter or oscillator Q-tuned filter.

exactly the same frequency, then the oscillator will inject noise at the image frequency. Thus, some small offset must be made to the oscillator operating frequency. This can be done by adding a small capacitor (a small percentage of the total resonator capacitance) to the oscillator resonator.

9.9 Frequency Tuning

The last step in making any of these circuits practical is the addition of some circuitry that will allow the frequency response of the filter to be adjusted, such as a phase-locked loop. The easiest way to do this is with the aid of the oscillator in the Q tuning loop. First, one or more of the capacitors in the slave filter and master oscillator must be replaced with a varactor of some kind. Then the master VCO that performs the Q tuning for the filter can be placed in the phase-locked loop that will set the frequency of the VCO through feedback and at the same time set the frequency of the notch. Alternatively, in a receiver application, the image frequency and hence the notch frequency are related to the local oscillator frequency by a constant offset. Because of this, the synthesizer that controls the LO can also be used to control the image reject filter [2].

9.10 Higher-Order Filters

So far, we have considered only second-order filters in this book. For many applications, higher order filters are required. In order to make higher order filters, many second-order filters must be cascaded together. There are many ways that LC networks can be cascaded together using the basic structures discussed here. Theory on how to do this can be found in several references devoted to filter design [3, 4].

Example 9.5 A Notch Filter for Image Rejection

Modify the LNA designed in Example 6.10 to include a notch at 7 GHz. Thus, when placed in a receiver with an LO frequency of 6 GHz, it will reject the image at 7 GHz. Further, modify the circuit so that it has a differential rather than a single-ended input.

Solution

Starting with the LNA that was previously designed, the first step is to make it differential. That means that everything must be mirrored around the horizontal axis and another buffer must be added. Then we will add a series-style LC resonator between the driver and cascode transistor with a $-G_m$ negative resistance circuit to provide the notching circuitry. Once all this is done, then the circuit will look like Figure 9.17. The values from the previous examples are used for all the components that are already sized. Thus, $L_b = 1$ nH, $L_e = 290$ pH, $L_T = 1$ nH, and C_T was adjusted to make the gain peak at 5 GHz, so it is now 375 fF. The loss R_T is twice what it was in the original example and is now assumed to be 628Ω. The buffers are also biased the same as before. The gain S_{11} and noise figure were simulated again in the differential configuration and were identical to the previous simulations except for the slight shift in passband frequency. Table 9.1 summarizes the performance of the circuit before the notch was added.

Next, the notch circuitry needed to be sized. The transistors Q_3 and Q_4 were kept small to avoid swamping the circuit with parasitic capacitance. They were chosen to be 5 μm arbitrarily. In simulation, if the noise from the resistors

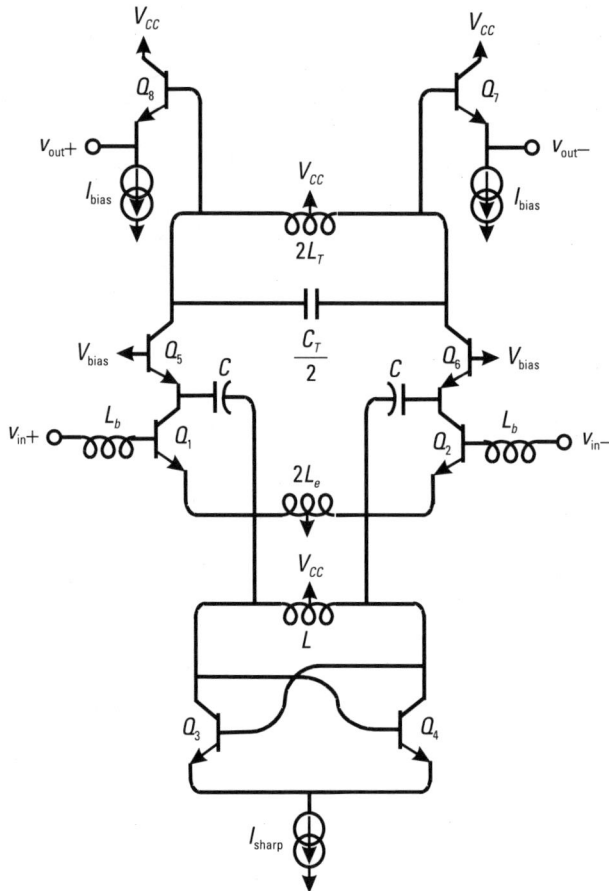

Figure 9.17 LNA modified and made differential with LC load and output buffers.

Table 9.1
Summary of LNA Parameters

Parameter	Value Before Notch Is Added	Value After Notch Is Added
Gain (dB)	28	28
NF at 5 GHz (dB)	1.74	2.7
S_{11} at 5 GHz (dB)	<−30	<−30
Image rejection (dB)	20	83

inside the transistor model starts to dominate, then these transistors will need to be made bigger. The inductor L was chosen so that $L = 1$ nH and the Q was assumed to be 10. This generated a parallel resistance across the resonator of 440Ω at 7 GHz. This meant that the capacitors were initially sized so that $C = 1.03$ pF for a center frequency of 7 GHz. In simulation, they were adjusted to be 825 fF to account for parasitic capacitance. Since the resonator had a loss from the inductor of 440Ω this meant that $2/g_m = 440Ω$. Thus, $I_C = 114$ μA and therefore $I_{sharp} = 227$ μA. Note that this simple analysis does not take into account transistor losses such as base resistance or output resistance. Through simulation, this current was refined to 337 μA. A plot of the gain with this current in the resonator is shown in Figure 9.18. This shows a very deep notch that reaches a depth of about −55 dB.

Of course, even with feedback, this current will never be able to be set with absolute precision. If the current varies by 5% either to 328.6 or 345.4 μA, then the notch depth drops to 28 dB, as shown in Figure 9.19. The noise figure of this design was also simulated and was found to rise to 2.7 dB. Table 9.1 also shows these numbers for direct comparison of the two designs.

Finally, the design was simulated with increasing input voltage to find out how large the signal could be made before the image rejection started to degrade. The current I_{sharp} was set to 345 μA and the voltage at the notch frequency was raised slowly in a transient simulation. As shown in Figure 9.20, image rejection in this design starts to degrade at about 1 mV. This is still a very small signal. Thus, work still needs to be done to increase the signal handling of this design. One thing that would help would be if the resonator

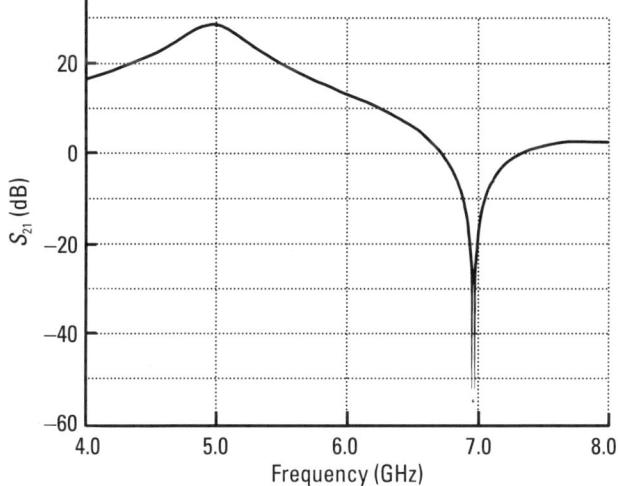

Figure 9.18 Simulated S_{21} for the LNA with notch filter.

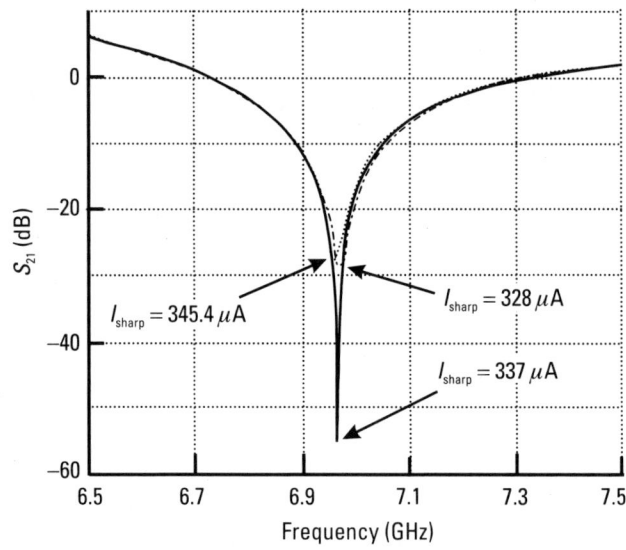

Figure 9.19 Simulated S_{21} for the LNA with notch filter showing the effect of imperfect adjustment of I_{sharp}.

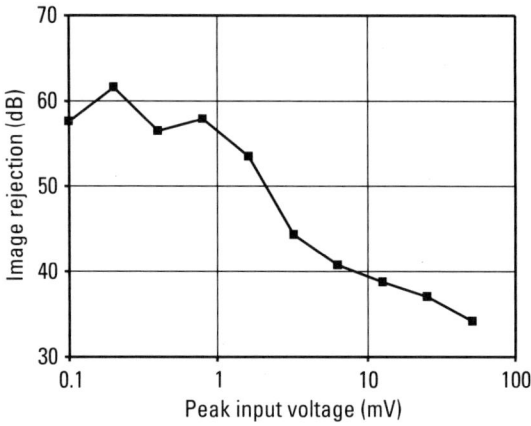

Figure 9.20 Simulated image rejection versus input voltage.

had a lower Q, then the ac current would become a smaller percentage of the total; also, as described previously, transformers could be used or degeneration resistors could be added to the resonator.

References

[1] Macedo, J., and M. A. Copeland, "A 1.9 GHz Silicon Receiver with Monolithic Image Filtering," *IEEE J. Solid-State Circuits,* Vol. 33, March 1998, pp. 378–386.

[2] Copeland, M. A., et al., "5-GHz SiGe HBT Monolithic Radio Transceiver with Tunable Filtering," *IEEE Trans. on Microwave Theory and Techniques,* Vol. 48, Feb. 2000, pp. 170–181.

[3] Schauman, R., M. S. Ghausi, and K. R. Laker, *Design of Analog Filters: Passive, Active RC, and Switched Capacitor,* Englewood Cliffs, NJ: Prentice Hall, 1990.

[4] Williams, A. B., and F. J. Taylor, *Electronic Filter Design Handbook: LC, Active, and Digital Filters,* New York: McGraw-Hill, 1988.

Selected Bibliography

Guo, C., et al., "A Fully Integrated 900-MHz CMOS Wireless Receiver with On-Chip RF and IF Filters and 79-dB Image Rejection," *IEEE J. Solid-State Circuits,* Vol. 37, Aug. 2002, pp. 1084–1089.

Li, D., and Y. Tsividis, "Design Techniques for Automatically Tuned Integrated Gigahertz-Range Active LC Filters," *IEEE J. Solid-State Circuits,* Vol. 37, Aug. 2002, pp. 967–977.

Pavan, S., and Y. Tsividis, *High Frequency Continuous Time Filters in Digital CMOS Processes,* Norwell, MA: Kluwer, 2000.

Pipilos, S., Y. Tsividis, and J. Fenk, "1.8 GHz Tunable Filter in Si Technology," *Proc. Custom Integrated Circuit Conference,* May 1996, pp. 189–191.

Rogers, J., J. Macedo, and C. Plett, "A Completely Integrated 1.9 GHz Receiver Front-End With Monolithic Image Reject Filter and VCO," *IEEE Trans. on Microwave Theory and Techniques,* Vol. 50, Jan. 2002, pp. 210–215.

Rogers, J., J. A. Macedo, and C. Plett, "A Completely Integrated Receiver Front-End with Monolithic Image Reject Filter and VCO," *Proc. IEEE RFIC Symposium,* June 2000, pp. 143–146.

Rogers, J., and C. Plett, "A 5 GHz Radio Front-End with Automatically Q Tuned Notch Filter," *Proc. Bipolar Circuits and Technology Meeting,* Sept. 2002, pp. 69–72.

Rogers, J., and C. Plett, "A Completely Integrated 1.8 Volt 5 GHz Tunable Image Reject Notch Filter," *Proc. Radio Frequency Integrated Circuits Symposium,* May 2001, pp. 75–78.

Rogers, J., D. Rahn, and C. Plett, "A Study of Digital and Analog Automatic-Amplitude Control Circuitry for Voltage-Controlled Oscillators," *IEEE J. Solid-State Circuits,* Vol. 38, Feb. 2003, pp. 352–356.

Samavati, H., H. R. Rategh, and T. H. Lee, "A 5-GHz CMOS Wireless LAN Receiver Front End," *IEEE J. Solid-State Circuits,* Vol. 35, May 2000, pp. 765–772.

Soorapanth, T., and S. S. Wong, "A 0-dB IL 2140 ± 30 MHz Bandpass Filter Utilizing Q-Enhanced Spiral Inductors in Standard CMOS," *IEEE J. Solid-State Circuits,* Vol. 37, May 2002, pp. 579–586.

Willingham, S. D., and K. Martin, *Integrated Video-Frequency Continuous-Time Filters,* Norwell, MA: Kluwer, 1995.

Yoshimasu, T., et al., "A Low-Current Ku-Band Monolithic Image Rejection Down Converter," *IEEE J. Solid-State Circuits,* Vol. 27, Oct. 1992, pp. 1448–1451.

10

Power Amplifiers

10.1 Introduction

Power amplifiers, also known as PAs, are used in the transmit side of RF circuits, typically to drive antennas. Power amplifiers typically trade off efficiency and linearity, and this tradeoff is very important in a fully monolithic implementation. Higher efficiency leads to extended battery life, and this is especially important in the realization of small, portable products. There are some additional challenges specifically related to being fully integrated. Integrated circuits typically have a limited power supply voltage to avoid breakdown, as well as a metal migration limit for current. Thus, simply achieving the desired output power can be a challenge. Power amplifiers dissipate power and generate heat, which has to be removed. Due to the small size of integrated circuits, this is a challenging exercise in design and packaging. Several recent overview presentations have highlighted the special problems with achieving high efficiency and linearity in fully integrated power amplifiers [1–3].

Power amplifiers are among the last circuits to be fully integrated. In many instances, there is no choice but to design them with discrete power transistors, or at least separately from the rest of the radio frequency front-end circuits. There is a lot of interest in discrete and semi-integrated power amplifier design, and for years the main reference was the classic book by Krauss, Bostian, and Raab [4]. Only recently has this book been complemented by a number of fine new books on the topic [5–7], showing that this subject matter is still of interest and may in fact be growing in importance.

10.2 Power Capability

One of the main goals of PA design is to deliver a given power to a load. This is determined to a large degree by the load resistor and the power supply. Given a particular power supply voltage V_{CC}, such as 3V, and a load resistance R_L, such as 50Ω, it is possible to determine the maximum power to be

$$P = \frac{V_{CC}^2}{2R} = \frac{3^2}{2 \times 50} = 90 \text{ mW} \Rightarrow 19.5 \text{ dBm} \qquad (10.1)$$

This assumes we have a tuned amplifier and an operating point of 3V, a peak negative swing down to 0V, and a peak positive swing up to 6V.

10.3 Efficiency Calculations

Efficiency η, sometimes also called *dc-to-RF efficiency*, is the measure of how effectively power from the supply is converted into output power and is given by

$$\eta = \frac{P_{out}}{P_{dc}} \qquad (10.2)$$

where P_{out} is the ac output power and, assuming voltage and current are in phase, is given by

$$P_{out} = \frac{i_1 v_1}{2} = \frac{i_1^2 R_L}{2} \qquad (10.3)$$

where i_1 and v_1 are the peak fundamental components of the current and voltage, respectively. These are determined from the actual current and voltage by Fourier analysis. P_{dc} is the power from the supply and is given by

$$P_{dc} = \frac{1}{T} \int_0^T V_{CC} i_C \, dt = \frac{V_{CC}}{T} \int_0^T i_C \, dt = V_{CC} I_{dc} \qquad (10.4)$$

where I_{dc} is the dc component of the current waveform.

Power-added efficiency (PAE) is the same as efficiency; however, it takes the gain of the amplifier into account as follows:

$$\text{PAE} = \frac{P_{out} - P_{in}}{P_{dc}} = \frac{P_{out} - P_{out}/G}{P_{dc}} = \eta\left(1 - \frac{1}{G}\right) \qquad (10.5)$$

where G is the power gain P_{out}/P_{in}. Thus, it can be seen that for high gain, power-added efficiency PAE is the same as dc-to-RF efficiency η.

Figure 10.1 shows the efficiency in comparison to power-added efficiency for a range of power gains. It can be seen that for gain higher than 10 dB, PAE is within 10% of the efficiency η. As the gain compresses, PAE decreases. For example, if the gain is 3 dB, the PAE is only half of the dc-to-RF efficiency.

A typical plot of output power and efficiency versus input power is shown in Figure 10.2. It can be seen that while efficiency keeps increasing for higher input power, as the amplifier compresses and gain decreases, the power-added efficiency also decreases. Thus, there is an optimal value of power-added efficiency and it typically occurs a few decibels beyond the 1-dB compression point.

10.4 Matching Considerations

In order to obtain maximum output power, typically the power amplifier is not conjugately matched. Instead, the load is designed such that the amplifier has the correct voltage and current to deliver the required power. We note that conjugate matching means that $\Gamma_S = S_{11}^*$ and $\Gamma_L = S_{22}^*$, as shown in Figure 10.3. In the figure, Γ_S is the source reflection coefficient and Γ_L is the load reflection coefficient.

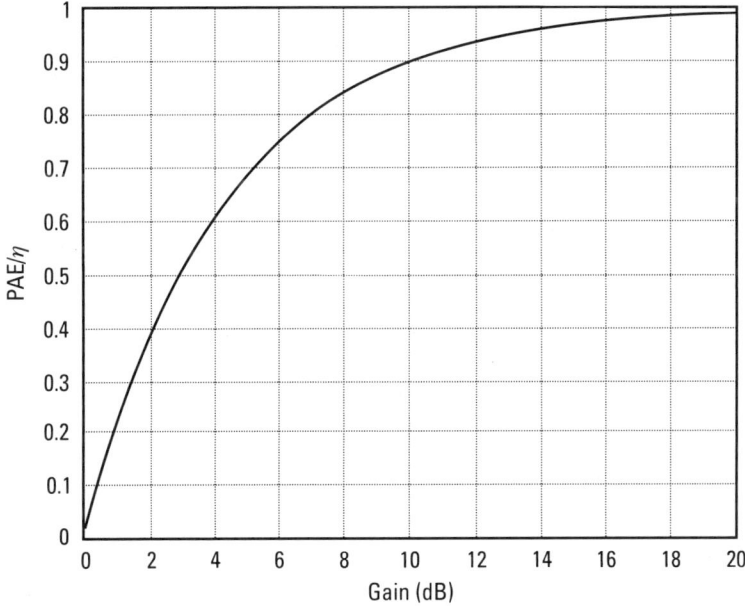

Figure 10.1 Normalized power-added efficiency versus gain.

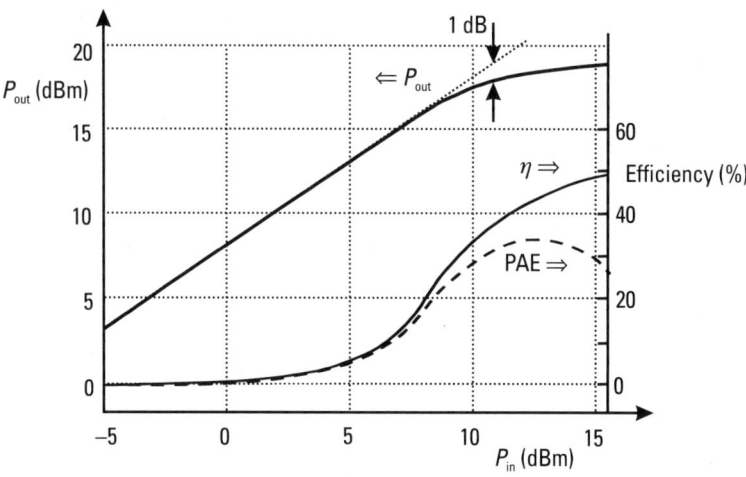

Figure 10.2 Output versus input power.

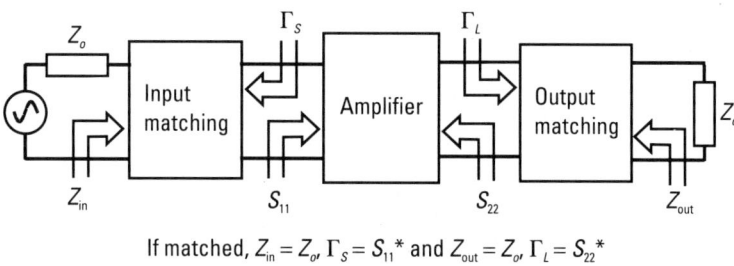

If matched, $Z_{in} = Z_o$, $\Gamma_S = S_{11}^*$ and $Z_{out} = Z_o$, $\Gamma_L = S_{22}^*$

Figure 10.3 Block diagram of amplifier and matching circuits.

10.4.1 Matching to S_{22}^* Versus Matching to Γ_{opt}

For low input power where the amplifier is linear, maximum output power is obtained with $\Gamma_L = S_{22}^*$. However, this value of S_{22} will not be the optimum load for high input power where the amplifier is nonlinear. Nonlinearities result in gain compression, the appearance of harmonics, and additional phase shift. The result can be a shift of the operating point and a shift in the optimal load impedance. For these reasons, for large-signal operation, tuning is done by determining the optimal load Γ_{opt}, typically by doing an exhaustive search called a *load pull*. The comparison between tuning for small signal and large signal is shown in Figure 10.4.

As illustrated, the small-signal tuning curve results in higher output power for small signals, while the large-signal tuning curve results in higher output power for larger signals. Typically, if operation is at the optimal PAE point (as

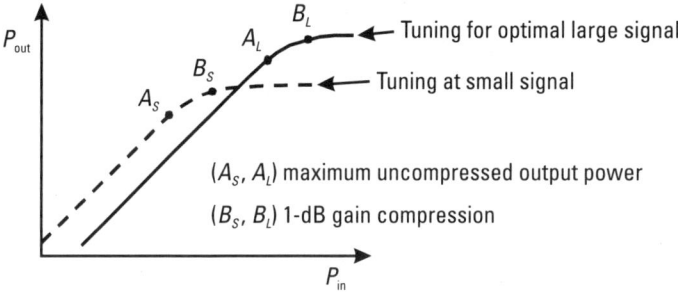

Figure 10.4 Optimal matching versus small-signal matching.

shown in Figure 10.2), optimal-power tuning produces about 1 to 3 dB of higher power. Gain is reduced (for small P_{in}) typically by a slightly smaller amount.

An estimate of the optimum impedance Γ_{opt} can be made by adjusting the load so that the transistor current and voltage go through their maximum excursion, as shown in Figure 10.5, with the output susceptance (typically capacitive susceptance) reactively matched.

10.5 Class A, B, and C Amplifiers

Power amplifiers are grouped into classes depending on the nature of their voltage and current waveforms. The first major classes to be considered are class A, B, and C amplifiers. Figure 10.6 shows a basic circuit that can be used for

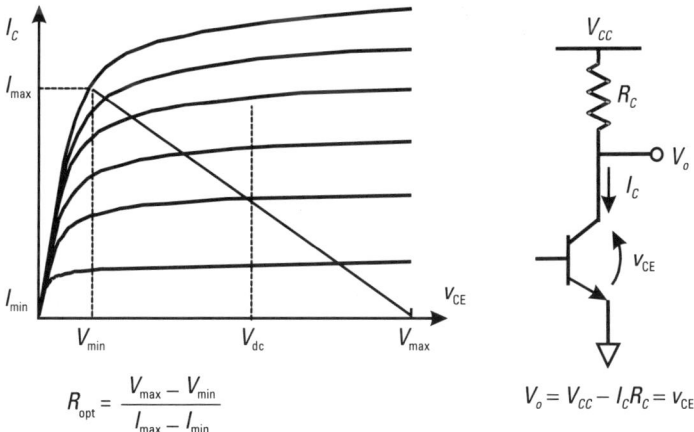

Figure 10.5 Current and voltage excursion of power amplifier.

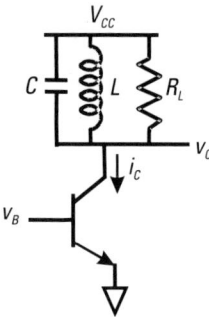

Figure 10.6 Power amplifier circuit with tuned load.

any of these classes. Waveforms for the base voltage v_B, collector voltage v_C, and collector current i_C are shown in Figure 10.7 for class A operation and in Figure 10.8 for class B and C operation. Class A amplifiers can be designed to have more gain than class B or class C amplifiers. However, as will be seen later, the achievable output power is nearly the same for a class A, class AB, or class B amplifier. For a class C amplifier, where the transistor conducts for a short part of the period, the output power is reduced.

The maximum sinusoidal collector voltage is shown from approximately 0V to $2V_{CC}$. The assumption that the collector voltage can swing down close to 0V is justified in that it simplifies the analysis and typically results in only a negligible error. While the collector voltage is assumed to be sinusoidal because of the filtering action of the tuned circuit, the collector current may be sinusoidal,

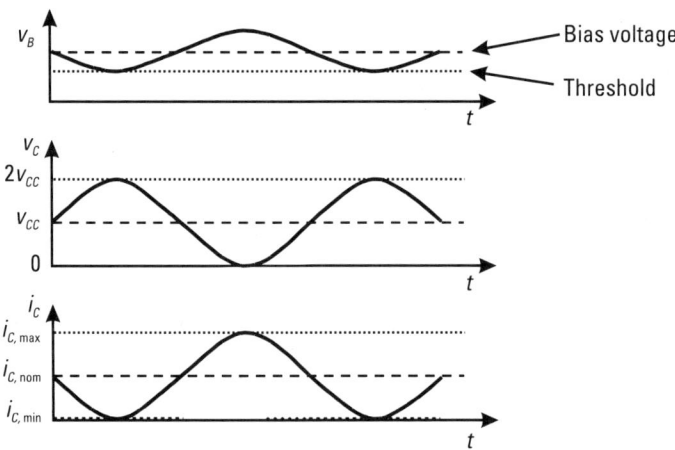

Figure 10.7 Waveforms for class A power amplifier.

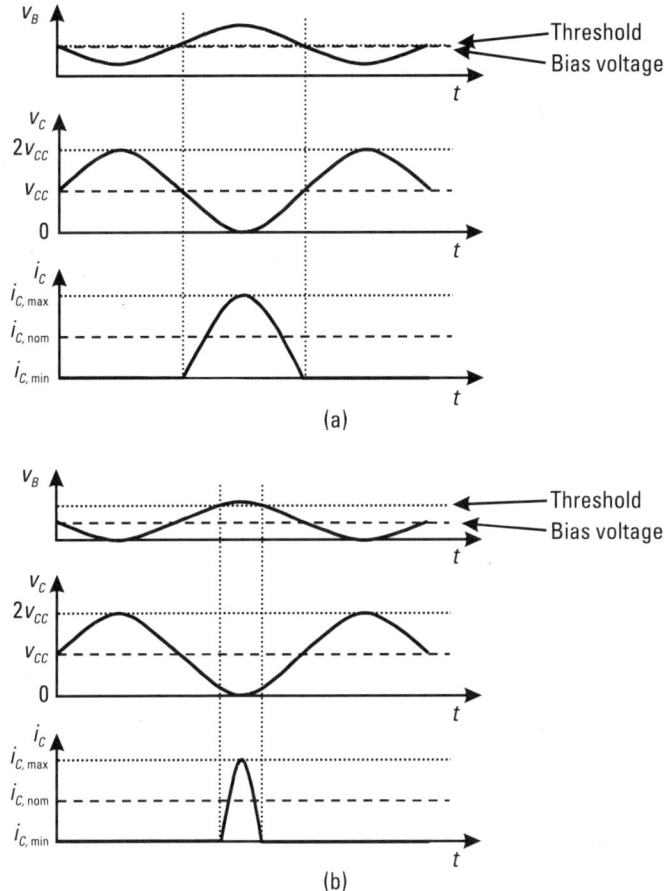

Figure 10.8 Power amplifier waveforms: (a) class B operation; and (b) class C operation.

as in class A operation, or may be nonsinusoidal, as in class B or C operation, which is determined mainly by how the transistor is biased.

The classification as A, AB, B, or C describes the fraction of the full cycle for which current is flowing in the driver transistor. Such a fraction can be described as a conduction angle, which is the number of degrees (out of 360°) for which current is flowing. If current is always flowing, the conduction angle is 360° and operation is class A. If current flows for exactly half of the time, the conduction angle is 180° and operation is class B. For conduction angles between 180° and 360°, operation is class AB. If current flows between 0° and 180°, the transistor is said to be operating as class C. It will be shown later that a higher conduction angle will result in better linearity but lower efficiency. A summary is shown in Table 10.1.

Table 10.1
PA Class A, AB, B, C Conduction Angle and Efficiency

Class	Conduction Angle (°)	Efficiency (max theoretical) (%)	Ouput Power (normalized)
A	360	50	1
AB	360–180	50–78.5	Nearly constant about 1 (theory: max of 1.15 at 240°)
B	180	78.5	1
C	180–0	78.5–100	1 at 180°, 0 at 0°

To obtain high efficiency, power loss in the transistor must be minimized, and this means that current should be minimum while voltage is high, and voltage should be minimum while current is high. It can be seen in Figures 10.7 and 10.8 that for all waveforms, maximum voltage is aligned with minimum current, and maximum current is aligned with minimum voltage. It can further be seen that for class B and class C, the current is set to zero for part of the cycle where the voltage is high. This leads to increased efficiency; however, there will still be some loss, since there is an overlap of nonzero voltage and current. Other classes of amplifiers, to be described in later sections, namely, classes D, E, F, and S, are designed such that the voltage across the transistor is also nonlinear, leading to higher efficiencies, in some cases up to 100%. A different way to improve efficiency, while potentially maintaining linearity, is to power a linear amplifier from a variable or switched power supply. This is the basis for class G and H designs. All of the above amplifiers will be discussed in more detail in Sections 10.6 to 10.9.

Figure 10.9 shows a simplified power amplifier and a plot of transistor current versus time for the various classes. The different classes can be obtained with the same circuit by adjusting the input bias circuit. For example, in class A, if the maximum current is I_{max}, the amplifier is set to have a nominal bias of half of I_{max} so that current swings from nearly 0 to I_{max}. For class B, the bias is set so that the transistor is nominally at the edge of conduction so that positive input swing will cause the transistor to conduct, while negative input will guarantee the transistor is off. Thus, the transistor will conduct half the time.

10.5.1 Class A, B, and C Analysis

Except for class A, the current through the transistor is not sinusoidal, but may be modeled as a biased sinusoid as shown in Figure 10.10.

The collector current can be expressed as

Power Amplifiers

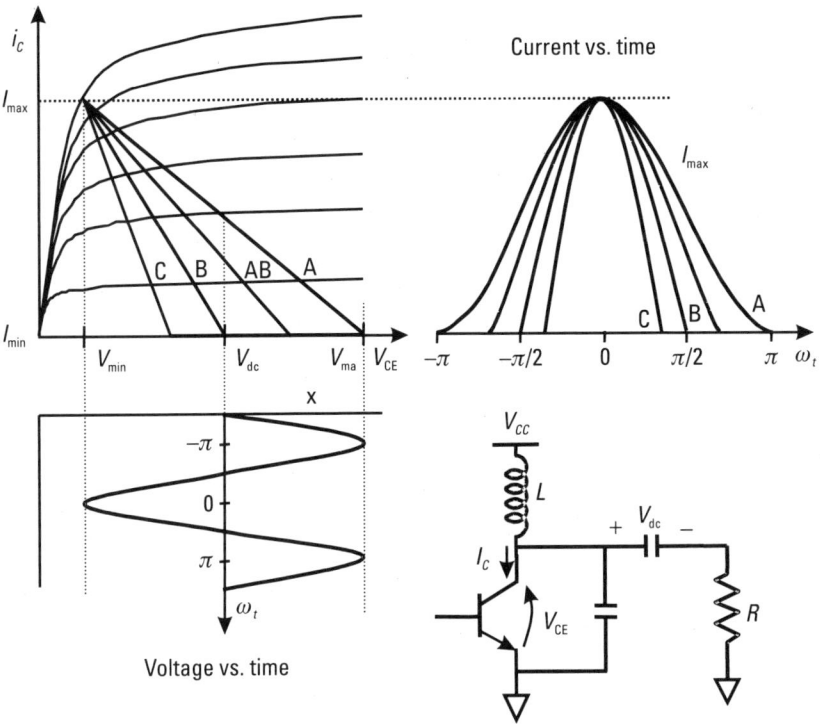

Figure 10.9 Current and voltage excursions for different classes of amplifiers.

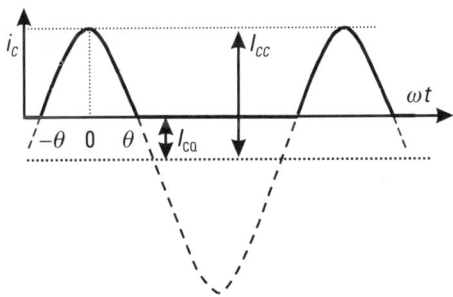

Figure 10.10 Waveform in analysis of classes A, B, and C.

$$i_C = I_{CC} \cos \omega t - I_{CQ} \quad (10.6)$$

This is valid from $-\theta < \omega t < \theta$. Here, I_{CQ} is given by

$$I_{CQ} = I_{CC} \cos \theta \quad (10.7)$$

We can find the dc component and the fundamental component by determining the Fourier series of this waveform. Note that the tuned circuit will give us the fundamental component (if tuned to f_0).

$$I_{dc} = \frac{1}{2\pi} \int_{-\theta}^{\theta} (I_{CC} \cos \omega t - I_{CQ}) d(\omega t)$$

$$= \frac{1}{\pi} \int_{0}^{\theta} [I_{CC}(\cos \omega t - \cos \theta)] d(\omega t) \qquad (10.8)$$

$$= \frac{I_{CC}}{\pi} [\sin \theta - \theta \cos \theta]$$

Power supplied is given by

$$P_{CC} = V_{CC} I_{dc} = \frac{V_{CC} I_{CC}}{\pi} (\sin \theta - \theta \cos \theta) \qquad (10.9)$$

The fundamental current i_1 given by

$$i_1 = \frac{4}{2\pi} \int_{0}^{\theta} (I_{CC} \cos \omega t - I_{CQ}) \cos \omega t \, d(\omega t) = \frac{I_{CC}}{2\pi} (2\theta - \sin 2\theta)$$

$$(10.10)$$

Output power is given by

$$P_{out} = \frac{i_1^2 R_L}{2} = \frac{v_{peak}}{\sqrt{2}} \cdot \frac{i_{peak}}{\sqrt{2}} \qquad (10.11)$$

The maximum possible v_{peak} is when the output swings from about 0V to $2V_{CC}$ or $v_{peak} = V_{CC}$. Thus,

$$P_{out, max} = \frac{V_{CC}}{\sqrt{2}} \cdot \frac{i_1}{\sqrt{2}} = \frac{V_{CC} I_{CC}}{4\pi} (2\theta - \sin 2\theta) \qquad (10.12)$$

Efficiency for this maximum possible voltage swing is given by

$$\eta_{max} = \frac{P_{out, max}}{P_{dc}} = \frac{2\theta - \sin 2\theta}{4(\sin \theta - \theta \cos \theta)} \qquad (10.13)$$

The efficiency is plotted in Figure 10.11.

The actual output power for an output peak voltage of V_{op} can be found as a function of θ:

$$P_{out} = \frac{V_{op} I_{CC}}{4\pi}(2\theta - \sin 2\theta) \qquad (10.14)$$

noting that to get maximum power, the load resistance has to be adjusted so that the maximum voltage $v_{o,max}$ is approximately $2V_{CC}$ and the minimum voltage $v_{o,min}$ is approximately zero, thus V_{op} is equal to V_{CC}.

The above equation will be more convenient to plot if we eliminate I_{CC}. Recall that I_{CC} is a measure of the peak current, not with respect to zero, but with respect to the center of the sine wave where the center for class C is less than zero, as shown in Figure 10.12.

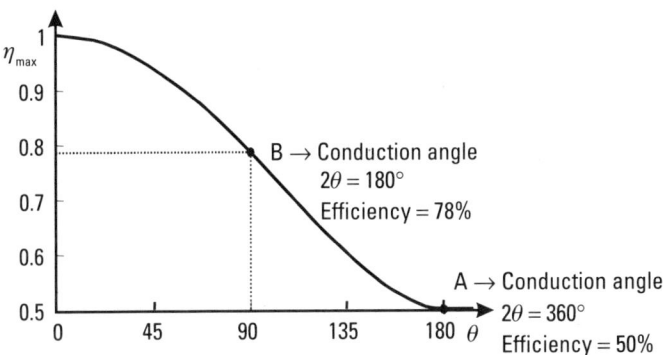

Figure 10.11 Maximum efficiency versus conduction angle.

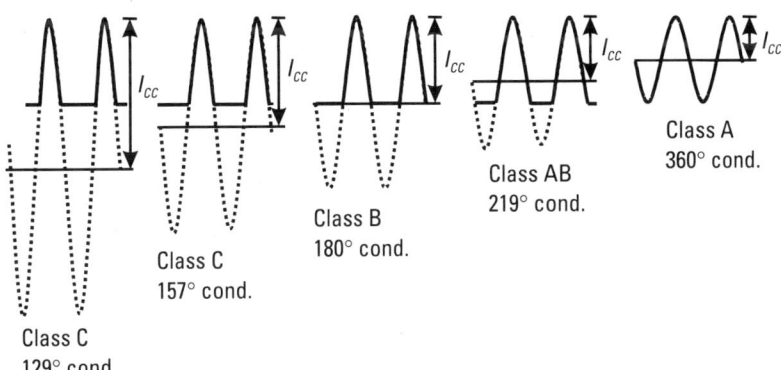

Figure 10.12 Time-domain waveforms for various conduction angles.

The peak current is

$$i_{peak} = I_{CC}(1 - \cos \theta) \quad (10.15)$$

For the same peak current, the maximum distorted, or class A, output power can be determined by noting that voltage goes from 0 to $2V_{CC}$ and the current goes from 0 to i_{peak}. Then converting peak-to-peak to rms, we obtain normalized output power $P_{o,norm}$ as

$$P_{o,norm} = \frac{(I_{max} - I_{min})(V_{max} - V_{min})}{8} = \frac{I_{CC} V_{CC}(1 - \cos \theta)}{4} \quad (10.16)$$

Then we could plot normalized, maximum output power as

$$P_{o,max,norm} = \frac{P_{o,max}}{P_{o,norm}} = \frac{1}{\pi} \frac{(2\theta - \sin 2\theta)}{1 - \cos \theta} \quad (10.17)$$

This has been plotted in Figure 10.13. It can be seen that at a conduction angle of 180° and 360° (or θ = 90° and 180°), the normalized maximum output power is 1. In between is a peak with a value of about 1.15 at a conduction angle of about 240°. For maximum output power, this might appear to be the optimum conduction angle; however, it can be noted that in real life, or in simulations with other models for the current (rather than the tip of a

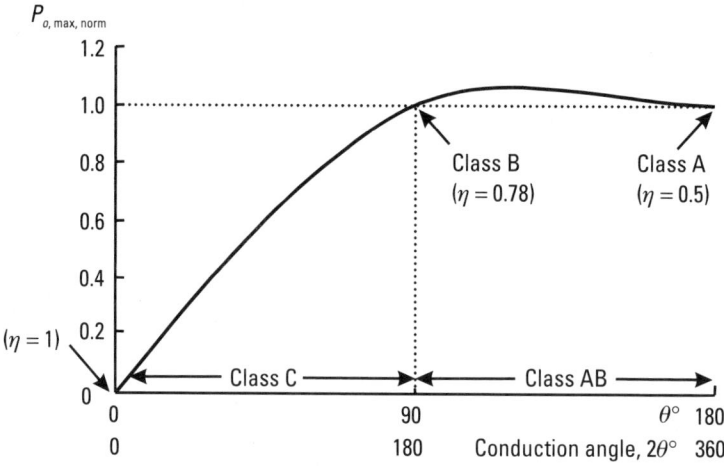

Figure 10.13 Maximum output power versus conduction angle.

sinusoid), this peak does not occur. However, overall, Figure 10.13 is a fairly good description of real life performance.

As an example, if $V_{CC} = 3V$, then V_{max} can go to 6V. If I_{max} is 1A, then maximum output power is given by $P_{out,max} = P_{out,norm} = (V_{max} \cdot I_{max})/8 = 0.75W$ at $\theta = 90°$ and at $\theta = 180°$.

An expression that is valid for $n = 2$ or higher can be found for i_n, the peak current of the nth harmonic:

$$i_n = \frac{2I_{CC}}{\pi} \left[\frac{\cos\theta \sin n\theta - n \sin\theta \cos n\theta}{n(n^2 - 1)} \right] \quad (10.18)$$

Figure 10.14 shows the current components normalized to the maximum current excursions in the transistor. The dc component is found from (10.8), the fundamental component is found from (10.10), and the other components from (10.18), with normalization done using (10.15).

We note that for class A ($\theta = 180°$ or conduction angle is $360°$) the collector current is perfectly sinusoidal and there are no harmonics. At lower conduction angles, the collector current is rich in harmonics. However, the tuned circuit load will filter out most of these, leaving only the fundamental to make it through to the output.

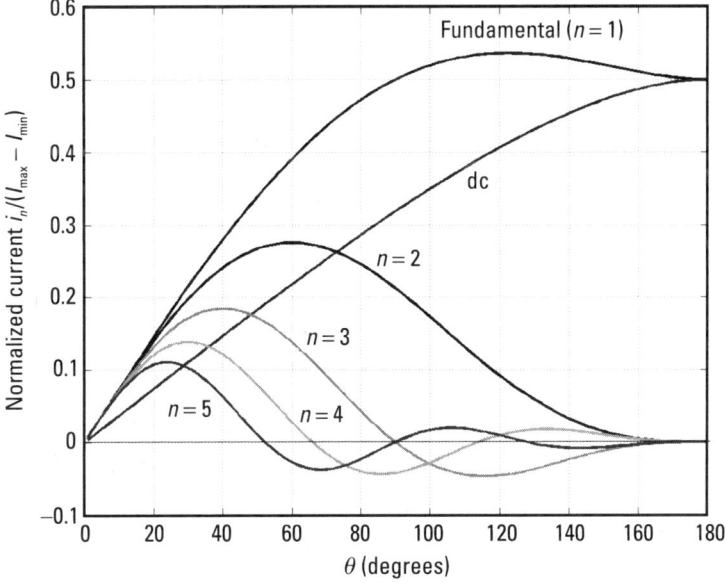

Figure 10.14 Fourier coefficients for constant transconductance.

At very low conduction angles, the current "pulse" is very narrow approaching the form of an impulse in which all harmonic components are of equal amplitude. Here efficiency can be high, but output power is lower.

10.5.2 Class B Push-Pull Arrangements

In the push-pull arrangement shown in Figure 10.15 with transformers or in Figure 10.16 with power combiners, each transistor is on for half the time. Thus, the two are on for the full time, resulting in the possibility of low distortion, yet with class B efficiency, with a theoretical maximum of 78%. The total output power is twice that of each individual transistor.

Mathematically, each transistor current waveform as shown in Figure 10.16 is described by the fundamental and the even harmonics as shown in (10.19).

$$i_A = \frac{I_P}{\pi} + \frac{I_P}{2} \cos \omega t + \frac{2I_P}{3\pi} \cos 2\omega t - \frac{2I_P}{15\pi} \cos 4\omega t + \ldots \quad (10.19)$$

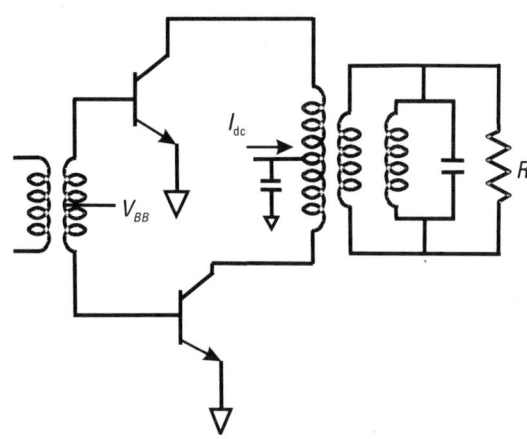

Figure 10.15 Push-pull class B amplifier.

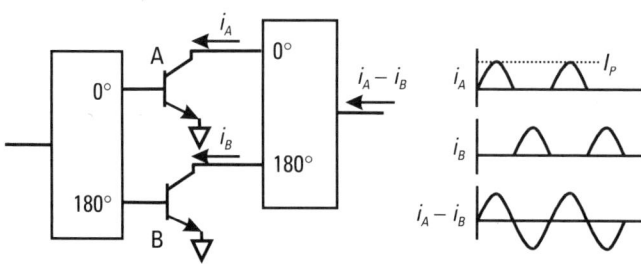

Figure 10.16 Class B amplifier with combiner.

Note that this agrees with Figure 10.14, which showed the third and fifth harmonic passing through 0 when $\theta = 90°$ (conduction angle is 180°). In the push-pull arrangement, the dc components and even harmonics cancel, but odd harmonics add, thus the output contains the fundamental only, as shown in (10.20):

$$i_A - i_B \approx I_P \cos \omega t \qquad (10.20)$$

Note that the cancellation of odd harmonics is only valid if the amplifier is not driven hard.

10.5.3 Models for Transconductance

There is an error in assuming current is the tip of a sine wave for a sinusoidal input. This model assumes that $i = g_m v_{in}$ with a constant value for g_m as long as v_{in} is larger than the threshold voltage. There are at least two other more realistic models for g_m. One is to assume that g_m is linearly related to the input voltage, as might be the case for an amplifier with emitter degeneration. Another model relates the transconductance g_m exponentially to the input voltage. While these models are more realistic, only the model with constant g_m results in easy analytical equations, but numerical results can be obtained for all cases. The resulting output powers and efficiencies are similar to the results for the simple constant g_m assumption (typically with somewhat reduced output power and efficiency). However, for high frequency, none of these models are completely accurate, so it is recommended that the simple constant g_m model be used for speedy hand calculations, and full simulations be used to continue the design.

Example 10.1 Class A Power Amplifier
Design a class A power amplifier that will drive 200 mW into a 50-Ω load at 1 GHz from a 3-V power supply. The transistor unit cell that is available has the f_T versus current relationship shown in Figure 10.17. Use as many of these in parallel as necessary.

Solution
Assuming the output voltage is centered at 3V and has a peak swing of 2.5V, the output resistance can be determined.

$$P_o = \frac{v_{rms}^2}{R} \text{ or } R = \frac{v_{rms}^2}{P_o} = \frac{2.5^2}{2 \times 0.2} = 15.6 \Omega$$

Thus, we estimate that we will need a 15.6-Ω load resistance for 200 mW of output power. This means we will need a matching circuit to convert between 15.6Ω and 50Ω.

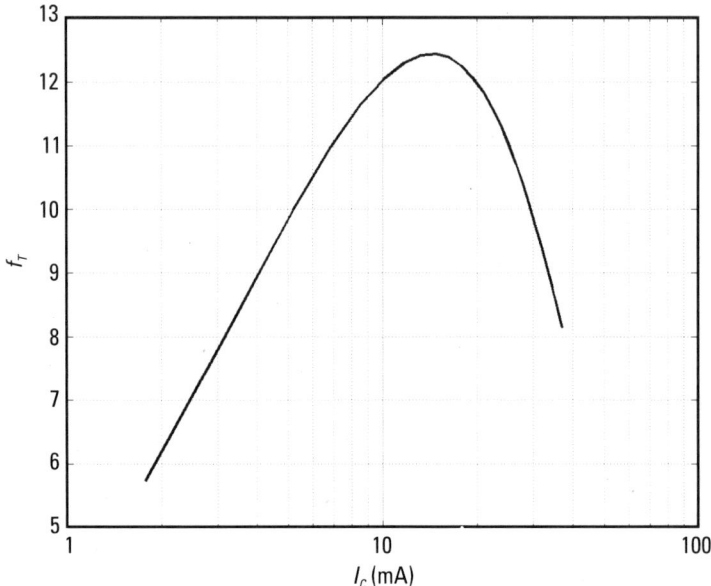

Figure 10.17 Power transistor f_T versus current.

We can also determine the current.

$$P_o = \frac{v_p \cdot i_p}{2}$$

or

$$i_p = \frac{2P_o}{v_p} = \frac{2 \cdot 0.200}{3} = 0.133 \text{ or } 133 \text{ mA}$$

Thus, for class A, the nominal current should be about 133 mA with peak excursion from about 0 to 266 mA. Thus, with the transistor as given, with peak f_T at about 15 mA, and recognizing that we will have extra losses, let us choose to use 10 of these units. Then with operation close to the peak f_T, the resulting simulated collector current is 147 mA, and with a dc current gain of about 85, the input current is set to about 1.75 mA.

It was noted that the circuit had a potential stability problem for low-impedance inputs. This problem was minimized with a 5Ω resistor in series with the input. The input impedance was then seen to be about $5.55 - j3.14$. Through simulations, the input was matched with 3 nH in series and 9 pF in parallel. Note that the finite Q of the input series inductor will help with stabilization. Then input power was swept to determine the power level at which the current went to zero. This power was used as a starting point for several iterations of sweeps of load pull and input power used to determine the optimal output load and the required input power. The transistor current crosses zero for an input power of about 8 dBm, as shown in Figure 10.18. The load pull shown in Figure 10.19 indicated the optimal load should be $9 + j7.6$. This is a little bit lower than the predicted 15.6Ω, explained largely by the reduced voltage swing compared to that predicted. The inductive portion of the load ($j7.6$) accounts largely for the transistor output capacitance. The sweeps of P_{out} and power-added efficiency versus P_{in} shown in Figure 10.20 shows that 1-dB compression occurs at an input power of about 9 or 10 dBm and that power-added efficiency is just over 30% at an input power of 8 dBm, rising to about 42% at 10 dBm. The output power is about 23 dBm as required.

Several differences can be seen between this simulation and simple theory. The simple equations were derived assuming that output voltage swings from 0 to 6V. This does not happen, and thus power is a little bit low. This also directly leads to a lower optimal load impedance than was initially calculated. In this example, ideal models were used for passives and packaging. Obviously, realistic models would have resulted in a reduction in efficiency.

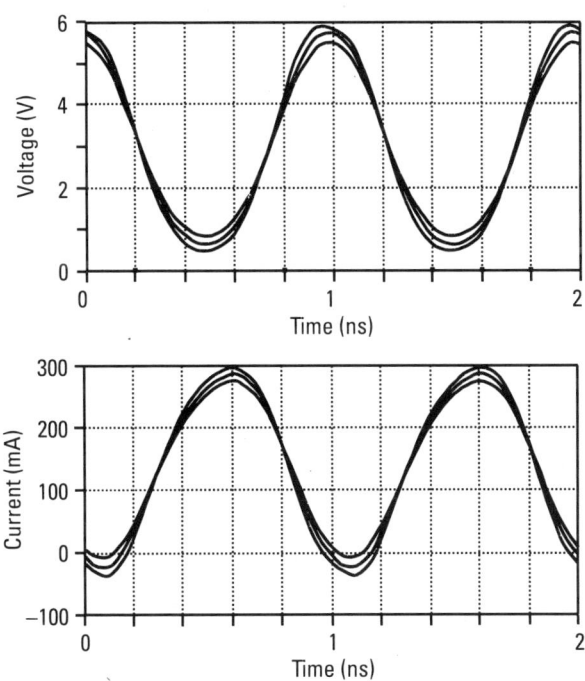

Figure 10.18 Voltage and current waveforms for input power levels of 8, 9, and 10 dBm.

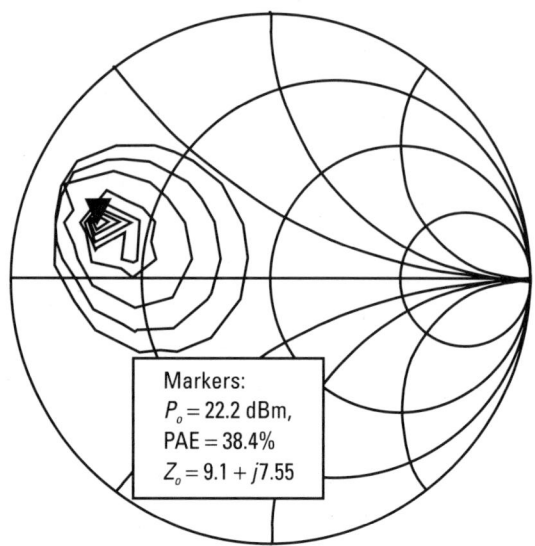

Figure 10.19 Load pull for an input power of 8.5 dBm.

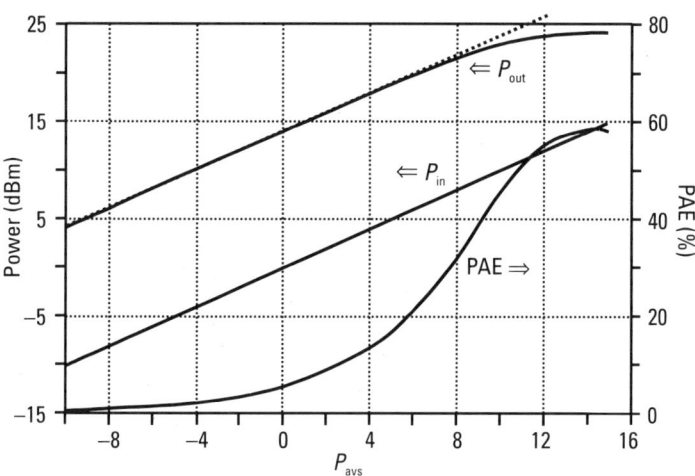

Figure 10.20 Gain and power-added efficiency versus input power.

Example 10.2 Class AB Power Amplifier

As a continuation of the previous example, design a class AB power amplifier that will drive 200 mW into a 50-Ω load at 1 GHz from a 3V power supply. As before, the transistor unit cell that is available had the f_T versus current relationship shown in Figure 10.17. Design for the optimal PAE.

Solution

The first thing to note is that the current extremes I_{max} and I_{min} will still be approximately the same as they were for the class A amplifier even though current will flow for a smaller percentage of the time. Examination of Figure 10.14 shows that the output fundamental current will be roughly constant for any conduction angle between 180° and 360°. Thus, it would seem that it should be possible to reduce the nominal current through this amplifier and achieve roughly the same output power by driving the amplifier into compression. In practice, with reduced bias, the achievable output power is reduced, as shown by simulation results summarized in Table 10.2.

It can be seen that by reducing the bias current to 105 mA, approximately the same results are obtained as for the class A amplifier. However, there are a few important differences. Because the amplifier is driven into compression, the efficiency is now 57.2% and the amplifier is now nonlinear. If instead the same amplifier is used as in Example 10.1, a bias current of 147 mA, output power is increased to 24.1 dBm and efficiency is increased to 58.9%. Time-domain waveforms for this case can be seen in Figure 10.21. It can be seen that waveforms do not match simple theory in that voltage is not sinusoidal and current goes negative due to transistor capacitance. By considering the positive portion of the current, the conduction angle for a 14-dBm input is estimated to be about 260° ($\theta = 130°$), and from Figure 10.11, efficiency is expected to be about 60%, which is close to the simulated value.

10.6 Class D Amplifiers

Two examples of class D amplifiers are shown in Figure 10.22. The two transistors alternately switch the output to ground or to V_{CC}. The output filter, consisting of L_o and C_o, is tuned to the fundamental frequency. This serves to remove the dc component and the harmonics, resulting in a sine wave at the output. While class D amplifiers can have high efficiency and have been

Table 10.2
Simulation Results for Class AB Power Amplifier Example

I_{bias} (mA)	Opt PAE (%)	P_{in} (opt PAE) (dBm)	P_{out} (opt PAE)	Compression (dB)	Γ_{opt}
63	55.8	12	20.7	2.5	11.5 + j18.7
105	57.2	13	22.6	3.2	12.2 + j14.3
147	58.9	14	24.1	3.7	13.1 + j10.3

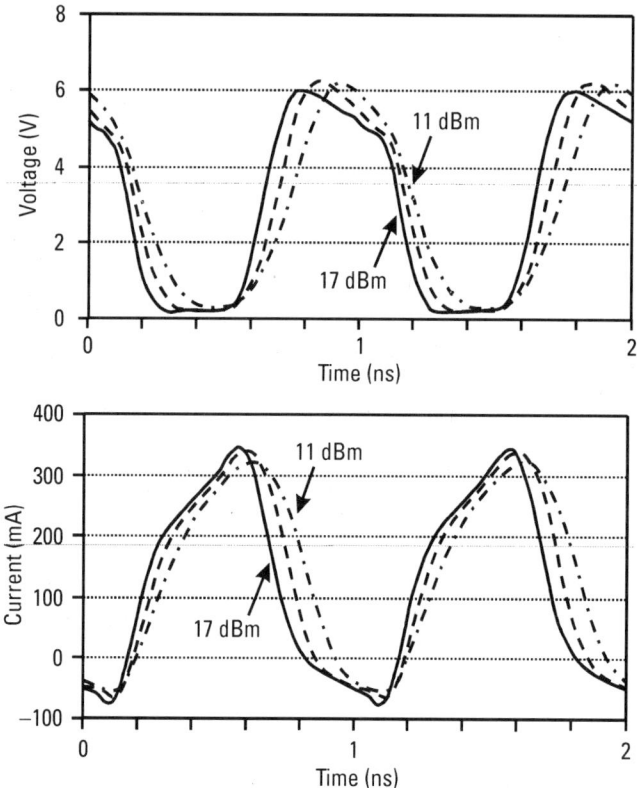

Figure 10.21 Voltage and current waveforms for bias current of 147 mA and input power of 11, 14, and 17 dBm.

demonstrated in the 10-MHz frequency range, they are not practical in the gigahertz range, especially when there is another type of switching amplifier (the class E amplifier) available which performs much better. The class E amplifier has the high efficiency of the class D amplifier without needing a push-pull structure or transformers, while being feasible at high frequencies. For this reason, the class D will not be discussed further. For the interested reader, more information on class D amplifiers can be found in [5, 6].

10.7 Class E Amplifiers

The class E amplifier is shown in Figure 10.23. It is designed to require a capacitor across the output of the transistor, which means that the capacitor C is the combination of the parasitic transistor output capacitor c_o and an actual

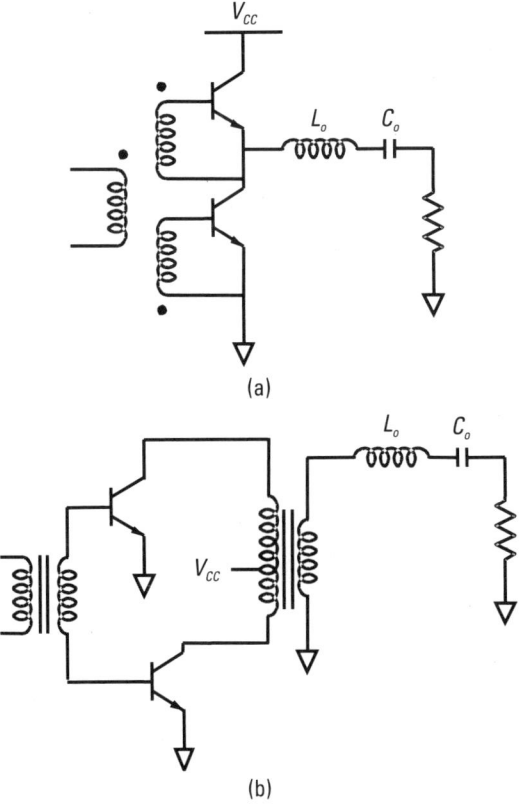

Figure 10.22 Class D amplifiers: (a) with a nonsymmetric driver, and (b) with a symmetric driver.

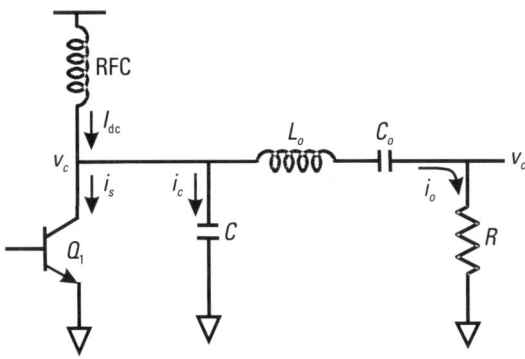

Figure 10.23 Class E amplifiers.

added capacitor C_A. Thus, it is possible to obtain close to 100% efficiency even in the presence of parasitics.

10.7.1 Analysis of Class E Amplifier

Several simplifying assumptions are typically made in the analysis [4]:

1. The *radio frequency choke* (RFC) is large, with the result that only dc current I_{dc} flows through it.
2. The Q of the output circuit consisting of L_o and C_o is high enough so that the output current i_o and output voltage v_o consist of only the fundamental component. That is, all harmonics are removed by this filter.
3. The transistor Q_1 behaves as a perfect switch. When it is on, the collector voltage is zero, and when it is off the collector current is zero.
4. The transistor output capacitance c_o, and hence C, is independent of voltage.

With the above approximations, the circuit can now be analyzed. Waveforms are shown in Figure 10.24. When the switch is on, the collector voltage is zero, and therefore the current i_C through the capacitor C is zero. In this case, the switch current $i_s = I_{dc} - i_o$. When the switch is off, $i_s = 0$. In this case, $i_C = I_{dc} - i_o$. This produces an increase of collector voltage v_C due to the charging of C. Due to resonance, this voltage will rise and then decrease again. To complete the cycle, as the switch turns on again, C is discharged and collector voltage goes back to zero again. If the component values are selected correctly, then the collector voltage will reach zero just at the instant the switch is closed, and as a result, there is no power dissipated in the transistor.

We cannot explicitly solve for voltage and current waveforms over the entire cycle. We can, however, determine the collector voltage waveform when the switch is off:

$$v_C(\theta) = \left[\frac{I_{dc}}{B}\left(y - \frac{\pi}{2}\right) + \frac{V_{om}}{BR}\sin(\phi - y) + \frac{I_{dc}}{B}\theta + \frac{V_{om}}{BR}\cos(\theta + \phi)\right]$$

(10.21)

where I_{dc} is the dc input current, I_{om} is the magnitude of the output current i_o, V_{om} is the magnitude of the output voltage and is given by the product of I_{om} and R, ϕ is the phase of v_o measured from the time the switch opens, $2y$ is the switch-off time in radians (e.g., $y = \pi/2$ for 50% duty cycle), and B is the admittance of the capacitance C.

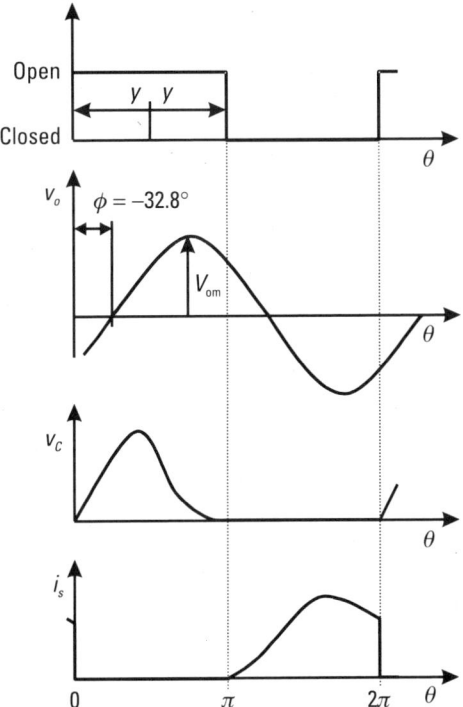

Figure 10.24 Class E waveforms.

The fundamental frequency component of $v_C(\theta)$ is $v_1(\theta)$. This is applied to $R + jX$ to determine output current, voltage, and power. Here jX is the residual impedance of the series combination of L_o and C_o, which are tuned to be slightly away from resonance at f_0.

For lossless components (as in the assumptions), the only loss is due to the discharge of C when the switch turns on. If the components are selected so that v_C just reaches zero as the switch turns on, no energy is lost and the efficiency is 100%. In practice, because the assumptions do not strictly hold and because components will not be ideal, the voltage will not be at zero and so energy will be lost. However, with careful design, efficiencies in the 80% range are feasible.

10.7.2 Class E Equations

It is necessary to choose B and X for the correct resonance to make sure $v_C = 0$ as the switch turns on, and to make sure that $d[v_C(\theta)]/d\theta = 0$ so that there is no current flowing into the capacitor.

Setting $v_C(\theta)$ and $d[v_C(\theta)]/d\theta$ to 0 at $\theta = \pi/2 + y$ results in

$$\phi = -32.48°$$

$$B = \frac{0.1936}{R} \tag{10.22}$$

$$X = \frac{1.152}{R}$$

It can be shown that

$$V_{om} = \frac{2}{\sqrt{1 + \pi^2/4}} V_{CC} \approx 1.074 V_{CC} \tag{10.23}$$

and that

$$P_o = \frac{2}{1 + \pi^2/4} \frac{V_{CC}^2}{R} \approx 0.577 \frac{V_{CC}^2}{R} \tag{10.24}$$

The dc current is given by

$$I_{dc} = \frac{V_{CC}}{1.734 R} \tag{10.25}$$

The peak transistor currents and voltages are given by

$$v_{C,peak} = 3.56 V_{CC} \tag{10.26}$$

$$i_{s,peak} = 2.86 I_{dc}$$

The resulting output power is 78% of class B, but the efficiency approaches 100%.

10.7.3 Class E Equations for Finite Output Q

If the Q of the output circuit is not infinity as initially assumed, but more typically less than 10, then some harmonic current will flow. This directly can result in the collector voltage not being zero at the instant the switch closes. Formulas for optimum operation in this case were shown by Sokal [8]:

$$X = \frac{1.110 Q}{Q - 0.67} R \tag{10.27}$$

$$B = \frac{0.1836}{R}\left(1 + \frac{0.81Q}{Q^2 + 4}\right) \quad (10.28)$$

$$Q \approx \frac{\omega_0 L}{R} \quad (10.29)$$

Then we may need to insert a filter between C_o and R to prevent excessive harmonic currents from reaching R.

10.7.4 Saturation Voltage and Resistance

Previously it was assumed that the output voltage would be zero when the transistor was on. In reality, it will be equal to the transistor saturation voltage V_{SAT}. As described in [4], this can be accounted for by replacing the power supply voltage with $V_{eff} = V_{CC} - V_{SAT}$ for all calculations, except power input. As for the transistor having nonzero on resistance R_{on}, this can be accounted for by changing the value of V_{eff} by $V_{eff} \approx R/(R + 1.365 R_{on}) V_{CC}$. This is valid for a 50% duty cycle.

10.7.5 Transition Time

Ideally, no power is dissipated during the transition between off and on. The turn-on transition for nonoptimum conditions can be approximated with a linear ramp of current. This produces a parabolic collector voltage waveform. As described in [4], the current and voltage waveforms can be integrated to determine dissipated power P_{dT}.

$$P_{dT} = \frac{1}{12} \theta_S^2 P_o \quad (10.30)$$

where θ_S is the transition time in radians and P_o is the output power. Then efficiency is given by

$$\eta = 1 - \frac{1}{12} \theta_S \quad (10.31)$$

The above losses due to saturation voltage on resistance and turn-on transient can be combined by summing dissipated power or by finding each efficiency by itself and then multiplying the efficiencies.

Further information and detailed examples of class E amplifier designs are shown by Cripps [5] and by Albulet [6].

Example 10.3 Class E Amplifier

Design a class E amplifier that delivers 200 mW from a 3-V power supply at 2.4 GHz. Assume an ideal transistor and aim for a Q of 3 for the output circuit. Specify device ratings and components.

Solution

Using $P_o \approx 0.577 V_{CC}^2 /R$ results in $R = 26\Omega$. The maximum transistor voltage is $v_{C,\max} = 3.56 \cdot 3 = 10.68\text{V}$. If this large voltage is not permissible (and it is quite likely that it is not), the power supply voltage may need to be reduced.

$$I_{dc} = \frac{V_{CC}}{1.734R} = \frac{3}{1.734 \cdot 26} = 66.6 \text{ mA}$$

$$i_{c,\text{peak}} = 2.861 I_{dc} = 190.6 \text{ mA}$$

From (10.28),

$$B = \frac{0.1836}{R}\left(1 + \frac{0.81Q}{Q^2 + 4}\right) = \frac{0.1836}{26}\left(\frac{1 + 0.81 \cdot 5}{3^2 + 4}\right) = 0.00755$$

Hence $C = 0.50$ pF. Since $Q = 3$, C_o has a reactance of 78Ω and is therefore 0.85 pF. Using (10.27),

$$X = \frac{1.10Q}{Q - 0.67}R = \frac{1.110 \cdot 3}{3 - 0.67}26 = 37.2\Omega$$

L_o therefore has a reactance of $37.2 + 78 = 115.2\Omega$ and is thus 7.64 nH. Ideally, the RFC should have a reactance of at least $10R$ and thus should be at least 17 nH, which would likely need to be an off-chip inductor.

This circuit was simulated using a process with f_T that is 25 times higher than the operating frequency. With numbers calculated as above, and choosing a transistor size that has optimal f_T at about 66 mA, the results in Figure 10.25 were obtained. For simplicity, the input was a ±1.5V pulse waveform through a 50-Ω source resistance. In a real circuit, a more realistic input waveform would have to be used.

It can be seen that the output transistor collector voltage has not gone down to zero when the transistor switches on. The problem is the parasitic output capacitance of the very large transistor. As a result, the output power is only about 77 mW and dc power is of the order of 100 mW. As a first-order compensation, the capacitor C can be reduced to compensate. With this done, the results are as shown in Figure 10.26.

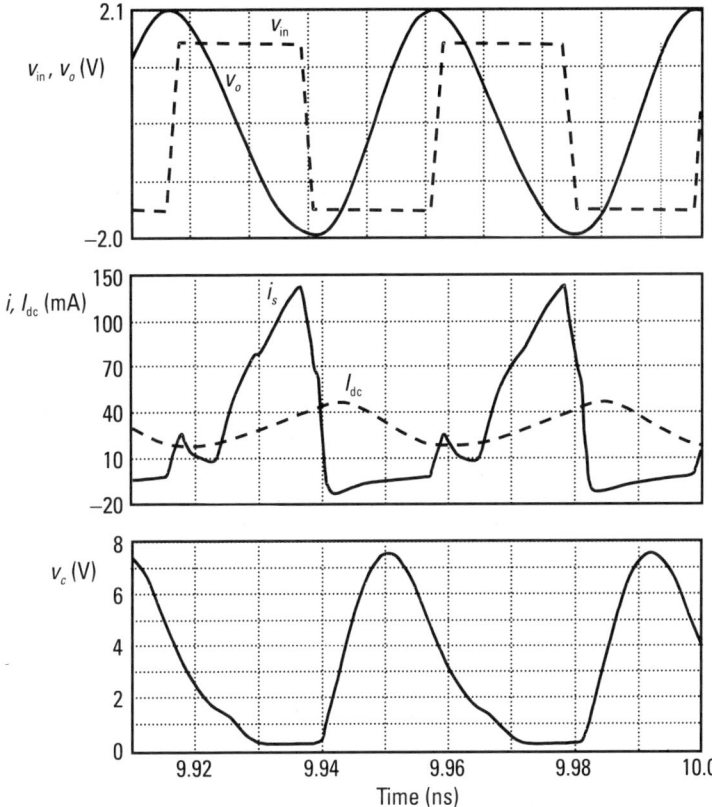

Figure 10.25 Initial simulated class E waveforms.

With this adjustment, the results are now close to the predicted values. The average current is about 60 mA; collector output voltage is just over 12V, a little bit more than predicted; the collector current peaks at 180 mA, close to the predicted value; and the output voltage is about 5.8V peak to peak, nearly the predicted value. The output power is 162 mW while the dc power is about 180 mW for a dc-to-RF efficiency of about 90%. However, with the unrealistic pulse input, a significant amount of power is fed into the input, so PAE will be lower; in this example, with an input current of nearly 20 mA, PAE is estimated to be about 75%.

10.8 Class F Amplifiers

In the class F amplifier shown in Figure 10.27, an additional resonator is used, with the result that an additional harmonic, typically the third harmonic, is

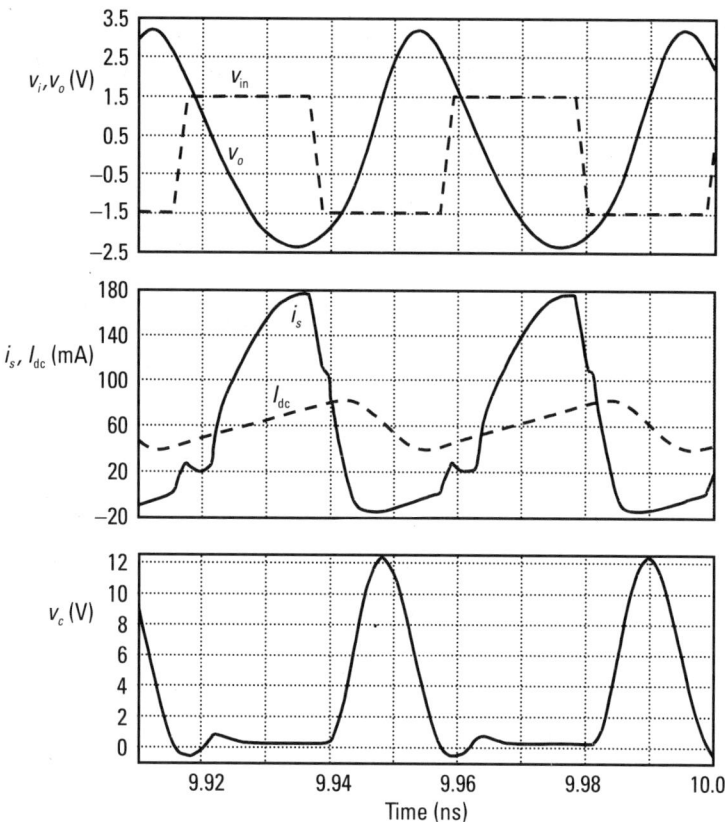

Figure 10.26 Simulated class E waveforms with reduced capacitance C.

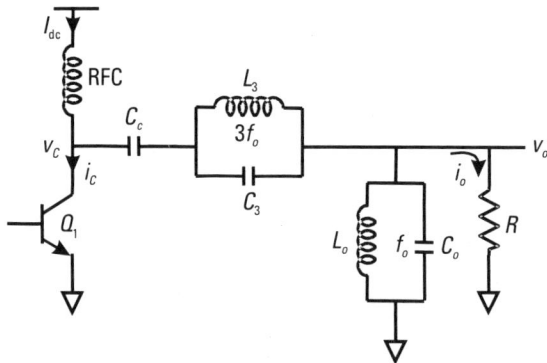

Figure 10.27 Class F amplifier.

added to the fundamental in order to produce a collector voltage more like a square wave. This means the collector voltage is lower while current is flowing, but higher while current is not flowing, so the overall efficiency is higher.

The typical waveforms for a class F amplifier are shown in Figure 10.28, with the collector voltage having a squared appearance while the output voltage is sinusoidal. Current only flows for half the time or less in order to ensure there are third-harmonic components and to maximize the efficiency (zero current while there is finite collector voltage).

The transistor behaves as a current source producing a half sinusoid of current $i_C(\theta)$, similar to class B operation. L_o and C_o make sure the output is a sinusoid. The third-harmonic resonator (L_3, C_3) causes a third-harmonic component in the collector voltage. At the correct amplitude and phase, this third-harmonic component produces a flattening of v_C as shown in Figure 10.29. This results in higher efficiency and higher output power.

If the amplitude of the fundamental component of the collector voltage is V_{cm} and the amplitude of the third harmonic is V_{cm3}, then it can be shown that maximum flatness is obtained when $V_{cm3} = V_{cm}/9$.

Thus, with

$$V_{cm3} = \frac{V_{cm}}{9} \qquad (10.32)$$

it can be seen from Figure 10.29 that the peak collector voltage is

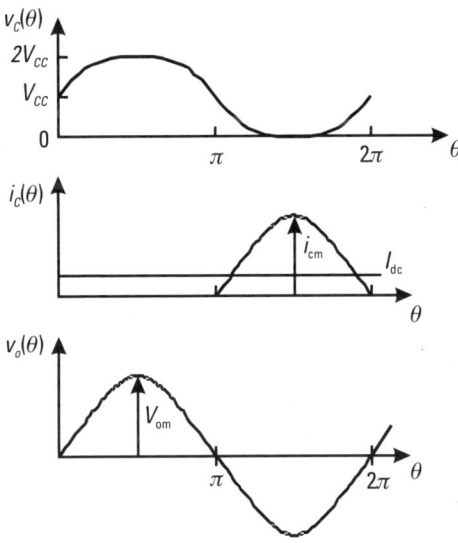

Figure 10.28 Class F waveforms.

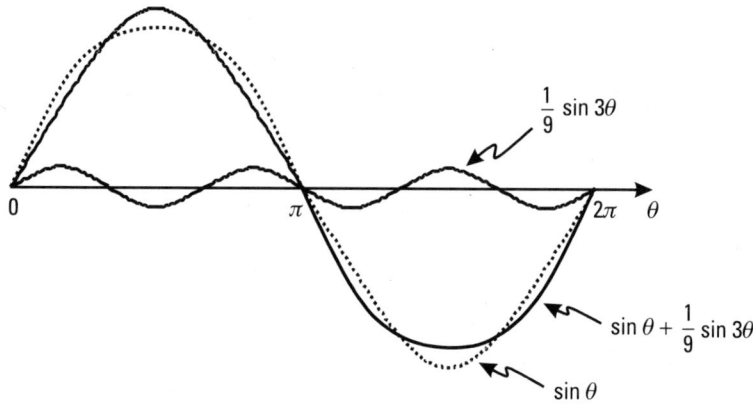

Figure 10.29 Class F frequency components of waveforms.

$$\frac{8}{9} V_{cm} = V_{CC} \therefore V_{cm} = \frac{9}{8} V_{CC} \quad (10.33)$$

As an aside, the Fourier series for the ideal square wave is

$$\sin\theta + \frac{1}{3}\sin 3\theta + \frac{1}{5}\sin 5\theta \ldots \quad (10.34)$$

However, choosing $V_{cm3} = 1/3 V_{om}$ would produce a nonflat waveform, as shown in Figure 10.30.

The efficiency can be calculated as P_o/P_{dc}. By taking a Fourier series of the $i_C(\theta)$ waveform with a peak value of i_{cm}, it can be shown that the dc value I_{dc} is equal to i_{cm}/π and the fundamental value of the output current I_o is equal to $i_{cm}/2$. As well, V_{om} is equal to V_{cm}. Thus, efficiency can be calculated as

$$\frac{P_o}{P_{dc}} = \frac{\frac{1}{2}(i_o \cdot V_{om})}{I_{dc} V_{CC}} = \frac{\frac{1}{2}\left(\frac{i_{cm}}{2} \cdot \frac{9}{8} V_{CC}\right)}{\frac{i_{cm}}{\pi} \cdot V_{CC}} = \frac{9}{8} \cdot \frac{\pi}{4} \Rightarrow 88.4\% \quad (10.35)$$

Figure 10.30 Fundamental and third harmonic to make square wave.

10.8.1 Variation on Class F: Second-Harmonic Peaking

A second resonator allows the introduction of a second-harmonic voltage into the collector voltage waveform, producing an approximation of a half sinusoid, as seen in Figure 10.31. It can be shown that the amplitude of the second-harmonic voltage should be a quarter of the fundamental.

It can be shown that the peak output voltage is given by

$$V_{om} = \frac{4}{3} V_{CC} \tag{10.36}$$

and the efficiency is given by

$$\eta = \frac{8}{3} \pi \approx 84.9\% \tag{10.37}$$

10.8.2 Variation on Class F: Quarter-Wave Transmission Line

A class F amplifier can also be built with a quarter-wave transmission line as shown in Figure 10.32 with waveforms shown in Figure 10.33.

A quarter-wavelength transmission line transforms an open circuit into a short circuit and a short circuit into an open circuit. At the center frequency, the tuned circuit (L_o and C_o) is an open circuit, but at all other frequencies, the impedance is close to zero. Thus, at the fundamental frequency the impedance into the transmission line is R_L. At even harmonics, the quarter-wave transmission line leaves the short circuit as a short circuit. At odd harmonics, the short circuit is transformed into an open circuit. This is equivalent to having a resonator at all odd harmonics, with the result that the collector voltage waveform is a square wave (assuming that the odd harmonics are at the right levels).

The collector current consists of the fundamental component (due to the load resistor) and all even harmonics. We note that there are no odd harmonics,

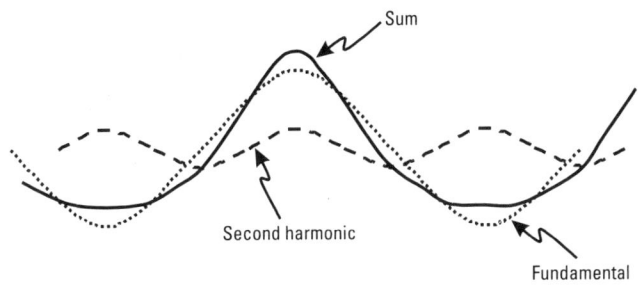

Figure 10.31 Second-harmonic peaking waveforms.

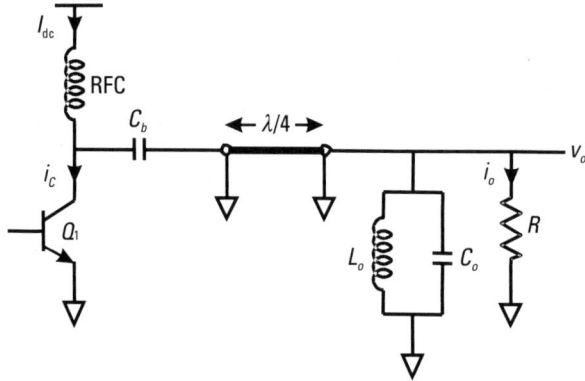

Figure 10.32 Transmission line in class F amplifier.

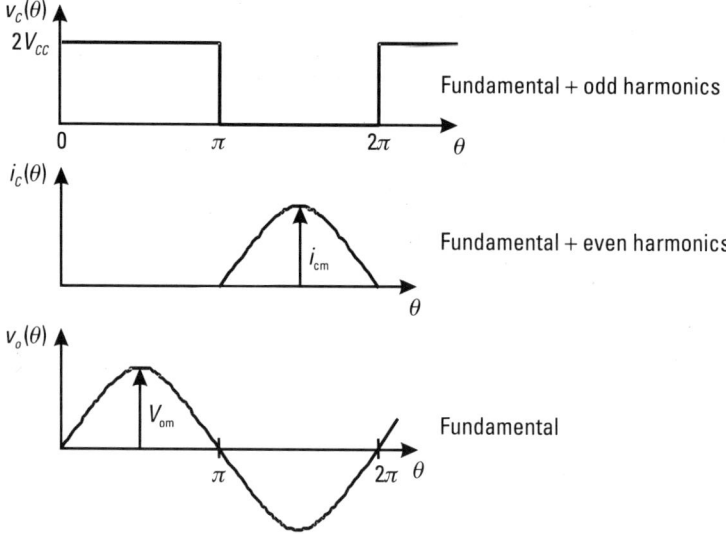

Figure 10.33 Waveforms for class F amplifier with transmission line.

since current cannot flow into an open circuit. This produces a half sinusoid of current.

Only the fundamental has both voltage and current; thus power is generated only at the fundamental. As a result, this circuit ideally has an efficiency of 100%.

Saturation voltage and on resistance of the transistor can be accounted for in the same way as for the class E amplifier.

Example 10.4 Class F Power Amplifier

Design a class F amplifier with third-harmonic peaking to deliver 200 mW from a 3-V supply.

Solution

The maximum output voltage can be determined.

$$V_{CC} = \frac{8}{9} V_{om} \Rightarrow V_{om} = \frac{9}{8} V_{CC} = 3.375 \text{V}$$

The required output resistance is found as

$$\frac{V_{om}^2}{2R} = 0.2 \Rightarrow R = \frac{3.375^2}{2 \cdot 0.2} = 28.5 \Omega$$

The maximum collector voltage swing, $V_{max} = 2V_{CC} = 6V$

Peak output current is $i_{o,peak} = V_{om}/R = 3.375/22 = 148$ mA

$I_{max} = 2 \cdot i_{o,peak} = 237$ mA $\cdot I_{dc} = I_{max}/\pi = 0.237/\pi = 75.5$ mA

Check powers, efficiencies $\Rightarrow P_{dc} = I_{dc} V_{CC} = 0.0755 \cdot 3 = 0.2264$ W

$$\eta = \frac{P_o}{P_{dc}} = \frac{0.20}{0.2264} \Rightarrow 88.4\%$$

Of course, in a real implementation, efficiency would be lower because of losses due to saturation voltage, on resistance of the transistor, finite inductor Q, imperfect RFC, and parasitics.

One useful application of a class F amplifier is as a driver for the class E amplifier, as shown in Figure 10.34. A class E amplifier is ideally driven by a square wave. Such a waveform is conveniently available on the collector of the class F amplifier, so this node is used to drive the class E amplifier.

10.9 Class G and H Amplifiers

The class G amplifier shown in Figure 10.35 amplifier has been used mainly for audio applications, although recently variations of this structure have been used up to 1 MHz for signals with high peak-to-average ratios (high crest factor), for example, in digital telephony applications.

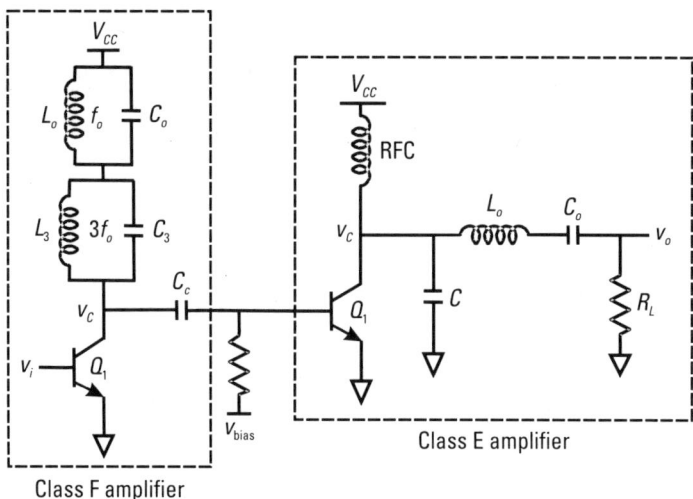

Figure 10.34 Class F amplifier driving a class E amplifier.

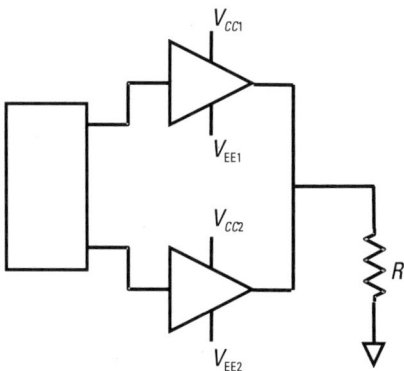

Figure 10.35 Class G amplifier.

This topology uses amplifiers powered from different supplies. For low-level signals, the lower supply is used and the other amplifier is disabled.

Class H uses a linear amplifier, such as a push-pull class B amplifier as shown in Figure 10.36, to amplify the signal. However, its power supplies track the input signal or the desired output signal. Thus, power dissipated is low, since the driver transistors are operated with a low-voltage V_{CE}. As a result, the efficiency can be much higher than for a class A amplifier.

The power supplies use a highly efficient switching amplifier, such as the class S shown later in Section 10.10. Noise (from switching) is minimized by the power supply rejection of the linear amplifier.

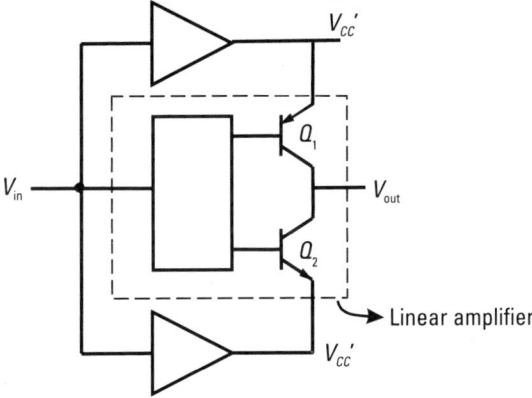

Figure 10.36 Class H amplifier.

As with the class G amplifier, this technique has mainly been used for lower frequencies. However, this technique can be modified so that the power supply follows the envelope of the signal rather than the signal itself. Discussion of such circuits for *code division multiple access* (CDMA) RF applications can be found in [2].

10.10 Class S Amplifiers

The class S amplifier, shown in Figure 10.37, has as an input a pulse-width modulated signal. This turns Q_1 and Q_2 on or off as switches with a switching frequency much higher than the signal frequency. L_o and C_o form a lowpass filter that turns the pulse-width modulated signal into an analog waveform. If only positive outputs are needed, only Q_1 and D_2 are required. For negative signals, only D_1 and Q_2 are necessary. Since the switching frequency must be

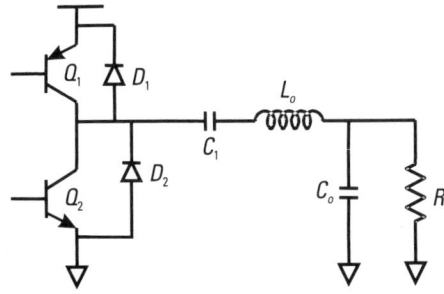

Figure 10.37 Class S amplifier.

significantly higher than the signal frequency, this technique is obviously limited in frequency capability and thus is not viable for amplification of signals in the gigahertz frequency range with the current state of process technology.

10.11 Summary of Amplifier Classes for RF Integrated Circuits

Classes D, G, H, and S are not appropriate for RF integrated circuits in the gigahertz, range so will not be included in this summary.

The main advantage of the class A amplifier is its linearity, but good linearity can also be achieved with class AB (push-pull class B) or if the power is backed off, thereby sacrificing efficiency for linearity. Thus, in spite of reduced efficiencies, these amplifiers are used for low-power applications, where efficiency is less important, or in applications requiring linearity, for example, in *quadrature amplitude modulation* (QAM), where the amplitude is not constant. Linearization techniques, to be discussed later, are not yet widely used for fully integrated power amplifiers.

For class A, AB, and B operation, the fundamental RF output power is approximately constant. However, class B has a theoretical maximum efficiency of 78%, while class A has a theoretical maximum efficiency of 50%. It should be noted that practical efficiencies are much lower in fully integrated power amplifiers due to a number of nonidealities such as finite inductor Q, saturation voltage in the transistors, and tuning errors due to process, voltage, or temperature variations. Efficiencies of half of the theoretical maximum would be considered extremely good, especially in a low-voltage process.

Class C amplifiers have theoretical maximum efficiencies higher than the class B, approaching 100% as the conduction angle decreases. However, this increase in efficiency is accompanied by a decrease in the output power. The output power approaches zero as the efficiency approaches 100%. Because of the difficulty of achieving low conduction angles on an integrated circuit, and because of other losses, completely integrated class C amplifiers are very rarely seen.

The class F amplifier can be seen as an improvement over class C or single-ended class B amplifiers in terms of output power (theoretically up to 27% higher than class B) and efficiency (theoretically 88% versus 78% for class B). However, the added resonant circuits have loss due to finite Q and result in an increase of complexity and larger chip area.

Class E can operate at radio frequencies because the resonant circuits are tuned to help the transistor switch. While efficiencies up in the 90% range can be achieved for hybrid designs, fully integrated designs have additional losses due to low-quality passives, such as inductors, so efficiencies of 60% are considered quite good. Output power is typically a few decibels lower than for similarly designed class AB amplifiers. While class AB amplifiers might have a maximum transistor voltage of about twice the power supply voltage, class E can have a swing higher than three times the power supply voltage. The maximum supply voltage and the breakdown voltage have been reduced for each new generation of process, with a typical value now being 1.8V or less. Thus, for class E amplifiers, the supply voltage must typically be set to less than the maximum supply voltage.

Other techniques exist for combining two amplifiers with different output power and different peak operating conditions, and then optimizing the combination for improved performance compared to a single amplifier. One possible optimization allows high efficiency over a broader range of input power [5]. While these techniques show promise, it still remains for someone to exploit these for integrated power amplifiers.

10.12 AC Load Line

The ac or dynamic load line shows the excursion of current versus voltage at the operating frequency. As shown in Figure 10.38, because of reactive impedance, voltage and current will be out of phase and the dynamic load line will no longer be a straight line, appearing instead as an ellipse. Figure 10.38 shows a simulated family of curves for increasing input amplitude, in this case from 25 to 200 mV peak. The amplifier is shown with an inductive output impedance resulting in current that lags voltage. In addition, for nonlinear circuits, with harmonics, the current versus voltage characteristics will no longer be a simple ellipse, and patterns that are more complex can be seen. An example is shown in Figure 10.39. In this example, the load is now tuned so that current and voltage are in phase. Inputs of 200 and 400 mV are applied, resulting in current and voltage, which are visibly nonlinear, resulting in dynamic load lines with loops in the characteristics.

10.13 Matching to Achieve Desired Power

Given a particular power supply voltage and resistance value, the achievable amount of power is limited by $P_o \approx V_{CC}^2/2R$. Obviously, R must be decreased to achieve higher P_o. This is achieved by an impedance transformation at the output, as illustrated in Figure 10.40. As an example, if $V_{CC} = 3$V and $P_o = $ 1W the required resistance R is approximately 4.5Ω. Similarly, for a P_o of 500 mW, R needs to be approximately 9Ω, and for a P_o of 200 mW, R needs to be about 22.5Ω.

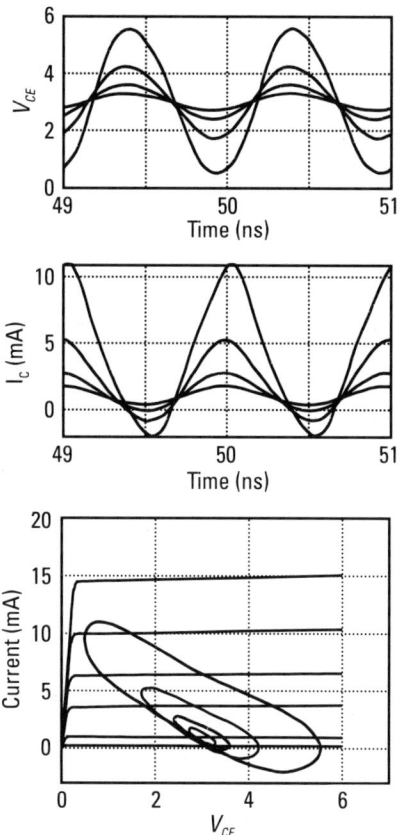

Figure 10.38 Time-domain waveforms and dynamic load line for reactive circuits.

Generally, the smaller R is, compared to R_L, the narrower will be the bandwidth of the circuit; that is, the amplifier will be able to produce useful output power over a narrower band of frequencies. It is possible to increase the bandwidth by using a higher order of matching network. For example, instead of an *ell* network, a *double ell* network can be used to convert first to an intermediate impedance, usually $R_{int} = \sqrt{RR_L}$, and then to the final value as shown in Figure 10.41. This choice of R_{int} maximizes the bandwidth. As an example, if $R = 4.5\Omega$ and $R_L = 50\Omega$, the optimal R_{int} is about 15Ω.

Sometimes higher Q is desired, and then an intermediate resistance higher than R_L might be used. A possible matching network to achieve this is illustrated in Figure 10.42.

Bond wire inductance can be used for realizing series inductance. Examples of series inductance can be found in impedance transformation networks or in the output of the class E amplifier. Bond wire inductance has the advantage of high power handling capability and high Q compared to integrated inductors.

Power Amplifiers 387

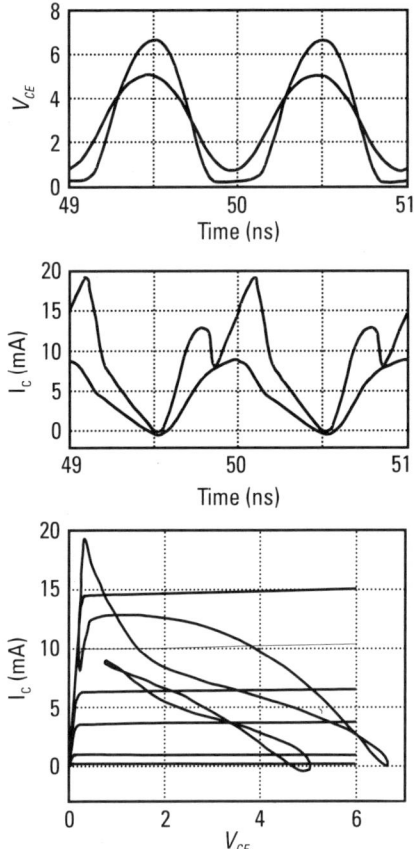

Figure 10.39 Time-domain waveforms and dynamic load line for large signals.

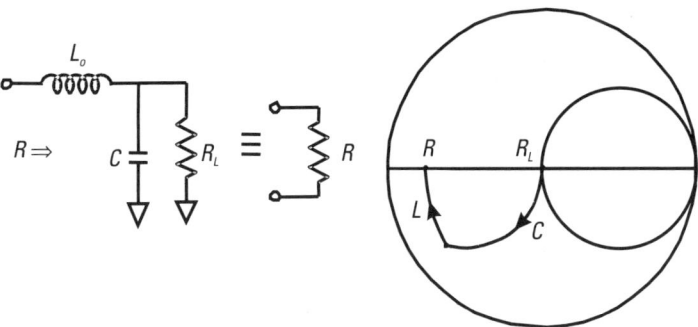

Figure 10.40 Matching example.

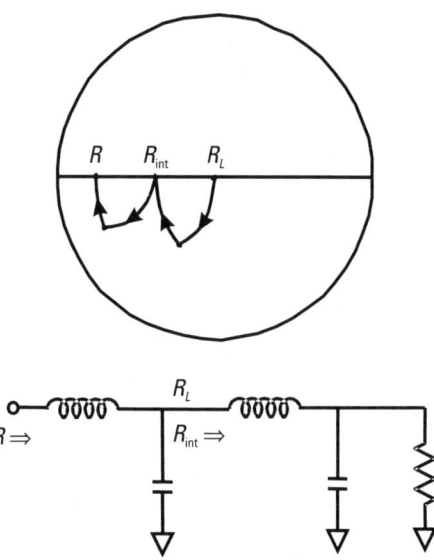

Figure 10.41 Broadband matching circuit.

10.14 Transistor Saturation

Efficiency increases rapidly with increasing input signal size until saturation of the input device occurs. After saturation, efficiency is fairly constant, but drops somewhat due to gain compression and the resulting increase in input power. Classes A, B, and C are usually operated just into saturation to maximize efficiency. Class E (and sometimes F) is operated as a switch, between saturation and cutoff. There may be difficulty in properly modeling the switching transistor, which increases the design difficulty. It can be noted that power series approximations are not particularly good for a transistor that is switching hard. An important design consideration for operation into saturation is that a proper base drive is required to remove stored charge to get a transistor out of saturation fast.

10.15 Current Limits

As described earlier, with a 3-V power supply, for an output power of 500 mW, the load resistance must be about 9Ω and the required current is 333 mA. Because of efficiency issues and because the transistor is on for some reduced time, the peak collector current can easily be over 1A. As a result, there is a requirement for huge transistors with very high current handling requirements. This requires the use of transistors with multiple emitter, base,

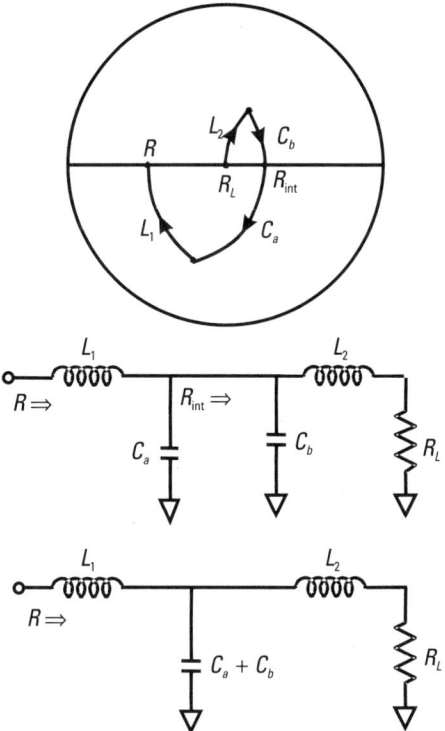

Figure 10.42 High-Q matching circuit.

and collector stripes (multiple fingers), as shown in Figure 10.43, as well as multiple transistors distributed to reduce the concentration of heat and to reduce the current density. However, the use of transistors with multiple fingers and multiple transistors introduces the new concern of making sure each finger and each transistor is treated the same as every other finger and transistor. This is important in order to avoid local hot spots or thermal runaway and to make sure that the connection to and from each transistor is exactly the same to avoid mismatch of phase shifts. Power combining will be discussed further in Section 10.17, and thermal runaway is discussed further in Section 10.18.

Also, for such high currents, metal lines have to be made wide to avoid problems with metal migration, as described in Section 5.6.

As for transistors, a large transistor cannot become too long or it will not be able to handle its own current. With transistors, the current handling capability is directly proportional to emitter area. Thus, as the emitter length is doubled, the current is doubled. However, if line width stays the same, the maximum current capability is the same. As an example, a transistor that has emitters that are 40 μm long and has 1-μm-wide lines can only handle about 1 mA. However,

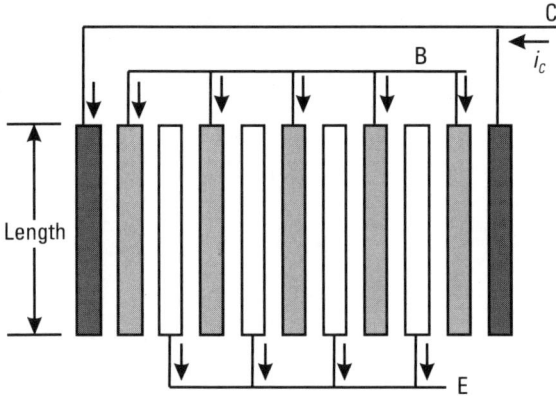

Figure 10.43 Transistor with multiple stripes.

to increase current capability, it is possible to use multiple metal layers, for example, metals 1, 2, and 3. The top metal is often thicker, resulting in a higher total current capability. In this example, the original 1 mA of current per emitter stripe might be increased to 4 mA/stripe for 1-μm-wide line. Another point to keep in mind is that since current flows from collector to emitter, the current density in the emitters is highest close to the external emitter contact, which for Figure 10.43 is on the bottom.

10.16 Current Limits in Integrated Inductors

Integrated inductors as used for LNAs and oscillators are typically 10 or 20 μm wide. This means they can probably handle no more than 20 to 40 mA of dc current, and maybe up to 80 mA or so of ac current. This obviously limits the ability to do on-chip tuning or matching for power amplifiers.

10.17 Power Combining

For high power, it is possible to combine multiple transistors at the output as shown in Figure 10.44. This distributes the heat and limits the current density in each transistor (compared to a single super-huge transistor).

However, with many transistors, the base drive to the outside transistors can be phase delayed compared to the shortest path, so it is important to keep the line lengths equal, as illustrated in Figure 10.44. Note also that as with all RF or microwave circuits, sharp bends are to be avoided. Line delay or phase shift can be determined by considering that the wavelength of a 1-GHz sine

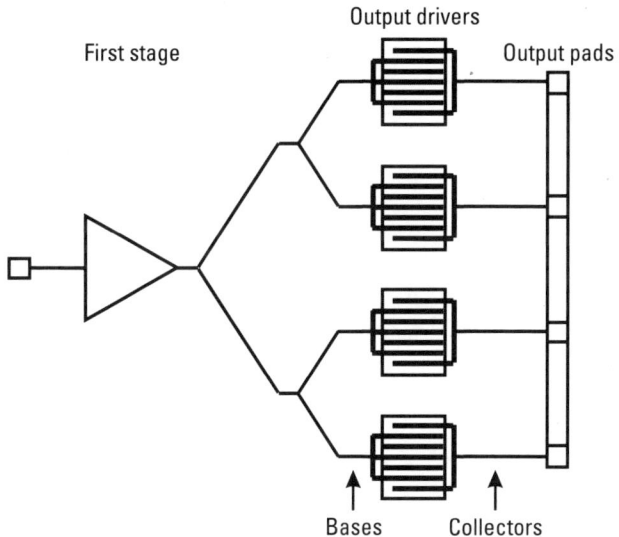

Figure 10.44 Multiple transistors.

wave in free space is 30 cm. This results in a phase shift of about 1.2°/mm. Wavelength is inversely proportional to frequency and inversely proportional to the square root of the dielectric constant ϵ_R. Thus, for SiO_2 with ϵ_R of about 4 and at 5 GHz, we now have phase shift of about 12°/mm. At 5 GHz, for a distance of 5 mm, we have a phase shift of about 60°. This is obviously of critical importance, especially when considering that sometimes an exact phase shift is required, such as the 32° of phase shift required for a class E amplifier.

Note that it is possible to use multiple output pads for parallel bond wires, or instead to have a "long" pad, shown in Figure 10.44, to connect more bond wires if desired. A typical requirement is that between 60 and 90 μm are required for each bond wire.

Power combining can also be done off-chip, using techniques shown in Figures 10.45 and 10.46, including backward wave couplers (stripline overlay, microstrip Lange) for octave bandwidth, or the branch-line, coupled amplifier. These are examples of quadrature combining.

Another way of combining differential signals is the "rat race" shown in Figure 10.47. This produces two outputs 180° out of phase (or can combine two inputs that are 180° out of phase). The rat race can replace a balun, but since it is dependent on the electrical delay along a path, it is only useful for small frequency deviations around f_0 (i.e., it is narrowband).

The push-pull arrangement, shown in Figure 10.48, is the same as for a class B push-pull amplifier discussed in Section 10.5.2.

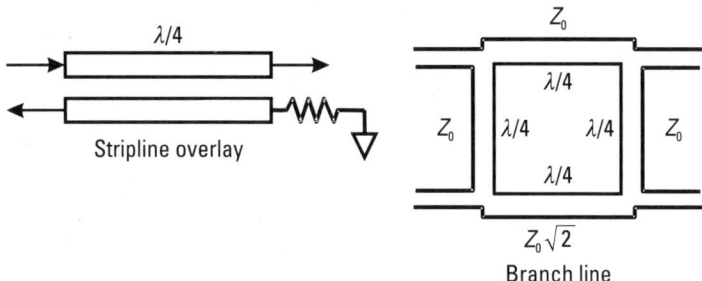

Figure 10.45 Stripline and branchline couplers.

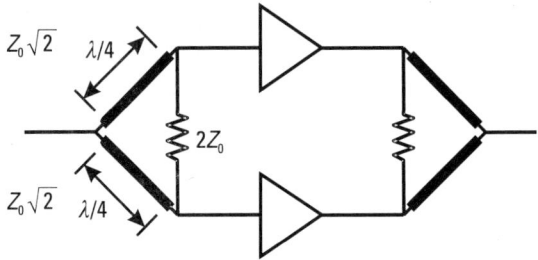

Figure 10.46 Coupled amplifiers.

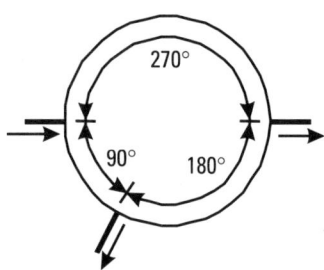

Figure 10.47 Rat race.

10.18 Thermal Runaway—Ballasting

Under high power, the temperature will increase. With constant base-emitter voltage, current increases with temperature. Equivalently, with constant current, base-emitter voltage decreases with temperature as

$$\left.\frac{\Delta V_{\text{BE}}}{\Delta T}\right|_{I=\text{constant}} \approx -2\,\frac{\text{mV}}{°\text{C}} \qquad (10.38)$$

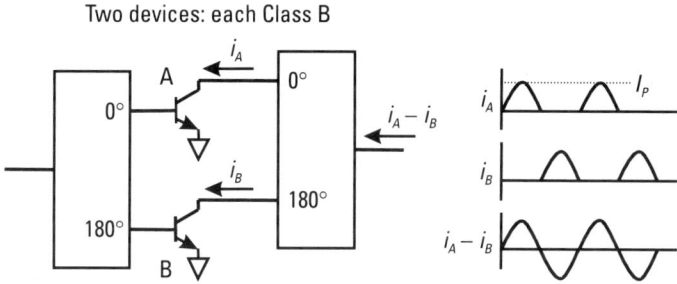

Figure 10.48 Combiners for push-pull operation.

Thus, if V_{BE} is held constant, if temperature increases, current increases, and as a result, more power is dissipated and temperature will increase even more. This phenomenon is known as *thermal runaway*. Furthermore, for unbalanced transistors, the transistor with the highest current will tend to be the warmest and hence will take an even higher proportion of the current. As a result, it is possible that the circuit will fail. Typically, ballast resistors are added in the emitters as a feedback to prevent such thermal runaway. With ballasting resistors, as shown in Figure 10.49, the input voltage is applied across the base-emitter junction and the series resistors, so as V_{BE} and current increase, there is a larger voltage across the resistor, limiting the increase of V_{BE}.

It is also possible to decrease input voltage as temperature increases, for example, by using a diode in the input circuit, using a current mirror, or using a more complex arrangement of thermal sensors and bandgap biasing circuits. An example of a thermal biasing circuit for a push-pull class B output stage is shown in Figure 10.50. In this example, all diodes and transistors are assumed to be at the same temperature. As temperature rises, V_D falls, reducing V_{BE} and keeping I constant.

10.19 Breakdown Voltage

Avalanche breakdown occurs when the electric field within the depletion layer provides sufficient energy for free carriers to knock off additional valence

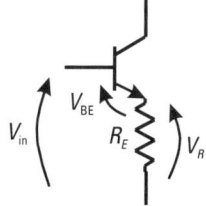

Figure 10.49 Balasting.

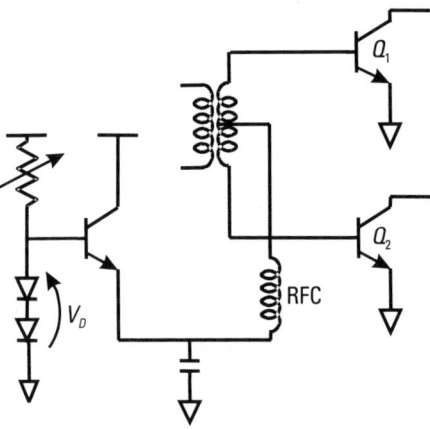

Figure 10.50 Temperature compensation.

electrons from the lattice atoms. These secondary electrons in turn generate more free carriers, resulting in avalanche multiplication. A measure of breakdown is $V_{CEO,max}$, which is the maximum allowable value of V_{CE} with the base open circuited. A typical value in a 3-V process might be 5V. However, under ac conditions and with the base matched, swings of V_{CE} past $2V_{CC}$ can typically be provided safely. In processes where this is not possible, it may be necessary to drop the supply voltage, use cascode devices, and possibly to make use of more complex biasing, which is adaptive or at least variable [9].

A complication is that many simulators are too simplistic and do not model the effects of breakdown. Clearly, while better simulation models are required, many designers depend on laboratory verifications, common sense, and experience.

10.20 Packaging

How does one remove heat from a power amplifier? One possible mechanism is thermal conduction through direct contact, for example, when the die is mounted directly on a metal backing. Another mechanism is through metal connections to the bond pads, for example, with wires to the package or directly to the printed circuit board (called *chip on board*). In flip-chip implementation, thermal conduction is through the solder bumps to the printed circuit board.

10.21 Effects and Implications of Nonlinearity

Linearity of the PA is important with certain modulation schemes. For example, filtered *quadrature phase shift keying* (QPSK) is often used largely because it can have very narrow bandwidth. However, quite linear power amplifiers are required to avoid spectral regrowth that will dump power into adjacent bands. Note that with *offset QPSK* (OQPSK) and $\pi/4$-QPSK, the drawback is less severe because of the smaller phase steps. *Minimum shift keying* (MSK) modulation is typically constant envelope and so allow the use of nonlinear power amplifiers; however, MSK requires wider bandwidth channels. FM and *frequency shift keying* (FSK) are two other constant envelope modulation schemes that can make use of high-efficiency power amplifiers.

In addition to spectral regrowth, nonlinearity in a dynamic system may lead to AM-PM conversion corrupting the phase of the carrier. Linearity is often checked using a two-tone test as previously described. However, this may not be realistic in predicting behavior when a real signal is applied. In such cases, it is possible to apply a modulated waveform and to measure the spectral regrowth.

10.21.1 Cross Modulation

Nonlinear power amplifiers can cause signals to be spread into adjacent channels, which can cause cross modulation. This is based on the same phenomena as third-order intermodulation for nonlinear amplifiers with two-tone inputs.

10.21.2 AM-to-PM Conversion

The phase response of an amplifier can change rapidly for signal amplitudes that result in gain compression, as illustrated in Figure 10.51. Thus, any amplitude variation (AM) in this region will result in phase variations (PM); hence, we can say there has been AM-to-PM conversion.

10.21.3 Spectral Regrowth

There can be additional problems that are worse for systems with varying envelopes. As an example, envelope variations can occur for modulation schemes

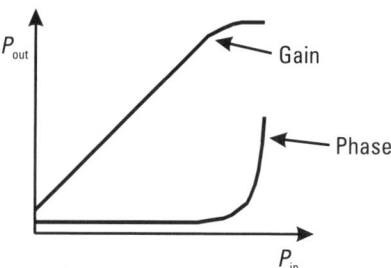

Figure 10.51 AM-to-PM conversion.

that have zero crossings, such as *binary phase shift keying* (BPSK) or QPSK. [We note that $\pi/4$ *differential quadrature phase shift keying* ($\pi/4$ DQPSK) or *Gaussian minimum shift keying* (GMSK) do not have zero crossings.] Due to band limiting, these zero crossings get converted into envelope variations. Any amplifier nonlinearities will cause spreading of frequencies into the adjacent channels, referred to as *spectral regrowth*. This is illustrated in Figure 10.52 for a QPSK signal with symbol period T_s.

10.21.4 Linearization Techniques

In applications requiring a linear power amplifier (filtered QPSK, $\pi/4$ QPSK, or systems carrying many channels, such as base station transmitters or cable television transmitters), one can use a class A power amplifier at 30% to 40% efficiency, or a higher efficiency power amplifier operating in a nonlinear manner, but then apply linearization techniques. The overall efficiency reduction can be minimal while still reducing the distortion.

Linearization techniques tend to be used in expensive complex RF and microwave systems and less in low-cost portable devices, often because of the inherent complexities, the need to adjust, and the problems with variability of device characteristic with operating conditions and temperature. However, some recent papers, such as [9, 10] have demonstrated a growing interest in techniques to achieve enhanced linearity for integrated applications.

10.21.5 Feedforward

The feedforward technique is shown in Figure 10.53. The amplifier output is $v_M = A_V v_{in} + v_D$, which consists of $A_V v_{in}$, the amplified input signal, and distortion components v_D, which we are trying to get rid of. This signal is attenuated to result in $v_N = v_{in} + v_D/A_V$. If this is compared to the original input signal, the result is $v_P = v_D/A_V$ and after amplification by A_V, results in $v_Q = v_D$. If this is subtracted from the output signal, the result is $v_{out} = A_V v_{in}$ as desired.

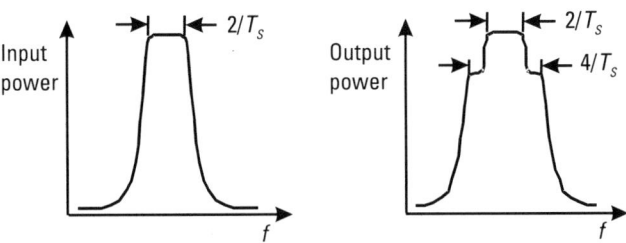

Figure 10.52 Spectral regrowth for QPSK signal with symbol period T_s.

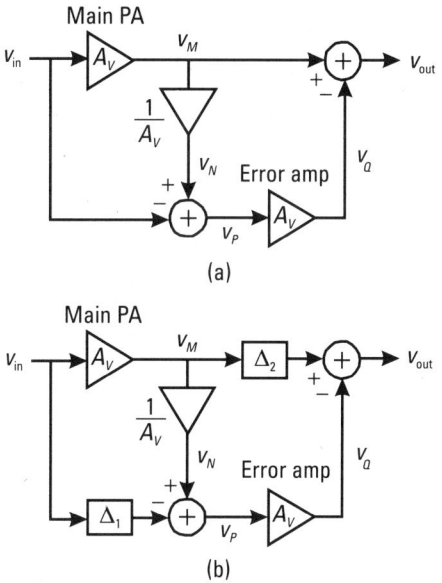

Figure 10.53 Feedforward linearization: (a) simple feedforward topology; and (b) addition of delay elements.

At high power, there is significant phase shift, and thus phase shift has to be added as shown in Figure 10.53(b).

A major advantage of feedforward over feedback is that it is inherently stable in spite of finite bandwidth and high phase shift in each block. There are also a number of difficulties with the feedforward technique. The delay can be hard to implement, as it must be the correct value and should ideally have no loss. The output subtractor should also be low loss.

Another potential problem is that linearization depends on gain and phase matching. For example, if $\Delta A/A = 5\%$ and $\Delta \phi = 5°$, then intermodulation products are attenuated by only 20 dB.

10.21.6 Feedback

Successful feedback requires high enough gain to reduce the distortion, but low enough phase shift to ensure stability. These conditions are essentially impossible to obtain in a PA at high frequency. However, since a PA is typically an upconverted signal, if the output of the PA is first downconverted, the result can be compared to the original input signal. At low frequency, the gain and phase problems are less severe. An example of using such a feedback technique is shown in Figure 10.54. There have been a number of variations on this technique,

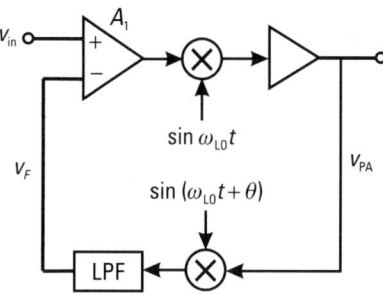

Figure 10.54 Feedback linearization techniques.

including techniques to remove envelope variations. The interested reader is referred to [7].

10.22 CMOS Power Amplifier Example

Recently, a number of CMOS power amplifiers have been published [9–16]. While CMOS is not likely to be the technology of choice for standalone power amplifiers, they are of interest in single-chip radios. Examples of CMOS power amplifiers include one that had an output power of over 2W in the 2.4-GHz region [15].

In Figure 10.55, a CMOS power amplifier in a 0.8-μm process at 900 MHz is shown [16]. This was designed for a standard that had constant amplitude waveforms, so linearity was of less importance. The cascode input stage operates

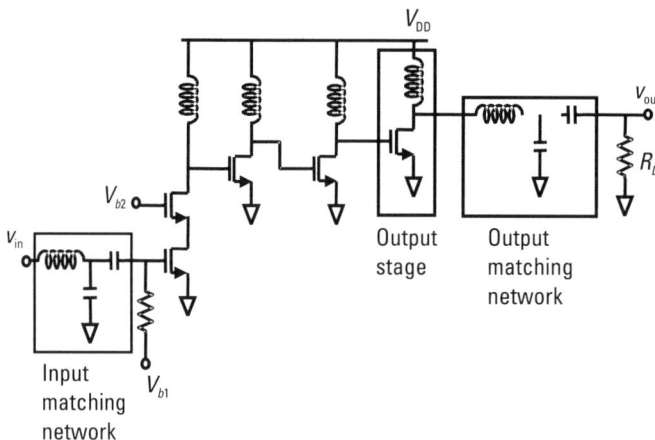

Figure 10.55 Power amplifier example.

in class A (input is +5 dBm), the second stage operates in class AB, and the last two stages operate as switching circuits to deliver substantial power with relatively high efficiency. (Note that class C amplifiers are also high efficiency, however, only at low conduction angles; thus they provide high efficiency only at low power levels.)

Measured results with a 2.5V power supply showed output power of 1W (30 dBm) with a power added efficiency of 42%.

References

[1] Fowler, T., et al., "Efficiency Improvement Techniques at Low Power Levels for Linear CDMA and WCDMA Power Amplifiers," *Proc. Radio Frequency Integrated Circuits Symposium*, Seattle, WA, May 2001, pp. 41–44.

[2] Staudinger, J., "An Overview of Efficiency Enhancements with Applications to Linear Handset Power Amplifiers," *Proc. Radio Frequency Integrated Circuits Symposium*, Seattle, WA, May 2001, pp. 45–48.

[3] Grebennikov, A. V., "Switched-Mode Tuned High-Efficiency Power Amplifiers: Historical Aspect and Future Prospect," *Proc. Radio Frequency Integrated Circuits Symposium*, Seattle, WA, May 2001, pp. 49–52.

[4] Krauss, H. L., C. W. Bostian, and F. H. Raab, *Solid State Radio Engineering*, New York: John Wiley & Sons, 1980.

[5] Cripps, S. C., *RF Power Amplifiers for Wireless Communications*, Norwood, MA: Artech House, 1999.

[6] Albulet, M., *RF Power Amplifiers*, Atlanta, GA: Noble Publishing, 2001.

[7] Kenington, P. B., *High Linearity RF Amplifier Design*, Norwood, MA: Artech House, 2000.

[8] Sokal, N. O., and A. D. Sokal, "Class E: A New Class of High Efficiency Tuned Single-Ended Power Amplifiers," *IEEE J. Solid-State Circuits*, SC-10, No. 3, June 1975, pp. 168–176.

[9] Sowlati, T., and D. Leenaerts, "A 2.4GHz, $0.18\mu m$ CMOS Self-Biased Cascode Power Amplifier with 23dBm Output Power," *Proc. International Solid-State Circuits Conference*, Feb. 2002, pp. 294–295.

[10] Shinjo, S., et al., "A 20mA Quiescent Current CV/CC Parallel Operation HBT Power Amplfier for W-CDMA Terminals," *Proc. Radio Frequency Integrated Circuits Symposium*, Seattle, May 2001, pp. 249–252.

[11] Pothecary, N., *Feedforward Linear Power Amplifiers*, Norwood, MA: Artech House, 1999.

[12] Yoo, C., and Q. Huang, "A Common-Gate Switched 0.9W Class E Power Amplifier with 41% PAE in $0.25\mu m$ CMOS," *IEEE J. Solid-State Circuits*, May 2001, pp. 823–830.

[13] Kuo, T., and B. Lusignan, "A 1.5W Class-F RF Power Amplifier in $0.25\mu m$ CMOS Technology," *Proc. International Solid-State Circuits Conference*, Feb. 2001, pp. 154–155.

[14] Su, D., et al., "A 5GHz CMOS Transceiver for IEEE 802.11a Wireless LAN," *Proc. ISSCC,* Feb. 2002, pp. 92–93.

[15] Aoki, I., et al., "A 2.4-GHz, 2.2-W, 2-V Fully Integrated CMOS Circular-Geometry Active-Transformer Power Amplifier," *Proc. Custom Integrated Circuits Conference,* May 2001, pp. 57–60.

[16] Su, D., and W. McFarland, "A 2.5-V, 1-W Monolithic CMOS RF Power Amplifier," *Proc. Custom Integrated Circuits Conference,* May 1997, pp. 189–192.

About the Authors

John Rogers received a B.Eng. in 1997, an M.Eng. in 1999, and a Ph.D. in 2002, all in electrical engineering from Carleton University, Ottawa, Canada. During his master's degree research, he was a resident researcher at Nortel Networks' Advanced Technology Access and Applications Group, where he did exploratory work on voltage-controlled oscillators and developed a copper interconnect technology for building high-quality passives for radio frequency (RF) applications. From 2000 to 2002, he collaborated with SiGe Semiconductor Ltd. while pursuing his Ph.D. on low-voltage RF integrated circuits (RFIC) for wireless applications. Concurrent with his Ph.D. research, Dr. Rogers worked as part of a design team that developed a cable modem integrated circuit for the DOCSIS standard. He is currently an assistant professor at Carleton University and collaborating with Cognio Canada Ltd. His research interests are in the areas of RFIC design for wireless and broadband applications.

Dr. Rogers was the recipient of an IEEE Solid-State Circuits Predoctoral Fellowship, and received the Bipolar/BiCMOS Circuits and Technology Meeting (BCTM) best student paper award in 1999. He holds one U.S. patent with three pending, and is a member of the Professional Engineers of Ontario.

Calvin Plett received a B.A.Sc. in electrical engineering from the University of Waterloo, Canada, in 1982, and an M.Eng. and a Ph.D. from Carleton University, Ottawa, Canada, in 1986 and 1991, respectively. From 1982 to 1984 he worked with Bell-Northern Research. In 1989 he joined the Department of Electronics at Carleton University, where he is now an associate professor. Since 1995 he has done consulting work for Nortel Networks in the area of RF and broadband integrated circuit design. He has also supervised numerous graduate students, often cooperatively with industrial partners, including Nortel Networks, Philsar, Conexant, Skyworks, IBM, and SiGe Semiconductors. His research interests are in the area of analog integrated circuit design including filters, radio frequency front-end components, and communications applications.

Index

1-dB compression point, 30–32, 40, 216, 240, 351

Additive phase noise, 283–91
Admittance, 89–93, 110, 135, 155
Alternating current, 47
Aluminum, 97, 102, 106
Amplifier circuit load line, 385–87
Amplifier circuit noise figure, 14–16
Amplitude mismatch, 224–27
Amplitude modulation noise, 283, 285
Amplitude modulation (AM) to phase modulation (PM) conversion, 395
Analog system design, 2–4
Antenna available power, 11–13
Antenna rules, 104
Audio amplifier, 381
Automatic-amplitude control (AAC), 302–13
Automatic gain control (AGC), 5
Available antenna power, 11–13
Available noise power, 11
Avalanche breakdown, 393–94

Back-end digital function, 57
Back-end processing, 95–97
Ballast resistor, 393
Bandgap reference generator, 187
Bandpass filter, 320, 327–29, 339
Bandpass LC filter, 321–22
Bandstop filter, 322–26

Bandstop filter with negative resistance, 329–33
Bandwidth, impedance transformation network, 83–84
Barkhausen criteria, 248, 249–50
Base bias current, 53
Base-collector depletion region, 46
Base-collector junction capacitance, 182
Base-emitter junction, 44, 45
Base pushout, 46
Base resistance, 169
Base resistance noise, 337–38
Base shot noise, 53, 55, 73–74, 161, 169–71, 337–38
Bessel function, 307
Bias current, 169–70
Bias current reduction, 232
Biasing, 44–45, 180, 233, 277
Bias network, 187–89
Bias resistor, 214, 234
Bias transistor, 340
Bipolar complementary metal oxide semiconductor (BiCMOS), 57
Bipolar radio frequency (RF) design, 1–2
Bipolar transistor, 43–46, 47, 197, 316
  design, 56–57
  noise, 53–54
  nonlinearity, 172–82
Bipolar transistor input-referred noise, 159–61

Blocker filtering, 39–41
Blocker rejection, 319
Blocking, 39
Boltzmann's constant, 45
Bonds pads, 233
Bond wire inductance, 386
Bottom noise, 214
Bottom-plate capacitance, 100–1, 104
Branchline coupler, 391–92
Breakdown voltage, 393–94
Broadband common-emitter amplifier, 154
Broadband linearity measures, 32–35
Broadband low-noise amplifier (LNA), 189–94

Capacitance, 51
Capacitive degeneration, 325
Capacitive feedback divider, 255–58
Capacitor, 69, 95
    metal insulator metal, 103–4
    tapped, 76–78
Capacitor ratios, 255–58
Cascaded circuit, 16–18
Cascaded noise figure, 21–22
Cascode low-noise amplifier (LNA), 141–42, 147–48, 152–54, 156, 165–66, 170, 184–85, 203, 209, 235, 322–23
Chip-on-board packaging, 139, 394
Circuit bandwidth, 84
Circular differential inductor, 107–8
Clipping, 232–33
Closed-loop feedback, 249, 252–55, 268–70
Code division multiple access (CDMA), 383
Collector-base junction, 44
Collector bias current, 53
Collector current dependence, 46, 48
Collector-emitter junction, 45
Collector shot noise, 53, 57, 59, 161, 169–72, 337–39
Colpitts circuit, bandstop filter, 329–33
Colpitts negative resistance circuit, 336–37
Colpitts oscillator, 250, 251–52, 255–58, 262–63, 270–72, 275–83, 287, 296–302
Colpitts oscillator with buffering, 270, 272
Common-base amplifier, 141, 142, 146–48, 251–52, 256, 257
Common-base oscillator, 271, 275–77, 280, 287, 295–97

Common-collector amplifier, 141, 142, 148–51
    linearity analysis, 182–83
    noise, 171–72
Common-collector oscillator, 251–52, 256, 270–72, 275, 276–80, 297–302
Common-controller buffer, 156
Common-emitter amplifier, 141–48
    differential pair, 183–84
    linearity analysis, 172–82
    noise figure, 161–63
    with series feedback, 152–54
    with shunt feedback, 154–58, 190
Common mode impedance, 131
Complementary metal oxide semiconductor (CMOS), 1–2, 43, 180, 316
    small-signal model, 58–60
    square law equations, 60–61
Complementary metal oxide semiconductor (CMOS) mixer, 242–44
Complementary metal oxide semiconductor (CMOS) power amplifier, 398–99
Complementary metal oxide semiconductor transistor, 57–61
Composite second-order beat, 33–35
Composite triple-order beat, 33–34
Compression, mixer design, 232–33
Compression point, 30–32, 40, 216, 232–33, 240, 351
Conductive plug, 97
Conjugate matching, 351
Contact layer, 97
Controlled transconductance mixer, 198–200
Coplanar waveguide, 129, 130, 131
Coplanar waveguide with ground, 131
Copper, 97, 103, 132
Correlation admittance, 15
Coupled amplifier, 391, 392
Coupled inductor, 78–81, 109, 111, 119–20, 122, 125
Coupled microstrip line, 131
Coupling-capacitor mixer, 230–31
Coupling network, mixer, 236
Cross-coupled double-balanced mixer, 197
Cross modulation, 395
Current effects, 51–53
Current handling, metal, 102–3
Current limits, 388–90
Current mirror, 187–89

Index

Damped resonator, 247–48
Dead zone, 99
De-embedding techniques, 134–39
Degeneration resistor, 192, 202, 228
Desired nonlinearity, 215
Differential amplifier, 183–84
Differential bandpass low-noise amplifier (LNA), 327–29
Differential impedance, 131
Differential inductor, 109, 116–17, 118
Differential oscillator, 270
Differential-pair amplifier, 183–84, 198–202, 204, 206
Differential-pair mixer, 208
Diffusion capacitance, 51
Diffusion resistance, 55, 103
Digital modulation, 5
Digital signal processing, 1, 57
Direct-conversion bias network, 187–89
Direct-conversion (dc) (homodyne) receiver, 37
Direct-conversion (dc) resistance, 113
Direct-conversion (dc)-to-radio-frequency (RF) efficiency, 350–51
Direct downconversion receiver, 54
Doping, 103
Doping region, 103
Double-balanced mixer, 200–2, 242–43
Double ell network, 386
Double-sideband noise figure, 207, 210, 214
Downconversion mixer, 202, 206, 207, 216, 218, 228
Dummy open, 135
Dummy short, 135
Dynamic load line, 385–87
Dynamic range, 35–36

Early voltage, 45
Edge effect, 131
Efficiency, amplifier, 350–51, 358–59, 378, 384–85
Electromagnetic simulator, 110
Electrostatic discharge, 291–92
Ell networks, 69–71, 87–88, 386
Emitter-base depletion region, 46
Emitter-coupled pair amplifier, 183–84
Emitter crowding, 46
Emitter degeneration, 137, 152–54, 164, 178–80, 190, 192, 206

Emitter-follower, 182. *See also* Common-collector amplifier
Equivalent impedance, 80
Equivalent inductance, 77
Equivalent noise model, 17
Equivalent source impedance, 16
Even-order impedance, 131
Excess noise, 54, 286
Excitation, inductor, 117
Exponential nonlinearity, 172–80

Fast Fourier transform (FFT), 194, 238
Feedback
    oscillator, 248–68, 325
    amplifiers with, 152–58
    See also Negative resistance
Feedback linearization, 397–98
Feedforward linearization, 396–97
Field-effect transistor, 197
Fifth harmonics, 206
Fifth-order nonlinearity, 32
Filtering, 105, 206, 209, 218, 221
    blockers, 39–41
    image signals, 37–39
    noise, 337–39
    overview, 319
    polyphase, 223–24, 239–41
    second-order, 319–20
    transceiver, 5–6
Finite input impedance, 14
First-order polyphase filter, 222–24
First-order roll-off, 49
First-order term, 24
Flicker noise, 54, 286, 286–87
Flip-chip packaging, 138–39, 394
Folded cascode, 184–85, 235
Forward active region, 44
Forward bias, 44
Fourier coefficient, 361
Fourier series, 203, 205, 378
FR4 material, 132
Frequency modulation (FM) noise, 283
Frequency shift keying (FSK), 395
Frequency synthesizer, 5, 6
Frequency tuning, 342–43
Fringing, 131
Fringing capacitance, 100–1
Fringing inductance, 102

Gain compression, 4, 25–26
Gallium arsenide, 44, 132

Gate resistance, 59–60
Gate voltage, 58
Gilbert cell, 197
Global Positioning System (GPS), 1
Global System Mobile (GSM), 39–40
Gold, 103, 139
Ground shield, 121–22, 123

Half thermally noise generation, 55
Harmonic distortion, 4
Harmonic filtering, 70, 319
Hartley architecture, 219–20
Hartley oscillator, 250, 251
HD2 terms, 29–30
Heterojunction bipolar transistor, 44
Higher-order filter, 343–46
High-frequency effects, transistor, 49–53
High-frequency measurement, passive circuit, 134–39
High-frequency nonlinearity, 182
High-linearity mixer, 234–38
Highpass filter, 221, 255, 256, 260
Highpass matching network, 70, 73, 74
Homodyne receiver, 37
HPADS, 164

Ideal circuit, 9
Ideal mixer, 207
Ideal oscillator, 283
Ideal transformer, 81
Image filter, 37
Image frequency, 39, 207, 208, 209
Image reject filter, 38–39, 208, 322, 333–35, 343–46
Image rejection, 38
Image reject mixer, 203, 217–27, 238–42
Impedance, 2–3, 49–50. *See also* Input impedance; Output impedance
Impedance matching, 63–65, 69–74
 one-step vs. two-step, 87–88
 using transformers, 81
Impedance mismatch, 20–21
Impedance parameter, 89–93, 115–16, 134–39
Impedance transformation network bandwidth, 83–84
Inductance ratio, 81
Inductive degeneration, 164, 325
Inductor, 69
 benefits, 95, 96, 106
 capacitor resonator, 83–88
 characterization, 115–17
 design, 106–8, 289–91
 isolation, 121
 lumped models, 108–9
 multilevel, 124–27
 on-chip spiral, 104–6, 110, 119–21
 quality factor, 111–15
 self-resonance, 110–11
 tapped, 76–78, 250
 using, 117–19, 122
Inductor-capacitor resonator, 247–51
Inductor-capacitor series circuit, 217–18
Inductor degeneration, 214, 229–30
Infinite impedance, 3
Input admittance, 155
Input frequency, 207
Input impedance, 2–3, 14, 49, 59, 69, 74, 142, 150, 154–58, 191, 324–25
Input matching, 163–69
Input preselection filter, 5–6
Input-referred noise model, 159–61
Input third-order intercept point, 28–29, 32
Integrated capacitor, 104, 105
Integrated circuit, 1, 95–97
Integrated inductor, 103, 111, 390
Intermediate frequency (IF), 5, 6, 197, 206, 207, 208, 214, 216, 228, 239
Intermodulation, 4, 206, 216
Interwinding capacitance, 113
Intrinsic transistor, 44, 45, 47
Isolation, mixer, 217

Junction capacitance, 51, 56, 57

Kelvin temperature, 45
Kirchoff's current law, 170
Kirk effect, 46

Large-signal nonlinearity, 275–77
Leeson's formula, 285, 294–95
Linearity, 23, 23–35, 30, 35
 amplifiers, 172–83, 356, 382, 396–98
 broadband measure, 32–35, 190–92
 mixer, 215–17, 234–38
 negative resistance circuits, 336–37
Linear phase noise, 283–91
Load line, 385–87
Load pull, 352
Load resistance, 232–33, 236

Local oscillator, 5
    frequency, 208, 209, 210, 214, 228, 233
    harmonics, 216
    quad switching, 202–6
    self-mixing, 37
Loop gain estimation, 260–62, 268–69
Lossless transmission line, 89
Low-frequency analog design, 2–4
Low-frequency noise, 291–92
Low-noise amplifier (LNA), 5, 21–22, 37, 40, 105, 141, 233
    broadband, 189–94
    differential bandpass, 327–29
    input matching, 163–69
    linearity, 172–83
    low-voltage, 184–87
    noise, 158–72, 338–39
    See also Cascode low-noise amplifier; Common-base amplifier; Common-collector amplifier; Common-emitter amplifier
Lowpass filter, 6, 217, 221, 383
Lowpass matching network, 70, 73, 74
Low-voltage low-noise amplifier, 184–87
Lumped components, 69
Lumped model, inductor, 108–9

Matching, power amplifier, 351–53, 385–88
Metal insulator metal capacitor, 103–4, 310
Metalization, 95–97
Metal migration, 102
Metal oxide semiconductor, 43, 54
Metal oxide semiconductor field-effect transistor, 43, 58, 180
Microstrip line, 129, 131–34
Microwave design, 2, 3–4
Microwave transistor, 91
Miller multiplication, 49–51, 142, 149, 152, 156, 163, 164, 182
Minimum shift keying, 395
Mixers, 5, 6, 37, 40, 197–98
    alternative designs, 227–31
    design, 231–42
    image reject/single-sideband, 217–27
    isolation, 217
    linearity, 215–17
    noise, 206–14
    See also Controlled transconductance mixer; Cross-coupled double-balanced mixer

Mixing components, 25
Mixing gain, 206
Moore mixer, 227–28, 229, 242
Multilevel inductor, 124–27
Multistage polyphase filter, 223–24
Multivibrator oscillator, 313–15
Mutual inductance, 78–81, 129, 136

Narrowband common-emitter amplifier, 152
Narrowband resistor, 75
Narrowband transformer model, 128–30
N-channel metal oxide semiconductor, 57–58, 243
N doping, 103
Near-far problem, 40
Negative resistance, 248–51, 262–68, 325
    bandstop filter, 329–33
    linearity, 336–37
    See also Feedback
Noise, 4, 9–22, 35, 203
    amplifiers, 158–72
    antenna power, 11–13
    bipolar transistor, 53–54
    broadband amplifier, 191–92
    CMOS small-signal model, 58–60
    filtering, 337–39
    impedance matching, 73–74
    low-frequency, 291–92
    mixers, 206–14, 236–38
    nonlinear, 292–95
    oscillator phase, 283–95
    thermal, 10–11
    transistor model, 55–56
Noise figure
    amplifier circuit, 14–16
    broadband amplifier, 192–94
    common-emitter amplifier, 161–63
    components in series, 16–22
    concept, 13–14
    low-noise amplifier, 164, 169–70
    mixers, 207, 209–14, 233
Noise floor, 12, 37
Noise matching, mixer, 229–30
Noise power, 10
Noise voltage, 10
Nonlinearity, 4, 23, 24, 26–27, 31–33, 40
    large-signal transistor, 275–77
    mixers, 215–17
    power amplifier, 172–83, 394–98
Nonlinear noise, 292–95

Nonlinear transfer function, 197–98
Notch filter, 320, 323–26, 330–31, 333, 339–46

Octagonal inductor, 110
Odd-order impedance, 131
Off-chip inductor, 107
Off-chip input transformer, 228
Off-chip passive filter, 319
Off-chip power combining, 391–92
Offset quadrature phase shift keying, 395
On-chip filter, 319
On-chip inductor, 327
On-chip input transformer, 228
On-chip passives, 134–39
On-chip spiral inductor, 104–6, 110–11, 119–21
On-chip transformer, 184–87
On-chip transmission line, 129–34
On-chip tuned circuit, 216
One-step impedance matching, 87–88
Open-circuit stub, 89
Open-loop feedback, 249, 252, 258–60, 268–70
Oscillator, 40, 245–46
    amplitude, 277–83
    filter tuning, 339–42, 339–43
Oscillator skirts, 246
Out-of-band signal, 5
Output buffer, 155–56
Output conductance, 58
Output filtering, 209
Output impedance, 2–3, 50, 69, 150–51, 155–58, 180–82
Output slope factor, 60
Output third-order intercept point, 28–29
Oxide capacitance, 112–13

Packaging
    amplifier, 394
    mixer design, 233
    passive circuit, 135–39
Parallel circuit negative resistance, 263–65
Parallel inductance, 69, 83
Parallel LC resonator, 247
Parallel-plate capacitor, 108–9
Parallel RC circuit, 217–18
Parallel resistance, 267, 288
Parallel resistor-capacitor network, 74–76
Parallel resistor-inductor network, 74–76

Parasitic capacitance, 45, 100–1, 105–6, 110–11
Parasitic inductance, 101–2
Parasitic resistance, 46
Parasitics effect, 274–75
Passband filter, 327, 333
Passive circuit, 95
P-channel metal oxide semiconductor, 57–58, 231, 242–43
P doping, 103
Peak power, 3
Peak-to-peak power, 3
Phase-locked loop, 287, 342
Phase mismatch, image rejection, 224–27
Phase noise, 41, 245–46
    oscillator, 283–95
Phase shifting, 130, 137, 218–24, 228, 239, 397
PN junction, 53, 55
Pole frequency, 143, 144, 146, 148
Poles, widely separated, 146
Poly capacitor, 104
Polyphase filter, 222–24, 239–41
Poly resistor, 103
Positive feedback oscillator, 250–52, 265–68, 270–74, 275, 280–83, 285–86, 287
Power-added efficiency (PAE), 350–51
Power amplifier (PA), 6, 340–53
    classes A, B, and C, 353–67
    class D, 367–68
    class E, 368–75
    class F, 375–81
    classes G and H, 381–83
    class S, 383–84
    class summary, 383–84
    nonlinearity, 394–98
Power combiner, 362, 390–93
Power matching, 229–30
Power series expansion, 23–27
Power spectral density, 54
Printed circuit board, 130, 394
Printed circuit board ground, 136–39
Process tolerance, 334–35
Push-pull amplifier, 362–63, 368, 382–84, 391, 393

Quadrature amplitude modulation (QAM), 12, 384

Quadrature phase shift keying (QPSK), 12, 394–95
Quad switching, 202–6, 214, 235
Quad transistor, 235–36
Quality factor
   capacitor resonator, 85–88
   inductors, 111–15
Quality measurement, 2
Quality tuning, 339–42
Quarter-wave transmission line, 379–81

Radio frequency (RF), 214
Radio frequency (RF) choke, 370
Radio frequency (RF) communications, 1
Radio frequency (RF) filter, 321–26
Radio frequency integrated circuit (RFIC), 1–2, 4–6
Radio frequency integrated circuit (RFIC) oscillator, 247
Rat race, 391, 392
Reactive matching circuit, 63, 69–71, 385–86
Receive side, 5
Reciprocal mixing, 40
Reflection coefficient, 66, 90–91, 351
Resistance. *See* Sheet resistance
Resistivity, 97
Resistor-capacitor network, 74–76, 217, 218, 220–22
Resistor-inductor network, 74–76
Resistor noise model, 11–13
Resonator with feedback, 248–51
Reverse bias, 44
Ring oscillator, 315–16
Root-mean-square (rms), 3

Saturation current, 45
Saturation voltage, 373, 380, 384, 388
Scattering, 89–93, 115, 127
Second-harmonic peaking, 379
Second harmonics, 25, 274, 277
Second-order bandpass filter, 320
Second-order beat tone, 35
Second-order filter, 319–20
Second-order intercept point, 29–30
Second-order intermodulation, 25, 34–35
Second-order transfer function, 83–84
Self-resonance, 110–11, 115–17, 120, 125
Series circuit negative resistance, 263–65
Series components noise figure, 16–22
Series feedback, 152–54

Series inductance, 63–64, 69, 386
Series resistance, 45–48, 60, 107, 119, 267
Series resistor-capacitor network, 74–76
Series resistor-inductor network, 74–76
Sheet resistance, 98–99, 111, 113
Shielded inductor, 122, 123
Short-channel device, 61
Short-circuit current gain, 48
Short-circuit stub, 89
Shot noise, 53, 55, 57, 59, 73–74, 161, 169–72, 337–39
Shunt feedback, 154–58, 190
Signal-to-noise ratio, 10, 12, 13, 14, 170, 208
Silicon, 103, 106, 122, 132
Silicon dioxide, 96
Silicon oxide, 132
Silicon substrate, 44
Simultaneous-noise-and-power-match mixer, 229–30
Single-balanced mixer, 242–43
Single-pole amplifier, 145
Single-sideband mixer, 217–27
Single-sideband noise figure, 207–8, 214, 237
Sinusoidal collector voltage, 354–55, 361, 377
Sinusoidal voltage source, 277–78
Skin depth, 98–99
Skin effect, 99, 111, 113
Small-signal model, 47–48, 56, 58–60, 73–74, 324
   amplifier, 142, 144, 146–47, 149, 153
   oscillator, 253, 259, 262–63, 265–66
Smith chart, 66–69, 130
Spectral regrowth, 395–96
SPICE, 164
Spiral inductor, 104–6, 110–11, 119–21
Square law equations, 60–61
Square spiral inductor, 107, 110, 111–13, 120–21
Stability, impedance matching, 70
Stripline coupler, 391–92
Substrate, inductor, 114, 116–17, 119
Superheterodyne receiver, 37–38, 208
Superheterodyne transceiver, 4–5
Switching modulator, 205–6
Switching quad, 200–3, 205–6, 216, 217, 230
Symmetric (differential) inductor, 109
Synthesizer spur, 41

Tapped amplifier, 250
Tapped capacitor, 76–78, 250
Tapped inductor, 76–78, 250
Taylor series, 283
Temperature effects, low-noise amplifier, 189
Thermal biasing circuit, 393, 394
Thermal conduction, passive circuit design, 139
Thermal noise, 10–11, 48, 53–55
Thermal noise spectral density, 10
Thermal runaway, 389, 392–93
Thermal voltage, 45
Thin quad flat pack, 137
Third harmonic, 25–26, 206, 375, 377, 378, 381
Third-harmonic resonator, 377
Third-order filter, 326
Third-order intercept point (IP3), 4, 27–29, 31–32, 34
Third-order intercept voltage, 174
Third-order intermodulation (IM3), 25–28, 33–34, 40, 178–80, 202
Third-order intermodulation (IM3) product, 216, 232–33, 240, 294
Third-order nonlinearity, 32, 33
Third-order terms, 24, 25–26, 34–35
Threshold voltage, 60
Tranconductance-controlled mixer, 198–200
Transceiver, 4–6
Transconductance, 48, 61, 363–67
Transformer-coupled negative resistance, 331–33
Transformer input, mixer with, 228–29
Transformers, 78–81
   application, 105–6
   characterizing, 127–29
   design, 106–7, 122–24
   matching, 81
   mutual inductance, 78–81
   noise, 339
   on-chip, 184–87
   tuning, 82–83
Transistor, 43, 340
   broadband amplifier, 191–92
   high-frequency effects, 49–53
   large-signal nonlinearity, 275–77
   multiple, 388-390
   noise sources, 55–56

   resistance, 373, 380
   sizing, 232
   See also Bipolar transistor; Intrinsic transistor
Transition time, amplifier, 373–75
Transmission coefficient, 90–91
Transmission line impedance, 88–89
Transmission lines
   matching, 130–31
   on-chip, 129–34
Transmit side, 5, 6, 349
Triode region, 61
Triple-beat products, 34
Tunable oscillator, 295–302
Tuned load, 217
Tuned output circuit, 233
Tungsten, 96–97, 102
Tuning
   amplifier, 352–53
   filter, 339–43
   frequency, 342–43
   quality, 339–42
   transformer, 82–83
Turns ratio, 81
Two-port network, 90–91
Two-step impedance matching, 87–88
Two-tone test, 24–25

Underpass, 113, 118, 119
Undesired nonlinearity, mixer, 215–17
Unity gain frequency, 143, 144–45
Upconversion mixer, 206, 218

Varactor, 295–302
Vias, 97, 125
Voltage-controlled oscillator (VCO), 5, 37, 106–7, 245–46
   automatic-amplitude control, 302–13
Voltage divider, 48, 50, 77, 81, 234, 261
Voltage gain, 141–42, 149, 206, 232
Volterra series, 23

Wafer processing, 104
Weaver architecture, 219–20
White noise, 11, 53
Wireless local-area network (WLAN), 1

Y Smith chart, 67–68

Zero impedance, 3
Z Smith chart, 67–68
ZY Smith chart, 67–68

Recent Titles in the Artech House Microwave Library

Advanced Techniques in RF Power Amplifier Design, Steve C. Cripps

Automated Smith Chart, Version 4.0: Software and User's Manual,
Leonard M. Schwab

Behavioral Modeling of Nonlinear RF and Microwave Devices,
Thomas R. Turlington

Computer-Aided Analysis of Nonlinear Microwave Circuits,
Paulo J. C. Rodrigues

Design of FET Frequency Multipliers and Harmonic Oscillators,
Edmar Camargo

Design of Linear RF Outphasing Power Amplifiers, Xuejun Zhang,
Lawrence E. Larson, and Peter M. Asbeck

Design of RF and Microwave Amplifiers and Oscillators,
Pieter L. D. Abrie

Distortion in RF Power Amplifiers, Joel Vuolevi and Timo Rahkonen

*EMPLAN: Electromagnetic Analysis of Printed Structures in Planarly
Layered Media, Software and User's Manual,* Noyan Kinayman
and M. I. Aksun

Feedforward Linear Power Amplifiers, Nick Pothecary

Generalized Filter Design by Computer Optimization,
Djuradj Budimir

High-Linearity RF Amplifier Design, Peter B. Kenington

Microwave Component Mechanics, Harri Eskelinen and
Pekka Eskelinen

Microwave Engineers' Handbook, Two Volumes,
Theodore Saad, editor

*Microwave Filters, Impedance-Matching Networks, and Coupling
Structures,* George L. Matthaei, Leo Young, and E.M.T. Jones

Microwave Materials and Fabrication Techniques, Third Edition,
Thomas S. Laverghetta

Microwave Mixers, Second Edition, Stephen A. Maas

Microwave Radio Transmission Design Guide, Trevor Manning

Microwaves and Wireless Simplified, Thomas S. Laverghetta

Neural Networks for RF and Microwave Design, Q. J. Zhang and K. C. Gupta

Nonlinear Microwave and RF Circuits, Second Edition, Stephen A. Maas

QMATCH: Lumped-Element Impedance Matching, Software and User's Guide, Pieter L. D. Abrie

Practical RF Circuit Design for Modern Wireless Systems, Volume I: Passive Circuits and Systems, Les Besser and Rowan Gilmore

Practical RF Circuit Design for Modern Wireless Systems, Volume II: Active Circuits and Systems, Rowan Gilmore and Les Besser

Radio Frequency Integrated Circuit Design, John Rogers and Calvin Plett

RF Design Guide: Systems, Circuits, and Equations, Peter Vizmuller

RF Measurements of Die and Packages, Scott A. Wartenberg

The RF and Microwave Circuit Design Handbook, Stephen A. Maas

RF and Microwave Coupled-Line Circuits, Rajesh Mongia, Inder Bahl, and Prakash Bhartia

RF and Microwave Oscillator Design, Michal Odyniec, editor

RF Power Amplifiers for Wireless Communications, Steve C. Cripps

RF Systems, Components, and Circuits Handbook, Ferril Losee

Stability Analysis of Nonlinear Microwave Circuits, Almudena Suárez and Raymond Quéré

TRAVIS 2.0: Transmission Line Visualization Software and User's Guide, Version 2.0, Robert G. Kaires and Barton T. Hickman

Understanding Microwave Heating Cavities, Tse V. Chow Ting Chan and Howard C. Reader

For further information on these and other Artech House titles, including previously considered out-of-print books now available through our In-Print-Forever® (IPF®) program, contact:

Artech House
685 Canton Street
Norwood, MA 02062
Phone: 781-769-9750
Fax: 781-769-6334
e-mail: artech@artechhouse.com

Artech House
46 Gillingham Street
London SW1V 1AH UK
Phone: +44 (0)20 7596-8750
Fax: +44 (0)20 7630 0166
e-mail: artech-uk@artechhouse.com

Find us on the World Wide Web at:
www.artechhouse.com